Ecological Implications of Livestock Herbivory in the West

Editors:
Martin Vavra
William A. Laycock
Rex D. Pieper

Published by the
Society for Range Management
1839 York Street
Denver, CO 80206

First Edition 1994

ISBN 1-884930-00-X

Library of Congress Catalog No.

Second Edition 1999

Printed in the United States

Table of Contents

Contributors

Steven Archer, Professor, Department of Rangeland Ecology and Management, Texas A&M University, College Station, TX.

David Briske, Professor, Department of Rangeland Ecology and Management, Texas A&M University, College Station, TX.

Jerrold Dodd, Professor, Department of Range Management, University of Wyoming, Laramie, WY.

Wayne Elmore, State Riparian Specialist, Oregon and Washington, Bureau of Land Management, Prineville, OR.

Richard Hart, Research Agronomist, High Plains Grassland Research Station, USDA- ARS, Cheyenne, WY.

Harold Heady, Professor Emeritus, Department of Forestry and Resource Management, University of California, Berkeley, CA.

Rodney Heitschmidt, Research Leader and Range Scientist, Fort Keogh Livestock & Range Experiment Station, USDA-ARS, Miles City, MT.

J. Boone Kauffman, Associate Professor, Department Rangeland Resources, Oregon State University, Corvallis, OR.

William Lauenroth, Professor, Department of Range Science and Natural Resource Ecology Laboratory, Colorado State University, Ft. Collins, CO.

William Laycock, Professor, Department of Range Management, University of Wyoming, Laramie, WY.

Daniel Milchunas, Research Associate, Department of Range Science, Colorado State University, Ft. Collins, CO.

Richard Miller, Professor, Eastern Oregon Agricultural Research Center, Oregon State University, Burns, OR.

Rex Pieper, Professor, Department of Animal and Range Sciences, New Mexico State University, Las Cruces, NM.

James Richards, Assistant Professor, Department of Land, Air and Water Resources, University of California, Davis, CA.

Larry Rittenhouse, Professor, Department of Range Science, Colorado State University, Ft. Collins, CO.

Kieth Severson, Supervisory Research Wildlife Biologist, Forestry Sciences Laboratory, School of Mines Campus, Rapid City, SD.

Tony Svejcar, Range Scientist, USDA-ARS, Eastern Oregon Agricultural Research Center, Burns, OR.

Phillip Urness, Professor, Department of Range Science, Utah State University and Research Biologist, Utah Division of Wildlife Resources, Logan, UT.

Martin Vavra, Professor, Eastern Oregon Agricultural Research Center, Oregon State University, Burns, OR.

Frederic Wagner, Associate Dean of Natural Resources and Director of Ecology Center, Utah State University, Logan, UT.

Neil West, Professor, Department Range Science, Utah State University, Logan, UT.

James Young, Range Scientist, USDA-ARS, Reno, NV.

Introduction

The following chapters resulted from a symposium presented at the forty-second annual meeting of the American Institute of Biological Sciences. The symposium was called to evaluate the credence of livestock grazing on western rangelands. Appreciation is expressed to Dr. Robert Zimbelman, Executive Vice President, American Society of Animal Science, for suggesting that action, and to Rex Cleary then President of the Society for Range Management for his interest and initiative in developing the symposium. Appreciation is also expressed to the American Society of Animal Science and the Society for Range Management for support of the symposium and this publication, and to the American Institute of Biological Sciences for providing the forum for the symposium.

The late 1980's saw the beginning of intense criticism of both livestock grazing and the ecological condition of public lands in the Western United States. Although to an extent warranted, the criticisms levied against grazing grew out of proportion to and in some cases contrary to the biological facts available. Grazing became the sole player in ecological decline of the West. The absurdity level was reached with the publishing of the book "Beyond Beef" by Jeremy Rifkin. Cattle grazing was blamed for most of the ills of the world. Perhaps even Hitler's politics were influenced by cows.

To understand present ecological conditions in the West we must remember that westward expansion of the United States was characterized by exploitation of natural resources. This exploitation was totally supported by the American public and legislated by the Federal government. Initially, one of the major players in that exploitation was the livestock industry. Vast public domain and a tremendous forage resource was available for consumption by the large herds of cattle and sheep that developed after the Civil War. The rapid expansion of the livestock industry coupled with the improper cultivation of vast acreages of nonarable land in the West through the Homestead Acts resulted in drastic changes in syn- and aut-ecological relationships of plants and plant communities. The concomitant unwitting introduction of exotic plants added to the ecological disarray that was occurring. Eventually, both the private livestock industry and government saw the resource damage being done and moved to alleviate the problem. The Taylor Grazing Act of 1934 was the first attempt to control grazing on the public domain and to put an end to the devastating series of Homestead Acts. However, not until the 1970's did the government or public consider uses other than extraction for Western lands. Management was for the sustainability of the single use of grazing.

When the public thinks of the West they seldom consider it to be more than one vast entity. In fact, the Western United States is made up of several distinct ecosystems delineated by precipitation amount and pattern. In this publication we have described three major systems, the Great Plains, Intermountain Sagebrush Region and Desert Southwest. These three major land types, their climates, plant communities and land uses; past and present; are discussed as examples of ecosystems impacted by settlement in the West, present ecological condition and potential for the future.

Recently, intensive ecological studies related to natural climate change, plant physiology, natural processes such as fire, and interspecific competition; and a rethinking of classic Clementsian ecology have changed the role of livestock as the sole disruption agent of the West to that of one piece of a complex matrix of several coincidental environmental and social events that acted together.

In the 57 years since the passage of the Taylor Grazing Act, subsequent legislation, Federal Agency guidelines and scientific advances in grazing management have drastically altered how both private and public lands are grazed by livestock. Recent changes in livestock grazing practices have accelerated geometrically as our knowledge of resource management and public awareness of our natural resources have increased. Perhaps no single group has suffered more blame for the past transgressions of an exploitive society than has the livestock industry of the West. Past abuses certainly occurred, and unfortunately, still continue. Since 1934 misuse in grazing has not necessarily been exploitation, but rather ignorance of proper grazing practices.

The knowledge of sustainable grazing is just now catching up to the practice of grazing. In this publication we hope to present the state-of-the-art of our knowledge of livestock grazing in the West; how we got here, where we are and where we can go. We will also explore other events such as climate change, fire suppression, weed invasion and competitive relationships that interact with grazing.

Not only is there a change in livestock grazing practices, but a change in how those practices and other natural resource decisions are made on public land. In the current period of environmental awareness a significant shift has occurred in how management decisions on public lands are developed. Citizen groups are often involved in the planning process. The role of the scientist is increasing as land managers and the public are asking for monitoring techniques, information on animal and plant species and keys to identify ecological change or potential for improvement.

It is our hope that this publication will serve as a benchmark for sound resource management decision making. We have attempted to explain the complexities of the ecosystems of the Western United States and the difficulties in management, with or without livestock grazing. Make no mistake, because of the ever increasing human population, the inherent resource demands and lack of a conservation ethic, all that remains are fragments of ecosystems. The only hope for sustained utilization with preservation of what is left is intensive science based management. In land management, due to past exploitive changes, increased human population and natural climate changes; it is true that "You can never go home again." The problem never was simply livestock grazing, but people.

Historical and Evolutionary Perspectives on Grazing of Western Rangelands

James A. Young

Abstract

Archaeologists can reach back over the millennia of human history and suggest with increasing precision when and where man began to *herd* instead of hunt certain large herbivores. Herdsmen gradually became animal husbandmen of domestic large herbivores, the difference being that man assumed a responsibility for the welfare of the animals. Cultures arose around the provision of seasonal forage needs for domestic herbivores and several of these transhumance cultures struggled to persist. From late 15th through the 19th century the new lands of the western hemisphere provided vast landscapes that could be exploited by domestic herbivores. These open landscapes became know as *ranges* over which livestock were free to roam to obtain forage and reproduce with minimum interference from animal husbandmen. The herdsmen's care and management of livestock became less intensive and the responsibility of animal husbandmen for their animals also diminished. Currently range is viewed as a type of environment where the grazing of livestock may be one of several relatively compatible uses of natural resources. The consequences of the sudden introduction of a large number of domestic livestock to new environments provide an in-depth view of the relation of primary productivity and utilization of rangelands. Concentrations of domestic livestock often truncate succession while service as agents for dispersal of seeds of plants adapted for dominant roles in secondary succession. Such plant species are often annuals that are also adapted to compete with herbaceous crops.

Introduction

This symposium is perhaps the first attempt to examine comprehensively the implications of livestock grazing on the western range since Senate Document No. 199 was compiled in 1936 (Anon. 1936). George Stewart's chapter in that document, History of Range Use, remains the classic historical document on range resources in the United States. The significance of Senate Document No. 199 lay in the national recognition its publication gave to problems that existed in the utilization of grazing resources on public and private rangelands. It was perhaps the most important document on grazing resources since Powell's report on the arid lands of the West in 1878 (Powell 1878). Time and experience have shown most of Powell's observations and suggestions for livestock production in semi-arid environments had merit, but virtually none of these became legislation or administrative policy of the United States. The movement to conserve natural resources and specifically to ensure sustained productivity of rangelands started at the end of the 19th century with the establishment of Forest Reserves. These in turn, became parts of our National Forest administered by the Forest Service, USDA (Rowley 1985). The first steps toward what became scientifically based grazing management were taken in investigations such as Coville's studies of sheep (*Ovis aries*) grazing in forest reserves (Coville 1898). By 1919, Jardine was publishing ideas on how ranges should be managed and Sampson was conducting studies on the effects of early season grazing (e.g. Jardine and Anderson 1919). Senate Document No. 199 was prepared a third of a century after the American conservation movement gained momentum, when a considerable body of literature, much of which was based on ecological theory, had accumulated.

Appreciating Senate Document No. 199 requires placing one's self in the context of 1936. Economically the country and the world were trying to crawl out of the greatest period of deflation and financial disruption of modern times. The early 1930s was a period of severe drought over vast regions of rangelands. Conditions in agriculture were so severe that large areas of marginal cropland were returned to public domain from abandoned homesteads and from Rural Resettlement Administration purchases. Most of this land eventually became parts of the National Grasslands, Forest Service, USDA. Things were so bad that, after years of Congressional deadlock, the Taylor Grazing Act was passed in 1934 and the President finally closed the remaining public domain to homesteading (Rowley 1985). Despite the climate of doom that prevailed during the period when Senate Document 199 was being prepared, it is noteworthy that a section was included on long-term successful examples of conservative use of grazing resources.

The authorship of Senate Document 199 was dominated by 17 professional foresters, most of whom held high administrative positions in the National Forest system and Forest and Range Experiment Stations with the Forest Service, USDA. This was very realistic, considering the contribution of the Forest Service to initiating and developing natural resource conservation in the United States. Perhaps the most significant portions of the report were authored by seven professional research (forest) ecologists. These were W.R. Chapline, M.W. Talbot, George Stewart, R.S. Campbell, David Costello, E.C. Crafts, and L.J. Palmer. This list includes major contributors to the development of the Ecological Society of America during the first half of this century. Only three authors were recognized as range managers (examiners).

The topic headings of Senate Document No. 199 have durable appropriateness. "Homes For Wildlife, Watershed Protection, Excessive Stocking Rates, Climatic Change, and Future Use and Ownership" may be dated in their wording, but 55 years later are still appropriate for content.

My purpose is to suggest perspectives that may prove useful as the talented and experience symposium participants reevaluate livestock herbivory on the western range.

Evolutionary Perspective

Perhaps this all began when some ancestor of vascular plants rose from the primeval seas to colonize terrestrial habitats. To optimize the capture of radiation by photosynthetic surfaces, without the buoyancy supplied by water, carbon compounds known as structural carbohydrates (predominately cellulose) evolved to strengthen cell walls and sheath vascular bundles (Bailey 1953). From these simple beginnings evolved a Pandora's box of complex chemicals (Young and Rowell 1986). Some of these compounds are so physically strong they serve as wooden structural timbers even after the plants are dead. The structures of a very few plants have so little complex cellulose structural carbohydrates that immature foliage or roots can be partially digested by the inefficient stomachs of humans, even without cooking.

Herbivory Mechanisms

For the vast majority of herbaceous plants, the roots and aerial portions are not digestible by humans, even after prolonged cooking. Herbivorous species have developed stomach or intestinal modifications which enable them to utilize cellulose and polysaccharides such as hemicellulose (Church 1969). Cellulose compounds are perhaps the most abundant organic compounds available to terrestrial animals. During evolutionary development, vertebrates failed to develop the capacity to produce enzymes in the gastrointestinal tract, capable of hydrolyzing the beta 1-4 glucosidic linkage which binds glucose molecules together to form cellulose.

Many bacteria, fungi and protozoa produce cellulolytic enzymes capable of hydrolyzing cellulose to cellobiose or glucose. Vertebrate herbivores and termites succeed in partially digesting cellulose through symbiotic relations with microorganisms that have the capacity to produce the necessary enzymes (Kirshna and Weesner 1970, Church 1969). The most successful of these symbiotic relations among large herbivores are those developed by the suborder Ruminantia of even-toed, three- or four-stomached mammals. The stomach modifications provide anaerobic fermentation chambers that house the activities of the micro-organisms. The four components of the ruminant stomach, consisting of the reticulum, rumen, omasum, and abomasum, represent specialized storage and function areas. The storage aspect is important in considering the ecological aspects for grazing. Because of the microbial breakdown of high cellulose, relatively low-nutrient-content forage, a bulk processing operation is required over extended periods. This is especially true for herbivores that ingest wads of herbage (cuds) that are later regurgitated for additional chewing. Animal behavior modes for large herbivore ruminants reflect variations of ingestion, resting, and cud chewing patterns that perhaps are unique for each species and/or habitat.

Ruminants take advantage of the products of anaerobic fermentation by locating their process before the true gastric system. A similar system has evolved in many other animals where the large intestine is enlarged in a post gastric fermentation chamber, known as caecum. Even non-herbivores have vestige structures similar to caeca, such as the human appendix. Animals trying to digest high-cellulose diets through a caecum have problems with absorption of the fermentation products (esp. nitrogen and vitamins). Rabbits partially solve this through coprophagy (ingestion of feces) (Church 1969). The significance of rumens and caeca is the degree of complex evolutionary specialization that is required to be a vertebrate herbivore. Functionally, the highly evolved ruminants are sufficiently more efficient as digesters of cellulose that they can be selective in their feeding. Dietary selection by herbivores brings into play the sensing of preferred forage by grazing animals.

Rumen Micro-organisms

The rumen micro-organisms is the fascinating portion of this symbiosis. It is easy to concentrate on the vertebrate that ingests and eventually defecates the remains of herbage and forget about the complex micro-organisms that are actually chemically breaking down the organic compounds that compose the plant materials in the diet. We consider a plant community containing perhaps a couple dozen higher plants to be complex. Add a moss or perhaps a lichen to this array and it becomes extremely complex. Stop and consider the complexity of the anaerobic vats that constitute the microbic digestive system. Vast numbers of different bacteria, fungi, and protozoa species constitute the micro communities of the rumen. Specific micro-organism functions are vaguely known. The relative composition of the micro-organisms change rather rapidly over time in response to age of the host and most importantly, diet.

Secondary Compounds In Foraging

The micro-organisms that actually break down cellulose and related compounds during the process of digestion do not function equally well on all types of forage or browse. As higher plants evolved they developed secondary compounds that influenced the activity of rumen micro-organisms. The secondary compounds became a major factor in predator-prey relations in herbivory. To play such a role in evolution, it is generally considered that the herbivore must be able to sense their presence, through visual, tactile, or olfactory sensing, and reject the non-digestible forage.

Such relationships are illustrated in the sagebrush (*Artemisia* sp.) of western North America. The various species that compose the section Tridentatae of the genus *Artemisia* are woody members of the Compositae that evolved from Arcto-Tertiary stock, perhaps as components of coniferous woodland communities (Axelrod 1950). These plants have been important components of the semi-arid regions for the Far West for a long period of recent geologic time. Currently the various species occupy, in varying density, perhaps 100 million hectares west of 102° longitude in North America (Beetle 1960). The various species, subspecies, and forms of sagebrush in the section Tridentatae have been shown to differ in the type, concentration, and distribution of phenol, sesquiterpene lactone and alkane compounds (Hanks et al. 1973). Grazing and browsing animals, both mammals and birds have been shown to exhibit strong preference or aversion for various species, subspecies, or highly specific ecotypes of sagebrush (Welch et al. 1981, Welch and

Pederson 1981). Domestic sheep on western rangelands seasonally include some sagebrush leaves in their diet depending on the species of sagebrush and other food availability, but they highly prefer browse from black sagebrush (*Artemisia nova* Nelson) on winter ranges (Hutchings and Stewart 1953). Cattle (*Bos* sp.) generally do not prefer sagebrush browse and especially not that of most forms of big sagebrush (*Artemisia tridentata* Nutt.). Mule deer (*Odocoileus hemionus*), with rumens apparently adapted by morphology and symbiotic relations to browsing, use much larger proportions of sagebrush, and especially of preferred forms on winter ranges (Welch et al. 1981). Pronghorns (*Antilocapra americana*) can consume relatively large portions of big sagebrush in their diet at all seasons of the year, yet historically they also existed on ranges that had no sagebrush (Young 1989).

Apparently all of these ruminants, except for pronghorns, suffer severe digestive problems that may cause mortality if the proportion of big sagebrush in the diet *becomes excessive* (Nagy et al. 1964). To keep this in perspective, you must remember that the grazing animal controls intake, except in starvation situations. In a starvation situation the fact that the herbage induces digestive problems does no alter the ultimate death of the animal. The important point is the chemical content of the herbage controls preference, making the ruminant more selective. The problem is a result of the chemical content of the big sagebrush herbage negatively impacting the rumen micro-organisms. Crude terpeniods have been indicated as the major interference with the rumen microbe activity (Striby et al. 1987). What are the consequences of vertebrate herbivory of big sagebrush? Individually our approach to this problem has been somewhat analogous to the blind man examin ing the elephant. Herbivory of big sagebrush results in poisoning. Preference unequivocally limits herbivory of big sagebrush. Herbivory of big sagebrush is essential in the diet of pronghorns. Such interpretations may be supported by an individual investigator's sample, but do the samples reflect the intrinsic and transient variability of populations?

The goal of this symposium can be clearly stated in terms of evaluating the consequences of herbivory. A very brief glance at the grazing animal reveals highly complex relations with symbiotic micro-organisms, and apparent evolutionary relations with secondary chemicals in plants that influence preference and in certain cases the function of rumen microbes. If we take this obviously complex consumer machine and superimpose its actions upon the host of plant communities from the equator to the arctic where animals graze, complexity suddenly becomes almost incomprehensible.

Temporal Perspective

This is an exciting time to be associated with environmental science because of the increasing awareness that relative short-term climatic change is both a reality and likely subject to influence by man's activities. Evaluating changes, such as the increase of woody plants on former grasslands, that have been attributed to excessive grazing, is very complex if concentration of atmospheric gases has also influenced such conversions.

Pleistocene Extinctions

It was not until the 1870's that Louis Agassiz's concept of Ice Ages in the recent geologic past was widely accepted by the scientific community. As the events of the close of the last ice age are known with increasing scientific detail it is worthwhile to remember that most of the really big, fierce, and wonderful mammals failed to survive the climatic change to post-Pleistocene conditions (Wallace 1876). Perhaps early man influenced this wave of extinctions (Martin 1967) as his modern counterpart's influence on habitats threatens many of the surviving mega-herbivores today (Owen-Smith 1988). Remember, we probably live towards the end of the Age of Mammals, certainly not the beginning. We know that modern mega-herbivores such as elephants (*Elephas* sp.) can severely impact woody vegetation on rangelands. When various species of mammoths (*Mammuthus* sp.) existed in abundance in North America what selective factors were leveraged on evolving plant communities?

We know that the Edentate mammals that once occurred in western North America were primary consumers of herbage not currently utilized by native or introduced herbivores in the region. What selective factors existed when junipers (*Juniperus* sp.) and Joshua trees (*Yucca brevifolia* Engelm.) were forage for the Shasta ground sloth (*Nothrotherium shastense*) (Houghton 1976)? Perhaps much of the influence of grazing animals on plants, in terms of co-evolution of secondary compounds that influence preference has already occurred, and was initiated by food niches of consumers that no longer exist?

Clonal Plants

In a temporal sense, past and present climatic change obviously influences grazing animal-plant relations. We are just beginning to realize that some plant species have persisted through extended periods of climatic change as vegetatively propagated clones. Creosote bush (*Larrea divaricata* Cav.) rings in the Mojave desert are a documented example (Barbour 1969). For this symposium blue grama (*Bouteloua gracilis [H.B.K.]*) is perhaps a more appropriate example. If clones have survived climatic change, but have reduced chances for seedling establishment in current or future climates, are they living links with the past with limited futures (Hyder et al. 1971)? Quaking aspen (*Populus tremuloides* Michx.) is also an appropriate example for this symposium. In the Great Basin, isolated clones of this species on arid mountains may be true Pleistocene relics. Senate Document 199 has references to the problem of damage to quaking aspen regeneration by range sheep operations. The range sheep industry has virtually disappeared from the Great Basin and the clonal populations of aspen still tremble in the breeze. Are these populations in more or less evolutionary jeopardy as they try to survive the perils of vegetative existence and wait for the return of a para-glacial climate to their altitude/latitude?

Domestication of Large Herbivores

Flannery's (1969) essays on the origins and ecological effects of domestication of plants and animals remain classic discussions of implications of the process for human culture and the *environment*. In relating man to environmental change, Flannery was perhaps 20 years ahead of his scientific time. In the course of archeological excavations in the Middle East, where Flannery was among the first to employ botanists in archeology, he determined that when ancient hearths were uncovered that contained bones of what could be identified as domesticated rather than wild herbivores, carbonized seeds of cheatgrass (*Bromus tectorum* L.), Russian thistle (current taxon *Salsola australis* R. Br.), and halogeton (*Halogeton glomeratus* C.E. Mey.) were associated with the bones. These species include some of the most successful alien colonizers of disturbed rangelands in the western United States. We often loosely group them as "Mediterranean" species because they found homes on the grossly overgrazed ranges of the Mediterranean basin. Their true center of origin is more correctly central and southwestern Asia where they became the shadow of herbivory and highly successful followers of the flag as civilization proceeded westward and eventually to the New World.

Is the success of these colonizing species due to preadaption to the abundant disturbed habitats that must have resulted from climatic change at the close of the Pleistocene and/or is their success as colonizers due to their inherent potential for rapid micro-evolution to take advantage of changing environments. "Studies Show Man Induced Changes In Atmospheric Gas Responsible For Plant Evolution On Rangelands", this news story has (to my knowledge) not yet appeared in *Science* or *Nature*, but if such a relationship is shown for epicular waxes or some aspect of gas exchange one would immediately look for the scientific name of one of the traditional grazing weeds.

Classic scholars have repeatedly related that ancient civilizations fell because they abused natural resources, especially forests and woodlands. This scenario usually includes excessive grazing by livestock of the denuded woodlands. In northern Europe conversion of woodlands to moors is an example. In more arid environments the supposed inevitable product is what has recently become known as desertification. In a noted essay, F.F. Darling (1956) suggested that pastoral societies were inherently counter to evolution of civilization because in semi-arid environments transhumantic movements were essential.

Livestock herbivory on the western range in the United States certainly involves seasonal movement of animals to take advantage of forage availability and quality. In mountainous and more arid areas such movement becomes more pronounced. Modern transportation systems alleviate most of the social problems attached with cyclic movements of livestock. Reorganizing such traditional patterns of grazing to more closely match the physiological requirements of plants may offer the potential of sustained production from rangelands.

Is Range Animal Husbandry Obsolete?

The production of red meat from rangelands will persist because of; a) dietary demand for quality protein, b) the inherent dispersal of animal waste and nutrient cycling with range operations, c) lack of direct or more efficient means of humans utilizing range forage, and d) the unrelenting growth in demand for food. It has been projected that we must double the world's food supply not too far into the

21st century. This will have to be accomplished with less cropland and greatly restricted irrigation water.

Any basic human nutrition text tells us that eight amino acids are essential for human development. They have to be supplied in the diet because humans cannot synthesize them. They are most easily obtained from consumption of meat and dairy products. It is possible to obtain these amino acids from a vegetarian diet if a wide variety of plant protein sources are included, especially legumes. By careful dietary planning an individual can have a balanced vegetarian diet. In practice it is difficult for a nation of vegetarians to have a balanced diet.

This is cruelly apparent in developing countries, especially in the tropics where essential amino acid deficient diseases such as kwashiorkor are apparent in young people who have higher requirements for protein for growth and development. Such countries can support unique habitats for the world's dwindling examples of wild large herbivores. Increasing meat production with domestic herbivores will impact negatively wild herbivores as well as their habitat. Moreover, such efforts at livestock production in third world countries can result in export sales for hard currency rather than domestic consumption. Despite all these negative connotations the bottom line remains the eight essential amino acids that man must receive from dietary sources. Unless this basic problem is addressed on a global scale apparently rational approaches to conservation biology are doomed to failure. Perhaps the logical conclusion to this scenario is that Americans must become self sufficient in quality protein and the western range must play a role in supplying this need.

Increasing scientific evidence suggests that excess animal fat in human diets is also detrimental to health. This should not be construed that declining per capita consumption of red meat will lead to rangelands free of domestic livestock in western North America. The type of animal grazed and the age and means through which it is marketed may dramatically change. Such changes will influence forage requirements and relations of livestock with native animals.

Western Range and Politics

Why does this symposium address livestock herbivory in the West and not the eastern United States? Politically, sub-humid and humid area pastures are largely located on private land. Environmental laws may limit management options on private lands, but certainly not to the extent they affect practices on publicly owned rangelands. The developed pastures of the eastern United States are usually artificially established and potentially could be renewed by such means. Most western rangelands are composed of significant portions if not absolute dominance of native plant species. Is it politically or biologically unacceptable to renew grazing lands west of the 102° longitude?

Historic Perspective

The forage characteristic of the western range that was most mystifying to the first livestock husbandmen to venture into these plains and mountains was self-curing of standing herbage. In the humid east standing herbage rapidly decayed when left exposed. Conservation of forage for winter feeding of livestock there required either curing or ensiling. John Clay (1922) relates the story that Wells Fargo freighters to northern plains military posts turned oxen loose on the range to winter by themselves and were shocked that the animals not only survived, but fattened during the winter on a diet of standing, sun-cured herbage. Hiram Latham (1871) exploited this characteristic as a pamphlet-writer entrepreneur exhorting ranching development along the Union Pacific railroad. The great open range boom of livestock production on the northern plains, which Dr. Latham helped promote, only lasted from the close of the civil war until the hard winter of 1886-87 (Webb 1931). Short-term climatic extremes, both drought and severe winters, ended the open ranging of cattle east of the Rocky Mountains in 1886-87 and west of the divide in 1889-90. During the scant two-decade period that unlimited range livestock production boomed, the abuse of vegetation resources by excessive grazing exacerbated the effects of the eventual climatic day of reckoning (Young and Sparks 1985). Out of these disastrous experiments with unlimited open ranging of livestock grew a system of livestock production that involved the conservation of forage for planned and/or emergency winter feeding of livestock. In the West very few areas outside of limited riparian areas were natural hay producing lands. Most areas for hay production were developed through the supplemental irrigation of riparian areas or through more elaborate engineering features that permitted the irrigation of upland areas (Young and Sparks 1985).

Herbivory of domestic livestock on the western range involves the utilization of native ranges, native and exotic hay production and crop aftermath. A consequence of this herbivory includes (certainly not exclusively) markedly changed stream seasonal flow patterns, wetlands often greatly expanded and changed in character, and changed water quality.

What a tangled web we trace from decomposition rates of standing herbage, through climatic extremes, to irrigated agriculture in search of the grail of understanding the consequences of herbivory on the western range.

Range Animal Husbandry Through Time

Range Animal Husbandry Cooperation

Note that the great boom in unlimited range livestock production in western North America lasted for only a couple of decades. Except for southern Texas, New Mexico and California, range livestock production in the western United States has only been carried out in any form for 100 to 130 years. This time period is a mere blip in the record of many areas in the world where livestock play a significant role in providing a protein source as meat, milk, or blood in a variety of cultures with great longevity. Perhaps it exists in another form in the anthropological literature, but what a great achievement it would be if the statesmen range ecologists participating in this symposium would prepare a compendium of the range livestock cultures they had visited around the world. Lapps herding reindeer (*Rangifer tarandus*) in the sub-arctic, yaks (*Bos grunniens*) in Tibet, Masai herdsmen on the Serengeti and goats (*Capra hircus*) on eroded Mediterranean hills would provide chapters in such a volume. Perhaps this data set would have such a temporal scale the consequences of herbivory would not be interpretable, but the dynamic equilibria attained in ancient husbandry systems and their reactions to 20th century perturbations would have fascinating implications toward herbivory on the western range.

Anthropocentric Versus the Green Revolution

In June 1955 a symposium was held at Princeton, New Jersey, to discuss "Man's Role In Changing The Face of the Earth" (Thomas 1956). Many of the leading biologists and anthropologists of the period participated in the symposium. The symposium proceedings have been a standard reference for many fields of science for years. In retrospect this conference may have been the high point of biological theory that placed man as the God-given dominant of all things and places. There has been a ground swell of awakening of ecological awareness in the American culture that has counterparts in most countries of the world. In the United States this ecological awareness has been translated into legislative action that has brought democracy to the management of natural resources through public review of environmental analysis and impact statements. Out of this movement has grown changing systems of values toward endangered species and man's inherent rights to modify the environment for economic gain. Perhaps some writer in the future will dub this man's ecological suicidal stage. One of the values of established civilization that has undergone scrutiny during this awakening has been the exploitation of domestic animals. Livestock production on rangelands is undergoing such scrutinies, as this symposium is apparently an example. These are important and necessary steps in evolving ecologically sound range management theory and practices.

Ecologists are being asked by a concerned society to join with traditional agriculturists to provide answers to the problem of adequate, sustainable supplies of food for the world's human population. In 1936 the scientific strength of the budding concept of range management was the use for the ecologic theory of climax vegetation in evaluating range condition and trend. As ecologists and ecologically based range managers join to evaluate the consequences of herbivory on the western range perhaps the greatest building block for future growth is application of evolving ecological theory to problem solving. As this process of scientific scrutiny is carried out, it is important to view the process of livestock herbivory from many-faceted perspectives.

Literature Cited

Anon. 1936. The Western Range Senate Document No. 199 74th Congress, 2nd Session. CPO, Washington, D.C.

Axelrod, D.I. 1950. Evolution of desert vegetation in western North America. Carnegie Inst. Publ. 590. Washington, D.C.

Bailey, W. 1953. Evolution of the tracheary tissue of land plants. Amer. J. Botany 40:4-8.

Barbour, M.G. 1969. Aged space distribution of the desert shrub *Larrea divaricats*. Ecology 50:679-685.

Beetle, A.A. 1960. A study of sagebrush. The section Tridentatae of Artemisia. Wyoming Agr. Exp. Sta. Bull. 368.

Church, D.C. 1969. Digestive Physiology and Nutrition of Ruminants, Vol. 1. D.C. Church Publ., Corvallis, OR.

Clay, J. 1922. My Life on the Range. Privately printed. Chicago, IL.

Coville, F.V. 1898. Forest growth and sheep grazing in the Cascade Mountains of Oregon. USDA Division of Forestry Bull. 15. CPO, Washington, D.C.

Darling, F.F. 1956a. Man's ecological dominance through domesticated animals on wildlands. p. 778-788. *In*: W.L. Thomas, Jr. (ed.) Man's Role in Changing the Face of the Earth. Univ. Chicago Press, Chicago, IL.

Flannery, K.V. 1969. Origins of ecological effects of early domestication in Iran and the near East. p. 73-100. *In*: P.J. Ucko and G.W. Dimbleby (eds.) The Domestication and Exploitation of Plants and Animals. Aldine Publ. Co., Chicago, IL.

Hanks, D.L., E.D. McArthur, Richard Stevens, and A.P. Plummer. 1973. Chromatographic characteristics and phylogenetic relations of Artemisia section tridentatae. Res. Paper INT-141. Intermountain Res. Station, Forest Service, USDA, Ogden, UT.

Houghton, S.G. 1976. A Trace of Desert Waters. The Arthur H. Clark Co., Glendale, CA.

Hutchings, S.S. and G. Stewart. 1953. Increasing forage yields and sheep production on Intermountain winter ranges. Circ. 925. Washington, D.C., USDA.

Hyder, D.N, A.C. Everson, and R.E. Bemont. 1971. Seedling morphology and seeding failures with blue grama. J. Range Manage. 24:287-292.

Jardine, J.T. and Mark Anderson. 1919. Range management on the National Forest. USDA Bull. No. 790. GPO, Washington, D.C.

Kirshna, R. and F.M. Weesner. 1970. Biology of Termites. Academic Press, NY.

Latham, H. 1871. Trans-Missouri Stock Raising: The Pasture Lands of North America: Winter Grazing. Nebraska Daily Herald Steam Printing House, Omaha, NE.

Martin, P.S. 1967. Pre-historic overkill. pp. 201-287. *In*: P.S. Martin and H.E. Wright, Jr., (ed.) Pleistocene Extinctions, the Search for a Cause. Vol. 6. Proc. of VII Congress of Inter. Assoc. for Quatenary Res. Yale Univ. Press. New Haven, CT.

Nagy, G., H.W. Steinhoff, and G.M. Ward. 1964. Effects of essential oils on sagebrush on deer rumen microbial function. J. Wildl. Manage. 28:788-790.

Owen-Smith, R.N. 1988. Mega herbivores: The influence of very large body size on ecology. Cambridge Univ. Press. New York, NY.

Powell, J.W. 1878. Report on the lands of the arid region of the United States House Doc. 73. 45th Congress, GPO, Washington, D.C.

Rowley, W.D. 1985. U.S. Forest Service Grazing and Rangelands. Texas A&M Univ. Press. College Station, TX.

Striby, D., C.L. Wambolt, R.G. Kelsey, and K.M. Hastad. 1987. Crude terpeniod influences on *in vitro* digestibility of sagebrush. J. Range Manage. 40:244-247.

Thomas, Jr., W.L. (ed.) 1956. Man's Role in Changing the Face of the Earth. Univ. Chicago Press. Chicago, IL.

Wallace, A.L. 1876. The geographical distribution of animals. Vol. I. McMillian. London, England.

Webb, W.P. 1931. The Great Plains. Ginn & Co. Boston, MA.

Welch, B. and J.C. Pederson. 1981. *In vitro* digestibility among accessions of big sagebrush by wild mule deer and its relationship to monoterpeniod content. J. Range Manage. 34:497-500.

Welch, B.L., E.D. McArthur and J.N. Davis. 1981. Differential preference of wintering mule deer for accessions of big sagebrush and for black sagebrush. J. Range Manage. 34:409-411.

Young, J.A. and B.A. Sparks. 1985. Cattle in the Cold Desert. Utah State Univ. Press. Logan, UT.

Young, J.A. 1989. Intermountain shrubsteppe plant communities -pristine and grazed. p. 3-13. B.G. Pendleton (ed.). Proc. Western Raptor Management Symposium and Workshop. National Wildlife Fed. Scientific and Tech. Serves No. 12. Washington, D.C.

Young, R.A. and R.M. Rowell (eds.) 1986. Cellulose: structure, modifications, and hydrolysis. Wiley-Interscience. New York, NY.

Woody Plant Encroachment into Southwestern Grasslands and Savannas: Rates, Patterns and Proximate Causes

Steven Archer

Abstract

Warm temperate grasslands and savannas which characterized many landscapes in southwestern North America at the time of European settlement have been replaced by shrublands and woodlands. These changes in plant life-form composition were coincident with the introduction of large numbers and high concentrations of livestock. While a cause-effect relationship is implied, it is difficult to demonstrate, since most evidence is based on anecdotal historical accounts or descriptions from localized long-term studies, many of which are conflicting. Case studies documenting the rate, pattern and extent of vegetation change are summarized and used to illustrate how historical inconsistencies might be resolved. Where vegetation history is reasonably known, causes for change are evaluated.

Explanations for the proliferation of woody plants and the associated decline of graminoids have typically centered around alterations in climatic, grazing and fire regimes. Each of these factors is addressed individually and in combination. It is argued that: (1) Atmospheric CO_2 enrichment and directional shifts in climate may have occurred, but have not been sufficient, to cause the vegetation changes observed to date; (2) Fire is not necessarily required to maintain grasslands or savannas; and (3) Although herbivory, lack of fire, atmospheric CO_2 enrichment and climate have interacted to produce recent vegetation change, selective grazing by large numbers and high concentrations of livestock has been the primary force in altering plant life-form interactions to favor unpalatable woody species over graminoids. Conceptual models illustrating the role of grazers in directing plant succession are presented in the context of ecosystem resilience, multiple steady states and positive feedbacks.

Introduction

In drylands around the world, including the North American southwest, desertification and/or the replacement of productive grasslands and savannas with shrub- and woodlands dominated by unpalatable species appears to have occurred since European settlement (Table 1). Explanations for these vegetation changes are the

Table 1. Documented instances of increased abundance of woody plants in arid and semi-arid ecosystems in recent history.

Geographic Location and Reference	Encroaching tree/shrub genera
North America	
Arizona	
Brown (1950)	*Prosopis*
Glendening (1952)	*Prosopis, Opuntia*
Humphrey and Mehrhoff (1958)	*Larrea, Prosopis, Aplopappus, Opuntia*
Hastings and Turner (1965)	*Larrea, Prosopis*, others
Johnson (1962)	*Juniperus*
Martin and Turner (1977)	*Prosopis*
Smith and Schmutz (1975)	*Prosopis*
California	
McBride and Heady (1968)	*Baccharis*
Hobbs and Mooney (1986)	*Baccharis*
Williams et al. (1987)	*Baccharis*
Young and Evans (1981)	*Juniperus*
Kansas	
Bragg and Hulbert (1976)	*Rhus, Cornus, Symphoricarpos, Ulmus, Quercus, Juniperus*
Nebraska	
Steinauer and Bragg (1987)	*Pinus*
New Mexico	
Buffington and Herbel (1965)	*Prosopis, Larrea, Flourensia*
Hennessy et al. (1983)	*Prosopis*
York and Dick-Peddie (1969)	*Larrea, Prosopis, Juniperus*
Montana	
Arno and Gruell (1986)	*Pseudotsuga*
Nevada	
Blackburn and Tueller (1970)	*Pinus, Juniperus*
North Dakota	
Potter and Green (1964)	*Pinus*
Oklahoma	
Snook, E.C. (1985)	*Juniperus*
South Dakota	
Progulske (1974)	*Pinus*
Tieszen and Archer (1990)	*Quercus, Celtis, Fraxinus, Tilia*

subject of some controversy, as retrogression and desertification can be natural (Haynes 1982), human-induced (Owen 1979, Gornitz and NASA 1985) or a combination of the two (Verstraete 1986). Anthropogenic activities may cause changes independent of climate, or they may reinforce, magnify, or accelerate changes instigated by natural processes. Retrogression associated with disturbance may be mitigated when climatic conditions for a given species are favorable or benign and magnified when conditions are unfavorable or stressful. In many cases, desertification and vegetation change has been coincident with the introduction of large numbers and high concentrations of livestock. However, while a cause-effect relationship is

Table 1 — continued

Geographic Location and Reference	Encroaching tree/shrub genera
North America (continued)	
Texas	
Archer (1990)	*Prosopis, Condalia, Zanthoxylum*
Bogusch (1952)	*Prosopis*, others
Ellis and Schuster (1968)	*Juniperus*
McPherson et al. (1988)	*Prosopis, Juniperus*
McPherson and Wright (1990)	*Juniperus*
Nelson and Beres (1987)	*Acacia, Larrea*
Smeins and Merrill (1988)	*Juniperus*
Scanlan and Archer (1991)	*Prosopis*, others
Wondzell (1984)	*Larrea*
Utah	
Madany and West (1983)	*Pinus, Juniperus, Quercus*
Sparks et al. (1990)	*Juniperus*
Other	
Humphrey (1958)	various species
Humphrey (1987)	various species
Robinson (1965)	*Tamarix*
Africa	
Acocks (1964)	*Acacia*
Trollope (1982)	various spp.
van Vegten (1983)	*Acacia, Dichrostachys, Grewia*
Skarpe (1990)	*Acacia, Grewia*
Australia	
Burrows et al. (1985)	*Eremophila, Dodonaea, Acacia, Cassia*
Harrington et al. (1979)	various shrub spp.
Booth and Barker (1981)	*Acacia, Cassia, Dodonaea*
Cunningham and Walker (1973)	*Acacia*
Lonsdale and Braithwaite (1988)	*Mimosa*
India	
Singh and Joshi (1979)	various spp.
Scandinavia	
Rosen (1988)	*Juniperus*
South America	
Adamoli et al. (1990)	various spp.
Schofield and Bucher (1986)	various spp.

implied, it is difficult to demonstrate, since most evidence is based on anecdotal historical accounts or descriptions from local, long-term studies, many of which are conflicting. Encroachment of woody plants into grasslands has been widely recognized, but the rates, patterns and dynamics have seldom been quantified.

Shifts in grass and woody plant abundance have broad implications for biodiversity, primary and secondary productivity, soil development and stability, livestock and wildlife composition and carrying capacity, recreational opportunities, water quality and water distribution. Changes from herbaceous to woody plant domination also constitute a potentially important global climate feedback by affecting carbon sequestering, non-methane hydrocarbon emissions and biophysical land surface-atmosphere interactions (albedo, evapotranspiration, surface roughness, boundary layer dynamics). As a result, an understanding of factors controlling the balance between grasses and woody plants is fundamental to developing natural resource management plans for sustained utilization of these globally extensive vegetation types.

This chapter will (1) provide a general overview of post-settlement vegetation change in the North American Southwest [see also Hastings and Turner 1965, Branson 1985, Grover and Musick 1990, Bahre 1991], emphasizing shifts from grassland or savanna to shrub- or woodland; (2) discuss why/how grass-woody plant ratios may have changed on some landscapes and not others by evaluating the role of atmospheric CO_2 enrichment, climate, soils, fire and grazing independently; and (3) evaluate proximate causes of shifts from grass-to-woody plant domination in the context of interacting factors.

Rate and Extent of Woody Plant Encroachment

Numerous historical accounts and photographic records indicate that shrublands, woodlands and forests of North America have expanded and replaced what were grasslands and savannas at the time of settlement (Fig. 1b, Table 1). Accounts of early settlers and travelers often describe expansive grasslands and savannas in what are now shrublands and woodlands (Table 2) (Inglis 1964, York and Dick-Peddie 1969, McCraw 1985). Although such historical accounts and historical-modern repeat photography are potentially limited in their information content (Forman and Russell 1983, Bahre 1991:14), they provide one of several types of information from which to gauge vegetation change. Despite some conflicting accounts of the extent of grasslands (Malin 1953) and the fact that woody vegetation may have characterized some landscapes at the time of settlement (Fig. 1a, c; Table 2), it seems unlikely that the livestock industry would have developed and flourished to the extent it did, were the acreages of grazable lands not substantial at the time of settlement. Subsequent studies of vegetation change, utilizing surveyor witness trees and notes, permanent plots, sequential ground and aerial photography, and stable carbon isotopes have documented shifts from grass-to-woody plant domination in recent history (Table 1). Such studies indicate that rates and patterns of this succession have been (1) rapid, with substantial changes occurring over 50- to 100-year time spans, (2) non-linear and accentuated by periodic drought, (3) influenced by topoedaphic factors, and (4) non-reversible over time frames relevant to management.

Factors Influencing Spatial/Temporal Patterns of Grass and Woody Plant Distribution and Abundance

Regional physiognomy and plant life-form distributions largely reflect the prevailing macroclimate (Box 1981). Yet substantial variation occurs across landscapes within a region, as soils, topography and elevation interact against a cli-

Figure 1. Paired photographs illustrating temporal patterns of grass-woody plant abundance. (A) Upper: creosotebush was well-established on this landscape in the Lower Sonoran Desert in 1893; lower: physiognomy has changed little through 1984 (Monument #158 in Humphrey [1987]).

Figure 1. Paired photographs illustrating temporal patterns of grass-woody plant abundance. (B) Succession from desert grassland in 1903 (upper) to *Prosopis* shrubland by 1941 (lower) at the Santa Rita Experimental Range in southern Arizona (from Martin, 1975:8).

Figure 1. Paired photographs illustrating temporal patterns of grass-woody plant abundance. (C) Meadows and ponderosa pine (*Pinus ponderosa*) savannas in the Black Hills of South Dakota photographed in 1874 (upper) gave way to closed-canopy pine forest by 1975 (lower) (Progulske and Shideler 1983:104-105).

Table 2. Historical observations indicating differences in local and regional distribution and abundance of woody plant abundance on Texas landscapes. Most landscapes in both regions are presently dominated by woody vegetation.

NORTH-CENTRAL TEXAS (Marcy Expedition of 1854 as seen in Malin, 1953):

"... nothing could be seen but one continuous mesquite flat, dotted here and there with small patches of open prairie. . . ."

"The country we are now passing is gently undulating and covered with mesquite trees."

"a broad level plain . . . covered with buffalo grass and mesquite tress, and extending as far as the eye could reach. . . ."

"[Mesquite often grew] upon the most elevated arid prairies, far from water courses. . . ."

SOUTHERN TEXAS (from Inglis, 1964):

Borders of creeks were well supplied with timber but there was "scarce a brush on uplands" — Lunday (1833), Jim Wells County

"The prairie was now dead level, the grass short...Not a bush or tree was to be seen . . . thousands of antelope. . . ." — Fremantle (1863), Nueces County

"high rolling prairie . . . covered with fine mesquite grass and interspersed with mesquite trees. . . ." — Michler (1849), Nueces County

"open country with mesquites" — Bollaert (1843), Frio County

matic background to influence vegetation patterns at smaller spatial scales. Disturbances and the utilization of plants by animals are superimposed on this background of topo-edaphic heterogeneity and climatic variability to further influence vegetation structure. As a result, plant species whose adaptations to prevailing climate and soil would make them the competitive dominants under one disturbance regime may assume subordinate roles or even face local extinction under other disturbance regimes. On grazed landscapes, it is often difficult to assess the extent to which herbivory influences ecosystem processes relative to abiotic factors (McNaughton 1983, Foran 1986). In some cases, comparisons of vegetation change between grazing systems have been shown to reflect differences in climatic conditions rather than differences in management (Herbel 1979, Chew 1982, Heitschmidt et al. 1982). In addition, frequent, small-scale perturbations (such as ant or termite mounds, rodent burrows, dung deposits, patch grazing) occur within larger-scale, less frequent disturbances (fire, floods) to produce a complex disturbance regime (Collins 1987, Coffin and Lauenroth 1990). To realistically understand cause-effect relationships in vegetation dynamics, environmental and disturbance variables must be dealt with simultaneously, in a spatially explicit manner and over a range of time scales. Given the lack of quantitative historical information and the short duration of most ecological investigations, few such studies exist.

The following sections review the independent role of atmospheric CO_2 enrichment, plant life history attributes, climate, soils, fire and grazing on grass and woody plant abundance in dryland ecosystems. The final section discusses interactions between these biotic and abiotic forces and attempts to evaluate conditions necessary and sufficient to elicit a change in vegetation from grassland or savanna to shrub- or woodland.

Atmospheric CO_2 Enrichment

Mayeux et al. (1991), Johnson et al. (1993), and Idso (1992) review an array of literature and data which support their hypothesis that increases in atmospheric CO_2 concentrations since the industrial revolution may have been a driving force for encroachment of woody plants into grasslands. Their arguments are based on observations which indicate: (1) woody plants typically possess the C_3 photosynthetic pathway, whereas the grasses they have replaced in the southwestern USA and tropical regions are primarily C_4; (2) increased atmospheric CO_2 may confer a significant advantage to C_3 species relative to C_4 species with respect to physiological activity, growth and competitive ability; (3) C_4 grasslands appear to have evolved at CO_2 concentrations below 200 ppm and thus at low CO_2/O_2 ratios; and (4) invasion of woody plants into C_4 grasslands has been accompanied by a 30% increase in atmospheric CO_2 over the past 200 years (from ca. 270 ppm to 350 ppm).

To what extent might historic increases in atmospheric CO_2 have contributed to the replacement of C_4 grasslands by C_3 shrublands or woodlands? A robust interpretation of the hypothesis suggests widespread replacement of C_4 grasses by C_3 grasses would also have occurred during this period, controlling for differences in land use and soils. Such does not appear to have been the case. This could reflect the fact that the predicted disparity in C_3 vs. C_4 plant response to increased CO_2 may be minimal. Indeed, recent modeling exercises (Hunt et al. 1991), controlled environment studies (Riechers and Strain 1988) and field investigations (Owensby et al. 1993) indicate C_4 grasses may be very responsive to CO_2 fertilization. The fact that C_3 woody plants have invaded some landscapes while C_4 grasses have persisted on others with similar soils and climate also argues against CO_2 enrichment as a proximate causal factor for community change. In addition, the differential effects of CO_2 enrichment on C_3 woody plants vs. C_4 grasses would not explain the invasion of woody plants into cold desert and temperate landscapes where dominant grasses were also C_3. The temporal correspondence between CO_2 enrichment and woody plant encroachment is also problematic. Although present-day atmospheric CO_2 levels are ca. 30% higher than those 200 YBP, enrichment has been exponential, with a prominent lag phase. In 1903-1925, CO_2 concentrations were ca. 300 ppm (Neftel et al. 1985), an 11% increase over that of the late 1700s. Yet, significant woody plant encroachment had occurred by this time on many sites (e.g., Fig. 1b). Would an 11% increase in atmospheric CO_2 have been sufficient to cause or contribute to woody plant encroachment into grassland? Finally, the extent to which benefits of increased CO_2 will be manifested in enhanced plant growth and competitive ability will be constrained by interactions of other environmental resources, stresses and higher-order ecological interactions (Bazzaz 1990, Mooney et al., 1991). Some data from tree rings indicate a significant CO_2 fertilizer effect on woody plants over the past century (LaMarche et al. 1984). However, others do not (Graumlich et al. 1989, Graumlich 1991), perhaps reflecting feedbacks to plant growth and alteration of biomass allocation patterns (Norby et al. 1992). In sum, these various exceptions limit the utility of the atmospheric CO_2 enrichment hypotheses as a robust explanation of the cause of woody plant encroachment into grasslands.

Life History Determinants

Different patterns of succession (or lack thereof) result from the interactions of species having various combinations of physiological and life history traits (Huston and Smith 1987). The rate, extent and dynamics of CO_2-, climate- or disturbance-induced shifts between grassland and shrub- or woodland would be strongly influenced by the life history characteristics of the species involved (e.g., potential growth rate, seed production, rates of seedling establishment, plant size, longevity, shade tolerance). Differences in longevity between grasses and woody plants may be particularly important in accounting for observed patterns of post-settlement succession from grass-to-woody plant domination. Grasses are relatively short-lived in comparison to many woody plants. Maximum life-expectancy of established perennial grass species ranges from 4 to 43 years, with a mean of 3 to 10 years (Canfield 1957, West et al. 1979, Wright and Van Dyne 1976). Longevity in woody plants varies with growth form. Limited evidence suggests a positive correlation between "woodiness" and longevity (e.g., suffrutescent < suffruticose < fruticose < arborescent) and ranges from 22 to >>400 years (West et al. 1979, Vasek 1980, Goldberg and Turner 1986, McAuliffe 1988, Young and Evans 1981). Populations of suffrutescents and small shrubs such as burroweed (*Haplopappus tenuisectus*), bursage (*Ambrosia deltoidea*), cholla cacti (*Opuntia* spp.), four-winged saltbush (*Atriplex canescens*) and catclaw acacia (*Acacia greggii*), may fluctuate markedly in response to amount and seasonality of precipitation, whereas large shrubs and small trees [e.g. mesquite (*Prosopis spp.*), *Acacia constricta*, creosotebush (*Larrea tridentata*), ocotillo (*Fouquieria splendens*), little leaf palo verde (*Cercidium microphyllum*)] with longer life-spans persist for extended periods once established (Shreve and Hinckley 1937, Shreve 1942, Martin and Turner 1977, Goldberg and Turner 1986). For woody plants capable of sprouting, ramets (individual stems or shoots) that succumb to stress or disturbance may be replaced vegetatively, thus perpetuating the genet (individual plant), sometimes for thousands of years (Vasek 1980).

Differences in mean and maximum plant longevities would influence the frequency of gap formation and hence the propensity for community change. Skarpe (1990) cites studies in southern Africa where encroaching small trees and shrubs formed thickets stable for 30-50 years, then died suddenly, followed by the re-establishment of grasses. These short-term (10-50 year), cyclic changes in vegetation associated with disturbance, climatic fluctuation and differences in plant longevities may be missed in widely spaced (50-100 year) matched photographs or inventories of permanent plots (e.g., Gehlbach 1981). Other factors held equal, the availability of gaps for establishment of new plants would be highest for short-lived species with high rates of turnover. Given the generally short life-span of grasses, opportunities for woody plant establishment in gaps vacated by grasses may be relatively abundant, whereas the frequency of gap formation and hence opportunity for grass establishment in a shrubland may be relatively low. Differences in rates and frequencies of gap formation associated with grass and woody plant life-history traits may explain why successional transitions between grass- and woody plant-dominated states are highly asymmetrical (i.e., probabilities (P) of transition from grassland to woody plant domination [P(g→w)] may be high, whereas P(w→g) is very low) (Scanlan and Archer 1991). The magnitude of gap formation discrepancies in grassland versus shrublands or woodlands would likely increase in response to stress and disturbance. For example, drought may increase mortality and hence gap formation in mesophytic grasses, but have relatively little impact on woody plants tolerant of moisture depravation. Preferential utilization by livestock may increase mortality rates and decrease seed production and seedling establishment of palatable grasses, while seed production, seedling establishment, growth rates and longevity of unpalatable woody plants remains constant or increases (see Livestock Grazing section).

Established individuals may tolerate the range of environmental stresses or disturbances likely to be encountered over a series of decades or even centuries. For example, in Texas the 1950s were characterized by the most severe, continuous droughts since 1698 AD (Stahle and Cleavland 1988). Even so, *Prosopis spp.* survival in southern Texas was observed to be >60% (Carter 1964). Although some subordinate shrubs died back to ground level, they were rarely killed and regenerated vegetatively. In the Edwards Plateau region of central Texas, *Quercus pungens var. vaseyana* (shin oak) mortality during the 1950s drought ranged from 7 to 47% depending on soils (Merrill and Young 1959). Shrubs such as *Diospyros, Acacia, Zanthoxylum, Eysenhardtia* and *Berberis (Mahonia)* were unaffected, despite *Quercus virginiana* (live oak) tree mortality of 38 to 68%. Older (=large= > 2 m height) *Juniperus* plants experienced about 90% mortality, but the drought induced no mortality in younger (=smaller= < 2 m height) plants. Chihuahuan desert populations of creosotebush experienced reductions in canopy area during the 1950s drought, but density was unaffected (Wondzell 1984). Rapid and large increases in shrub cover occurred in the post-drought period. Thus, severe as it was, the drought of the 1950s does not appear to have produced significant compositional changes or local extinctions of established woody plant populations. Disturbance or climate may therefore trigger episodes of establishment of woody plants, but not necessarily their subsequent extinction (Neilson and Wullstein 1985).

For woody plants with potentially long life-spans and low post-establishment mortality rates, the seedling recruitment phase would be the most critical in dictating the long-term dynamics of community structure. When conditions conducive to seedling establishment occur infrequently and sporadically, identification of determinants of community composition is problematic. Correlations between the distribution and abundance of adult plants and environmental factors may be misleading, since factors relevant and readily studied for mature plants may be quite different from those which determine the fate of seeds or seedlings (Harper 1977:112).

Species that endure chronically or periodically adverse conditions may persist long enough vegetatively to encounter rare or infrequent windows of opportunity for successful seed production, germination and seedling establishment. Recruitment of new individuals during these unpredictable windows may be sufficient to maintain the population and partially or fully compensate for mortality since the last recruitment episode and/or enable a species to increase in density by establishing in gaps left by other species. Jameson (1987) suggested for pinyon-juniper communities that "if during the life span of trees there are several weather events leading to regeneration, a continuing tree stand should develop. On the other hand, if a stand of trees dies before the appropriate weather events leading to regeneration occur, the area of land will be occupied by trees intermittently."

Rates and patterns of woody plant establishment in grasslands would be influenced by physiological attributes of the woody species involved. *Prosopis* and *Juniperus* represent contrasting growth forms whose abilities to compete with grasses should vary markedly. *Prosopis* is a deciduous arborescent capable of N_2-fixation (Johnson and Mayeux 1990) whose seeds are widely and effectively dispersed by livestock (Brown and Archer 1987). *Prosopis* grows rapidly, has high rates of photosynthesis (Sharifi et al. 1982), and produces abundant seed. Seedlings exhibit rapid tap root development and effectively partition soil moisture with grasses soon after germination (Brown and Archer 1989, 1990, Bush and Van Auken 1991a). Within two weeks of germination, *Prosopis* seedlings can resprout following shoot disturbance (Weltzin 1990). A high proportion (>40%) of seedlings 2- to 3-years of age can tolerate temperatures associated with intense prairie fires (Wright et al. 1976). Not unexpectedly, *Prosopis* is an aggressive invader of grasslands not easily eliminated once established. In contrast, junipers are evergreens with slow growth rates, low rates of photosynthesis and possess many other features characteristic of stress tolerant plants. Many species of *Juniperus* cannot regenerate vegetatively and are highly susceptible to fire. However, despite these apparent limitations, it too has successfully invaded grasslands and savannas in recent history (Table 1). This suggests (1) woody plants possess common features that make them aggressive invaders of grass-dominated systems and/or (2) beyond certain thresholds of herbaceous retrogression or disturbance, establishment of contrasting woody growth forms is high and primarily regulated by abiotic factors.

Climatic Determinants

Ecosystem processes are ultimately constrained by climatic variables. Plant growth, microbial activity and soil development are directly influenced by radiation, temperature and precipitation. Climatic limits to the adequacy of particular combinations of leaf and plant size, architecture and longevity, as measured by positive water and energy balances, are among the most important mechanisms of environmental limitation to plant distribution and abundance (Box 1981). Climate also influences ecosystems indirectly, by causing or creating conditions conducive to natural disturbances such as fire (Clark 1988, Baisan and Swetnam 1990, Swetnam and Betancourt 1990), floods, wind-throw, and drought. On world maps, the boundaries of natural vegetation zones, soil types and climate roughly coincide (Eyre 1963, Trewartha 1968, Kuchler 1978, Walter 1979, Box 1981). Regional and global patterns of plant productivity (Rosenzweig 1968, Leith 1972, Sala et al. 1988) and soil organic matter content (Post et al. 1982) have also been related to climate. So strong are the correlations between vegetation and climate, that one is often used to infer the other.

In climate-vegetation models (e.g., Holdridge [1947, 1964] and Whittaker [1975]) grasslands (steppe) and savannas are precariously situated between shrublands, woodlands and forest with respect to mean annual temperature, rainfall and evapotranspiration. Changes or fluctuations in one or more of these variables would shift the balance between grasses and woody plants and the areal extent of their respective domination. Regions where growth forms are near their climatic fringes would be particularly sensitive to climate change or fluctuation. However, in arid lands, episodic climatic events may mask, confound, override, reinforce or negate changes ostensibly attributed to changes in mean conditions (Beatley 1974, MacMahon 1980, Chew 1982, Griffin and Friedel 1985, Lonard and Judd 1985, Harrington 1991). Changes in the frequency and/or magnitude of extreme events (either favorable or unfavorable for plant establishment) may thus be more important than gradual shifts in mean values (Katz and Brown 1992).

Prehistoric Perspective.

Changes in grass and woody plant abundance since European settlement should be evaluated relative to changes which have occurred over longer time spans, if their significance is to be appropriately gauged and their potential causes understood. Evaluating causes of vegetation change in relation to anthropogenic activities since settlement requires an assessment of the extent to which vegetation was in equilibrium with climate. Were plant communities encountered by early settlers stable and in equilibrium with their environment? When one considers the paleoecological record, the answer would be "probably not." The fossil record, glacial ice cores and historical observations suggest appreciable climatic and vegetation instability on time scales of decades to centuries to millennia. Fossil pollen and pack rat midden data indicate that Late-Wisconsin landscapes of the North American Southwest were characterized by a cool, moist climate and sagebrush and chaparral shrublands, pygmy conifer and oak woodlands, extensive pine parklands and spruce, fir and pine forests (see Martin and Mehringer 1965, Van Devender and Spaulding 1979, Smeins 1984, Betancourt et al. 1990). The warmer, drier climates of the Holocene produced our modern landscapes whose potential natural vegetation was dominated by xerophytic, hot desert shrubs, warm-temperate grasslands and savannas, subtropical thorn scrub and, at higher elevations, oak and conifer woodlands and montane and subalpine forests (Kuchler 1964). However, this overall trend towards increasing aridity was apparently epicyclic rather than gradual and linear, with woodlands, desert grasslands and shrublands alternately expanding and contracting in response to numerous climatic shifts over the past several thousand years (Van Devender and Spaulding 1979, Van Devender 1980). Shifts between grassland and shrub- or woodland communities similar to those observed in historic times thus appear to have happened naturally, apparently in response to climatic fluctuations.

Vegetation Inertia.

The extent to which shifts in vegetation structure lag behind climatic changes that drive them, and the extent to which vegetation can ever be said to be in equilibrium with climate are not easily determined (Davis 1982). Vegetation established from seed under one climatic regime may survive under a subsequent regime in a vegetative state. This phenomenon, whereby perennial plants persist for periods of tens to hundreds of years under conditions very different from those under which they initially became established, represents biological inertia (Cole 1985, Lewin 1985). In these instances, correlations between recent changes in vegetation and short-term climatic variables or disturbance regimes may be spurious or low. If the present climatic conditions are such that the dominant plants cannot successfully reestablish from seed with sufficient frequency to maintain the population, community composition is destined to change. Neilson (1986) hypothesized that "the pristine vegetation of the Chihuahuan desert recorded 100 years ago, was a vegetation established under and adapted to 300 years of 'little ice age' and is only marginally supported under the present climate." If this hypothesis is correct, climate-driven succession from desert grassland to shrubland may have been in progress at the time of settlement, and augmented by anthropogenic alteration of grazing and fire regimes.

Evidence and Proposed Scenarios for Climate-Induced Vegetation Change.
Studies that utilize long-term records of plant cover and density provide perspectives into vegetation dynamics of long-lived plants not apparent from short-term experimental manipulations. Yet because such data are descriptive and typically lack a "control" it may be difficult to determine causal mechanisms behind change and to clearly sort effects of climate from those of disturbances such as fire and livestock grazing. However, in cases where livestock grazing or fire have been eliminated or experimentally accounted for, climatic influences over grass and woody plant abundance are apparent:

- Branson (1985:16) presents data from Weaver and Albertson that indicate changes in basal area of shortgrass steppe plants associated with light, moderate and heavy livestock grazing regimes were relatively minor in comparison to reductions in cover caused by the drought of the 1930s. Clarkson and Lee (1988) made similar observations in Australia. Rainfall (current and the previous year's) accounted for 69% to 93% of the variance in plant biomass in variously grazed ranges in western New South Wales, Australia (Robertson 1988).
- In the absence of fire, savannas in the Edwards Plateau of central Texas experienced a 2- to 4-fold increase in woody plant cover between 1949 and 1983, with the greatest increases occurring on pastures protected from livestock grazing (Table 3). The relative proportions of woody growth forms changed, with oak (*Quercus spp.*) decreasing and *Juniperus* increasing on all sites (Smeins and Merrill [1988]).

Table 3. Changes in relative cover (%) of *Quercus* and *Juniperus* spp. and changes in total absolute wood plant cover (%) under different grazing regimes on the Edwards Plateau, Texas (from Smeins and Merrill 1988).

Grazing Regime:	Continuous		Rotational		Exclosure	
Year:	1949	1983	1949	1983	1949	1983
Relative Cover:						
Quercus spp.	89	41	90	50	93	41
Juniperus spp.	7	40	4	39	3	32
Total Woody Cover	14	30	10	30	8	35

- Turner (1990) charted two *Prosopis* plants on permanent plots at a Sonoran Desert site inaccessible to domestic livestock and woodcutters in 1960. By 1982 the number of *Prosopis* plants had increased to 186. Most of the recruitment occurred in the early 1970s in conjunction with unusually large rainfall events. In other plots on the site, creosote bush populations declined 50-90% during the first half of this century with little or no recruitment since.

- In the absence of fire, woody plant encroachment occurred in grasslands characterized by a wide range of herbaceous standing crop, composition, density and/or a wide range of livestock grazing histories, suggesting that climate and soils are capable of supporting woody vegetation including *Juniperus, Artemisia* and *Prosopis* (Burkhardt and Tisdale 1969, Bragg and Hulbert 1976, Tisdale and Hironaka 1981, Young and Evans 1981, Johnson 1987, Smeins and Merrill 1988, Brown and Archer 1989 and references therein).

- Increased abundance of woody plants in grasslands and savannas has been geographically widespread (Table 1) and relatively synchronous, suggesting climate as a causal or contributing factor.

What climate-based scenarios might lead to the demise of grasses and the rise of shrubs such as *Larrea, Juniperus, Acacia* and *Prosopis*? Some combination of the following might elicit a shift from grassland to shrub- or woodland:

- *Increased rainfall.* Walter (1979:92) proposes a scenario for the role of annual precipitation in controlling the mixture of grasses vs. woody plants (Fig. 2). In addition, discrete pulses of high rainfall may trigger episodes of woody plant establishment. In years when surficial soil moisture is abundant, grasses may not competitively exclude woody plant seedlings and an opportunity for root development to exploit deeper and more abundant soil resources may be presented. Once established, woody plants may coexist with grasses in a dynamic equilibrium (tree or shrub savanna) by partitioning soil moisture (Knoop and Walker 1985, Stuart-Hill and Tainton 1989, Sala et al. 1989, Brown and Archer 1990). Alternatively, established woody plants may redirect successional processes and initiate a series of positive feedbacks (allelopathy, shade, soil resource depletion, decreased fire frequency) which eliminate grasses and enable development of shrublands or woodlands (Petranka and McPherson 1979, Archer et al. 1988, McPherson et al. 1988).

  Increased rainfall may also induce shifts from woody plant to herbaceous domination. In the Great Basin of the western United States, extensive die-offs of suffrutescent salt desert shrubs such as shadscale (*Atriplex confertifolia*) have occurred during and following successive years of above-average precipitation (Pyke and Dobrowolski 1989). Factors potentially contributing to shrub mortality included increased soil salinity and anaerobiosis in the root zone, iron deficiencies associated with increased soil bicarbonate concentrations, outbreaks of parasitic dodder (*Cuscuta spp.)* and soil-borne disease organisms (McArthur et al. 1990). Extended periods of inundation in lowlands of North American Coastal Prairie may periodically eliminate stands of *Aca-*

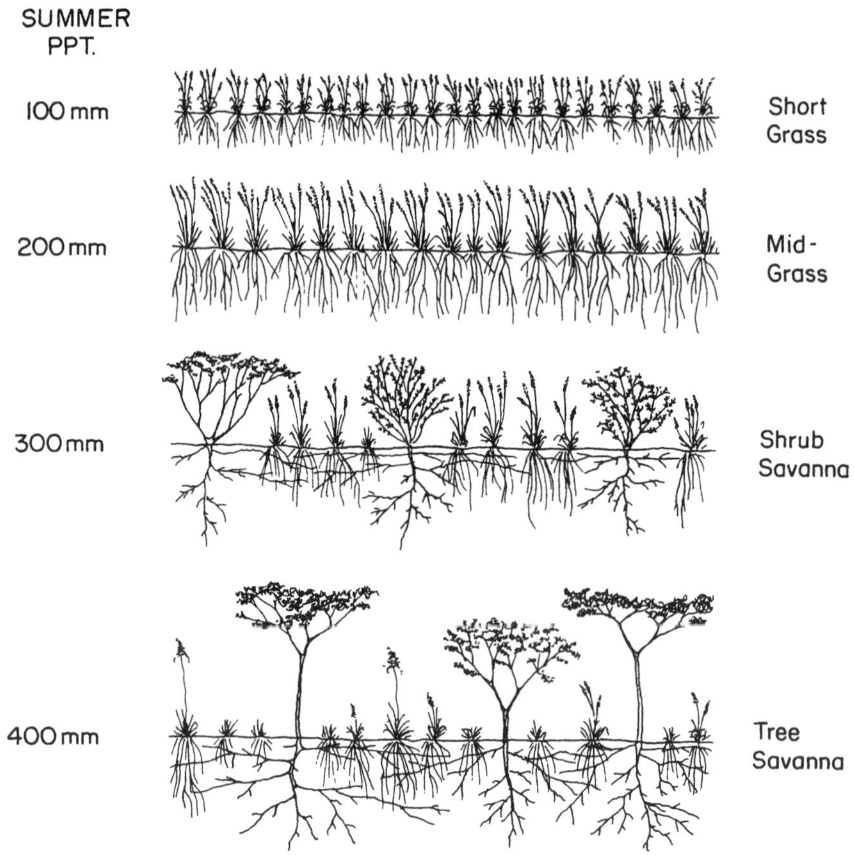

Figure 2. Schematic transformation from grassland to savanna to dry woodland with increasing annual rainfall (from Walter 1979).

cia, Prosopis and Quercus spp. (Scifres and Mutz 1975). In the Amboseli Basin of East Africa, Western and Van Praet (1973) attributed widespread mortality of *Acacia zanthophloea* to salinity associated with elevation of the water table over a series of high rainfall years.

- *Periodic drought.* During drought periods, grass cover, biomass and density decline, reducing the probability of fire and creating unoccupied gaps whose size and abundance would vary with soils and the duration and magnitude of drought (Nelson 1934, Paulsen and Ares 1962, Herbel et al. 1972). Woody plants may preferentially establish in some proportion of these gaps in subsequent pluvial periods (a function of seed bank and seed rain) and persist. Periodic drought may thus set the stage for subsequent episodes of woody plant establishment and canopy development (Fig. 3) (Herbel et al. 1972, Archer et al. 1988, Harrington 1991).

- *Decreased rainfall.* If conditions become chronically drier, grass mortality may increase while seed production and seedling establishment decrease and

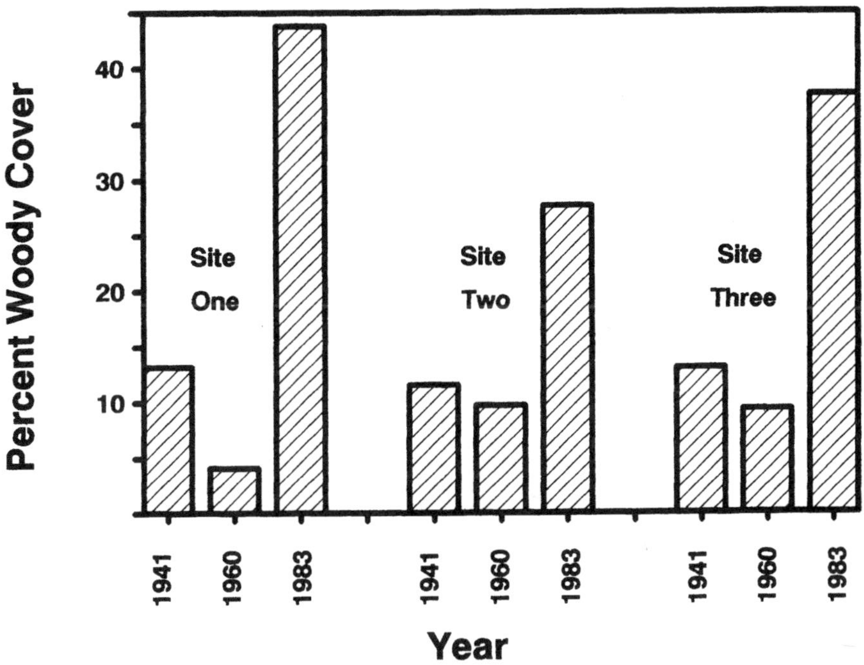

Figure 3. Changes in woody plant cover, 1941-1983 in southern Texas (from Archer et al. 1988). The 1941-1960 period was characterized by the severe drought of the 1950s; the 1960-1983 period received normal to above-normal annual rainfall.

the seed bank deteriorates. Resulting gaps may be colonized by woody plants that establish during moist periods and persist by tolerating or avoiding drought stress.

- *Shift in seasonality of rainfall.* Growth of woody plants in many arid and semi-arid systems is favored by cool season moisture, whereas growth and establishment of grasses (especially C_4's) is favored by warm season rainfall (Fritts 1976, Soriano and Sala 1983, Neilson 1986). Similarly, summer droughts may be more detrimental to grasses than woody plants, especially in years where there has been winter rain. When precipitation coincides with temperatures suitable for physiological activity, grasses may utilize soil moisture in upper horizons that might otherwise percolate to deeper depths for use by more deeply rooted woody plants; when precipitation occurs during the dormant or cool season, the soil wetting front may penetrate to deeper depths, and effectively uncouple woody plants with deep root systems from soil moisture conditions in upper horizons. Shifts in seasonality of rainfall and temperature over the past century may have contributed to the expansion of shrublands into southwestern grasslands (Hastings and Turner 1965, Neilson 1986). Shifts from summer to winter precipitation predicted to occur as a result of atmospheric CO_2 enrichment (Mitchell and Warrilow 1987) may increase susceptibility of existing grasslands to woody plant encroachment.

- *Shift in size class distribution of precipitation events.* When rainfall occurs as small, frequent events, grasses, with their opportunistic growth habit would be favored, particularly on fine-textured soils (Sala and Lauenroth 1982, Soriano and Sala 1983, Sala et al. 1989, Brown and Archer 1990). More deeply rooted woody plant growth forms would be favored by larger events, especially on coarse textured soils, where moisture may percolate to deeper depths and accumulate beyond the rooting zone of grasses. In addition, soils associated with woody plants typically have higher infiltration rates than soils associated with bunchgrasses and shortgrasses of interstitial zones (Pressland 1973, Thurow 1991) and deep percolation near shrubs may be facilitated via channels created by death of coarse roots. Larger rainfall events may therefore enhance moisture content of soils associated with woody plants, by influencing the lateral distribution of water between mosaic constituents (i.e., surface runoff and interflow from interstitial patches = runon to woody plant patches w/ greater infiltration and deep percolation) (Parsons et al. 1992).
- *Increased temperature.* Seasonal heat stress would be expected to increase, in conjunction with increased potential evapotranspiration and decreased plant water use efficiency. Stress tolerant shrubs would be more likely to persist under such conditions and potentially expand to occupy sites vacated by mesophytic grasses. Emanuel et al. (1985a, b) have compared contemporary maps of the distribution of Holdridge Life-Zones with maps resulting from changes in mean annual temperature increments predicted by models of climate under elevated CO_2. Their exercise provides an initial test of the sensitivity of the distribution of major vegetation associations to temperature change when other environmental factors were held constant. In tropical and subtropical grasslands, acreages of forest and woodland associations were predicted to increase 13 to 38%, while desert bush acreages in tropical, subtropical and cool temperate regions increased 30 to 67% (Table 4).

There is some evidence of a slight tendency toward increased winter precipitation, higher summer temperatures and an overall warming trend in the Northern Hemisphere from the late 1800s through the 1940s and extended regional droughts in the 1770s, 1860s, 1930s and 1950s (Hastings and Turner 1965, Bryson 1974, Stockton and Meko 1975, Norwine 1978, Mitchell 1980, Neilson 1986, Stahle and Cleavland 1988). These climate-related factors may have contributed to changes in distribution and abundance of grasses and woody plants on some landscapes in the last 100-200 years.

Edaphic Determinants

Lowe and Brown (1982), in mapping the biotic communities of the Southwest, regarded desertland and grassland as the two "base" formations, encompassing more than 70% of the land area. They further noted that while montane and piedmont communities were subject to strong edaphic control, Great Basin and Chihuahuan desertscrub and semi-desert grassland communities were difficult to delineate as they often intergraded over broad areas. Slight changes in elevation, slope exposure, available soil moisture may therefore influence the local patterns

Table 4. Predicted changes in the areal extent of woody plant associations within tropical and subtropical grasslands and desert shrublands associated with changes in mean annual temperature anticipated to result from atmospheric CO_2 enrichment (from Emanuel et al., 1985a, b).

	AREAL EXTENT (X 10^6 KM^2)		
LIFE ZONE	Base	Predicted	% CHANGE
Tropical Grasslands			
Very Dry Forest	4.7	6.3	+34
Thorn Woodland	2.4	3.3	+38
Subtropical Grassland			
Thorn Woodland	1.7	2.1	+24
Thorn Steppe	5.2	5.9	+13
Desert Bush			
Tropical	2.3	3.0	+30
Subtropical	1.5	2.5	+67
Warm Temperate	4.9	4.7	–4
Cool Temperate	3.6	5.0	+39
Boreal	1.3	2.5	+93
Total	27.6	35.3	+28

of grass versus woody plant distributions. Annual amounts of precipitation and seasonality of rainfall interact with soil texture and depth to influence the relative abundance of grasses vs. woody plants on a site (see Belsky 1991).

The spatial pattern and areal extent of life-form composition shifts from grass- to woody plant-domination (Table 1) is difficult to estimate without knowing something of the pre-settlement distributions. The major woody elements encroaching into grasslands in the North American southwest (e.g., *Prosopis, Larrea, Juniperus, Pinus, Opuntia*) have been present throughout the Holocene and historical accounts (see York and Dick-Peddie 1969, Gross and Dick-Peddie 1979, Malin 1953, Inglis 1964), photographic evidence (see Hastings and Turner 1965, Martin and Turner 1977, Humphrey 1987, Nelson and Beres 1987) and long-term records from permanent plots (Goldberg and Turner 1986, Turner 1990, Wondzell 1984) suggest that shrub and woodland communities may have characterized certain landscape units at the time of settlement (e.g., Fig. 1a, c; Table 2). It seems reasonable to expect that (1) patchy, discontinuous mosaics of grass vs. shrub-or tree-dominated vegetation existed at the time of European settlement, their distributions dictated primarily by topography, soils and fire; and (2) these patches served as sources of population expansion subsequent to changes in climate and/or disturbance regimes. For example, Johnston (1963) argues that the geographic range of honey mesquite (*Prosopis glandulosa*) has changed little over the past 300-500 years and that this plant has merely increased in stature and abundance within its historic range. Some landscapes of southern Texas presently dominated by *Prosopis*-thorn woodlands were described in the early 1800s as open prairies with woody plants confined to areas adjacent to watercourses, suggesting the spread of woody vegetation has been lateral from lowlands and drainages to uplands; in other regions of Texas, *Prosopis* may have been a natural, historical component of upland landscapes (Table 2)(Bogusch 1952, Inglis 1964, Tharp 1926), its density within those landscapes increasing in recent decades. Woody xerophytes such as *Larrea* and *Juniperus* may have historically inhabited shallow, rocky, calcareous erosional sites such as hilltops and ridges prone to runoff, while transport and depositional sites with deeper soils receiving runoff may have supported mesophytic grassland vegetation (Gardner 1951, Hallmark and Allen 1975, Stein and Ludwig 1979, McCraw 1985).

Topography, Texture and Geomorphology

Topography and soil texture, depth, fertility and chemistry can influence patterns of grass and woody plant distribution. Moisture may accumulate and depth to water table may be reduced on intermittent drainages, arroyos and other landscape "run-on" sites, thus favoring woody vegetation. Coarse-textured soils that permit greater infiltration and deeper percolation may enable water to accumulate beyond depths effectively exploited by grasses. These deeper stores of moisture would enable woody plants, with their generally deeper and more extensive root systems, to co-exist with grasses or dominate such sites. Finer-textured soils that retain moisture and nutrients in the upper soil layers generally favor grasses with high root densities in upper horizons by limiting water penetration to lower horizons (see Streng and Harcombe 1982, Johnson and Tothill 1985, Davis and Mooney 1985, Knoop and Walker 1985, Liang et al. 1989, Brown and Archer 1990).

Rates and patterns of woody plant encroachment into grasslands is influenced by topography and texture (e.g., Bragg and Hulbert 1976, McPherson and Wright 1988). Soil texture can also influence patterns of grass and woody plant survival. During the extended and severe drought of the 1950s, Carter (1964) observed highest rates of *Prosopis* mortality on fine-textured soils and lowest mortality on gravelly sites in southern Texas. During this same period, Herbel et al. (1972) observed reductions in grass cover to be greatest on deep, sandy soils and least on soils where cemented caliche layers were relatively shallow. Where topoedaphic conditions are not conducive to development of herbaceous vegetation (extremely shallow and/or rocky, gravelly soils), fine fuels required to initiate or carry a fire would also be lacking. Woody plants inhabit such sites by exploiting fissures and crevasses, thereby accessing resources accumulating beneath such barriers.

Depth to structural barriers such as ironstone or caliche layers can also influence the distribution and abundance of woody vs. graminoid life-forms. Grasses may monopolize resources when soils are present but relatively shallow, with woody plants dominating sites where geomorphic barriers are relatively deeper or where their roots exploit fissures, crevasses or subterranean catchments were water and nutrients might accumulate (Morison et al. 1948, Walter 1979, San Jose and Farinas 1983). Many upland landscapes in central and southern Texas support discrete clusters (mottes) of woody vegetation embedded within a matrix of grasses (Whittaker et al. 1979, Amos and Gehlbach 1988). It appears that these striking contrasts in life-form distributions over small spatial scales are related to subterranean characteristics. For example, field investigations on cluster-soil relationships in southern Texas have revealed that laterally extensive argillic horizons contain inclusions of coarse-textured, non-argillic soils. Where the argillic horizon was present, the vegetation was dominated by grasses interspersed with small woody plant clusters, typically with a single *Prosopis* plant and several subordinate shrubs. Vegetation of non-argillic inclusions consisted of large groves with numerous *Prosopis* trees and subordinate shrubs (Fig. 4c). The favorability of these non-argillic inclusions for woody plants is demonstrated by the fact that *Prosopis* plants in groves were larger and had mean radial growth rates more than twice that of *Prosopis* plants growing on argillic soils (Table 5). In addition, the size and shape of groves coincided with that of the non-argillic inclusions suggesting edaphic constraints to shrub cluster expansion and coalescence (Loomis 1989).

Soil Nutrients

Grasses generally have high densities of fine roots concentrated in upper soil horizons, whereas many dryland trees and shrubs have lower root densities and coarser, deeper root systems. These contrasting rooting patterns suggest grasses may have a competitive advantage for acquisition of surficial soil resources. However, even

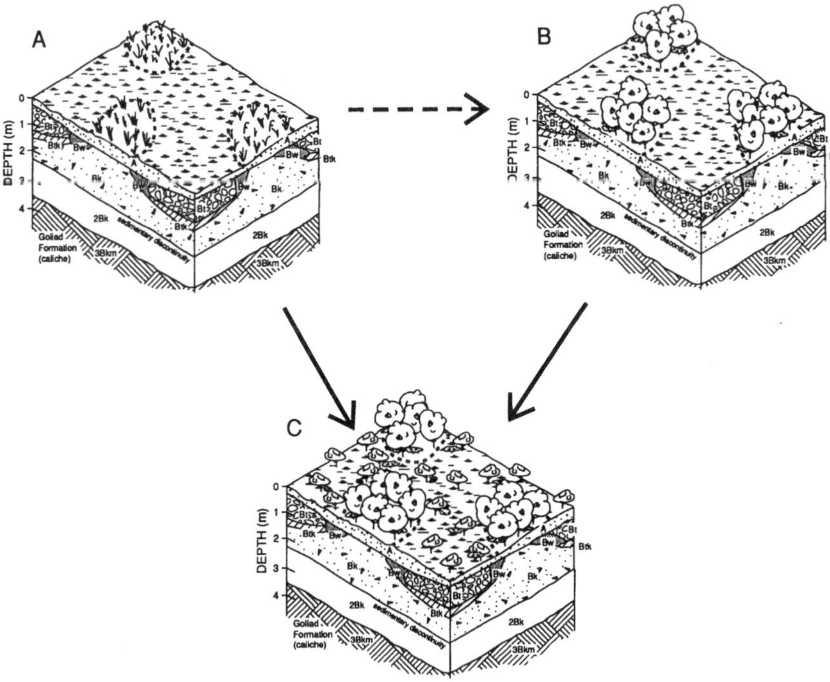

Figure 4. Hypothesized role of subsurface soil texture and grazing in regulating tree (*Prosopis*)-grass distribution in a savanna parkland Texas (Archer, unpublished). The argillic (B_t) horizon on uplands is laterally extensive except where groves occur (Loomis 1989). The savanna parkland physiognomy described at the time of settlement (A or B, Table 2) may have been edaphically regulated. With advent of livestock grazing transpirational grass leaf area and root biomass would have decreased enabling shrubs to establish. In addition, *Prosopis* seed dispersal would have increased and fire frequency would have decreased. Available data suggest mottes and groves began forming about the same time, with plants on the non-argillic inclusions developing faster (Table 5).

Table 5. Mean (+ SD) size of *Prosopis glandulosa* stems on upland landscapes in southern Texas. *Prosopis* plants in discrete clusters occurred on soils underlain with a distinct argillic (clay) horizon. This horizon was absent within groves (Fig. 4) (Archer and Flinn, unpublished data).

	Discrete Clusters	Groves
Canopy Diameter (m)	6.5 (1.1)	7.2 (2.2)
Basal Diameter (cm)	15.7 (2.4)	24.1 (6.2)
Height (m)	3.7 (0.6)	5.8 (0.7)
Radial Trunk Growth[a]		
(mm/month)	0.04 (.03)	0.10 (.07)
Trunk Age[b]		
Mean	49.3 (11.9)	74.6 (5.5)
Max	64	81

[a] From dendrometer bands over a 2 year period
[b] From annual ring counts (Archer and Flinn 1991, Flinn et al. 1991). Ages *not* adjusted for time required for stems to produce rings at height of cutting; ages are thus underestimated by ca. 5-10 years.

grasses that share similar characteristics in terms of canopy architecture, shoot phenology and root distribution may differ substantially in their ability to compete with woody plants for water and nutrients (Fig. 5) (Eissenstat and Caldwell 1988). Higher cation exchange capacities of shrub roots (mean = 33.5 meq/100g based on survey of 11 species from 9 genera) in comparison to grass roots (mean = 14.7 meq/100g based on survey of 9 species from 7 genera) (Woodward et al. 1984) may also reduce differences in nutrient absorption that might otherwise be expected from differences in growth form root densities. N_2-fixation may also enable Leguminous or Rhamnaceous trees and shrubs to exploit low fertility sites more successfully than grasses (Klemmedson 1979, de Faria et al. 1989, Hogberg 1989, Johnson and Mayeux 1990). Adaptations for nutrient conservation and integrated nutrient use efficiency by evergreen woody plants (slow growth rates, low nutrient demands, low maintenance costs, sclerophylly, low leachability of foliage, low rates of leaf turnover, effective resorption of nutrients from senescing leaves) (Gray 1983, Gray and Schlesinger 1983, Merino et al. 1984, Aerts 1989) may give these growth forms an advantage over deciduous growth forms (Chapin and Shaver 1989, Schlesinger et al. 1990) on nutrient poor sites. Relative to grasses, woody growth forms may thus be better suited to shallow or low fertility sites, or to sites where erosion associated with grazing or other disturbances has reduced fertility. However, other factors held equal, these adaptations to nutrient and water stress tolerance may limit the ability of evergreen trees and shrubs to effectively compete with grasses and other deciduous growth forms on more mesic and fertile sites (Fig. 6) (Grime 1977, 1979; Chapin 1980, Leps et al. 1982, McGraw and Chapin 1989).

Some studies suggest growthform responses to nutrient availability are inversely related to rooting depth and plant longevity (Gray and Schlesinger 1983, Witowski 1989). In tundra communities, the growth response to nutrient additions is generally in the following order: graminoids > deciduous shrubs > evergreen shrubs (Lechowicz and Shaver 1982, Henry et al. 1986). Herbaceous growth forms, with their intercalary and basal meristems, are more plastic in their morphological

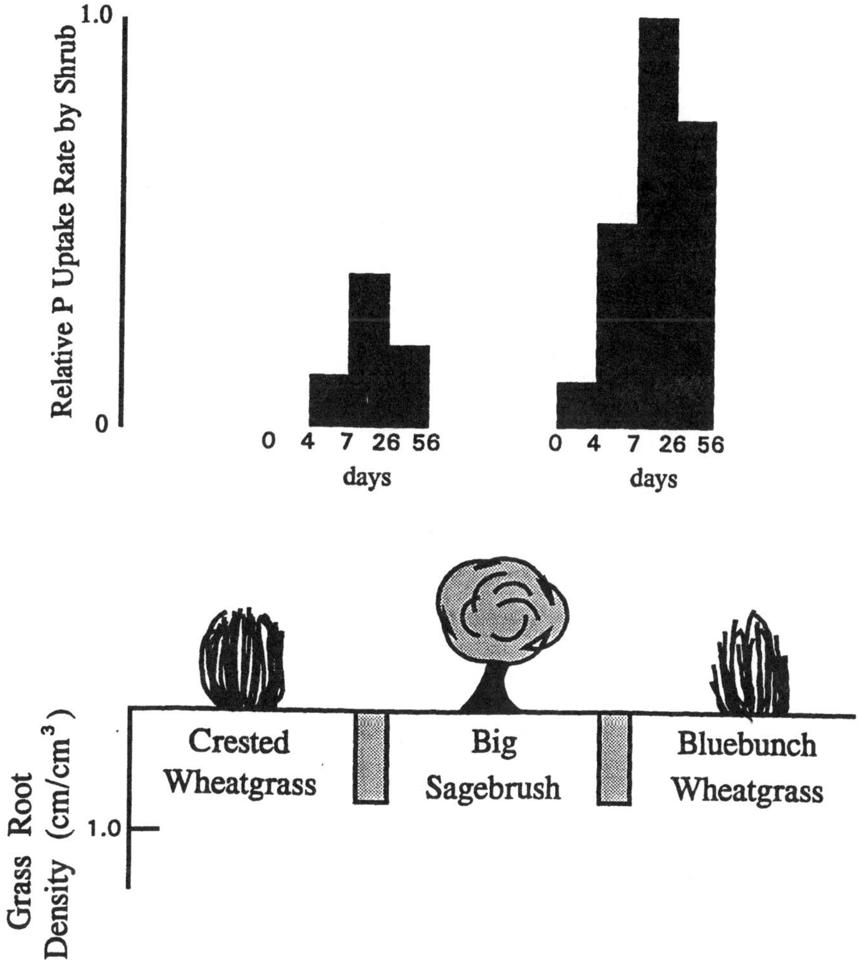

Figure 5. Phosphorus uptake by big sagebrush growing in association with crested wheatgrass vs. bluebunch wheatgrass (from Caldwell et al. 1985). Sagebrush plants acquired significantly more P from bluebunch wheatgrass neighborhoods than from crested wheatgrass neighborhoods, even though these grass species were strikingly similar with respect to root density, shoot phenology and canopy architecture.

growth responses to nutrient and moisture additions than are shrubs and arborescents (Sala and Lauenroth 1982, Soriano and Sala 1983, Knoop and Walker 1985, Aerts 1989, Witowski 1989). Differential life-form growth responses to nutrient availability would ultimately be expressed in community composition. Nutrient additions to Australian heathlands over a period of 20 years produced a shift in species and growth form composition from predominantly sclerophyllous evergreen shrubs to herbaceous species (Heddle and Sprecht 1975, Sprecht et al. 1977). Application of N and N+P to Californian chaparral resulted in increased proportions of annuals and graminoids (McMaster et al. 1982).

Given the differences in plant growth form response to nutrient availability, the kinds of woody plants (e.g. evergreen vs. deciduous) that replace grasses may

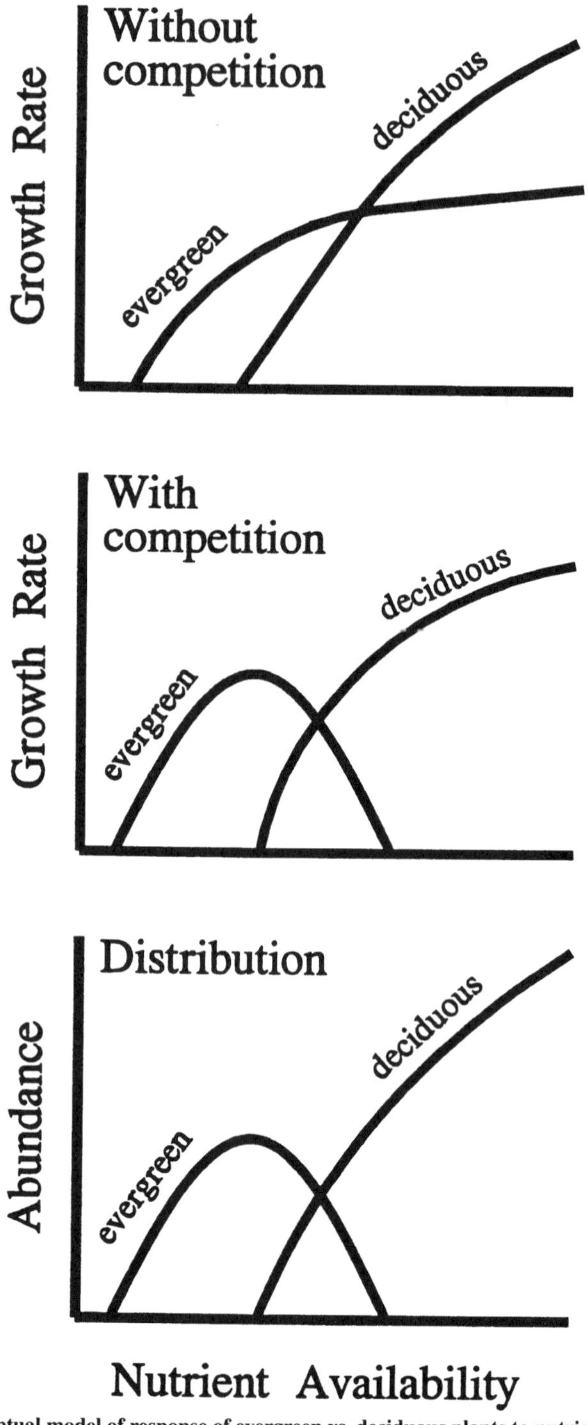

Figure 6. Conceptual model of response of evergreen vs. deciduous plants to nutrient availability (from Goldberg 1982).

depend upon site fertility characteristics (Fig. 6). In addition, the likelihood of succession from grassland to shrub- or woodland, or the reverse, may depend upon the extent to which site fertility and chemistry has been affected by grazing or by other disturbances.

Pyric Determinants

It is commonly assumed that fire is a primary determinant in the creation and/or maintenance of grasslands and savannas. Fires caused by lightning are a potentially frequent occurrence and the active and accidental uses of fire by aboriginals is well-known. However, for such fires to occur, continuity of fine fuels (grasses) must exist, i.e., in order to have a prairie fire, you must first have a prairie (Gleason 1913). This argues for climatic and/or edaphic controls over physiognomy in some systems, with fire playing a secondary role in shaping community structure (Walter 1979:76, 91; Wright and Bailey 1982:82). The extent to which fire occurred in southwestern grasslands is geographically variable and related to seasonal and annual rainfall, elevation, slope aspect and inclination. Fire may have been rare in desert grasslands and limited in areal extent, owing to erratic topography and low biomass and continuity of fine fuels (York and Dick-Peddie 1969, Hastings and Turner 1965). In more mesic grassland and savanna systems where fire was a prevalent and recurring force, pre-historic frequency and intensity appear to have been regionally synchronized by climatic conditions (Swetnam and Betancourt 1990). However, fire frequencies have decreased since settlement, a result of fine-fuel removal by livestock grazing, cessation of ignition by Native Americans and active fire suppression (Young and Evans 1981, Madany and West 1983, Arno and Gruell 1986, Baison and Swetnam 1990, Savage and Swetnam 1990). Reductions in fire frequency may have enabled fire-sensitive woody species to invade and establish in grasslands and savannas and/or may have enabled fire-tolerant, but suppressed woody species to assume greater dominance and subsequently redirect successional processes by altering soils and microclimate. In cases where woody plants suppress herbaceous production and/or create discontinuities in fine fuel distribution, a positive feedback may develop, whereby the propensity for fire is reduced as additional woody plants establish and assert dominance. On many shrub- and woodlands, the potential for ignition and spread of fire is low, owing to woody plant-induced discontinuity in fine fuel (e.g., McLaughlin and Bowers 1982).

Temporal patterns of fire and soil moisture are the primary factors influencing woody plant-grass biomass ratios in the semi-arid woodlands of Australia (Harrington and Hodgkinson 1986) . Under the influence of above-average rainfall in critical seasons, shrub seedling establishment is high (Harrington 1991). However, because such rainfall events also produce an unusually large grass biomass, fire becomes a possibility which, if realized, kills most juvenile shrubs. Grazing may increase the frequency of woody plant seedling establishment by reducing the soil moisture utilized by grasses and by reducing fire frequency. These two factors combine to favor vegetation dominated by woody plants.

With a sufficient fire-free interval, woody plants can eventually overtop grasses and persist despite subsequent fires (e.g., Bragg and Hulbert 1976). The length of the fire free period required for a tree or shrub to reach sufficient size or age to tolerate fire varies among species. Western juniper (*Juniperus occidentalis*) plants less than 50 years old are quite susceptible to wildfires (Burkhardt and Tisdale 1969). In contrast, survival of 2- and 3-year old honey mesquite seedlings exposed to flame temperatures approximating that of hot grass fires exceeded 60 and 90%, respectively (Wright et al. 1976). Only short fire-free intervals would thus be required for its establishment. Recurring fire may have kept *Prosopis* plants low in stature in pre-settlement grasslands, but would not necessarily have eliminated them from a site. Plants thus suppressed might persist for long periods, then enter a rapid phase of aboveground growth (Fig. 7) and overtop the herbaceous vegetation and achieve heights less sensitive to subsequent fire. Trunks and crowns of taller plants may incur minimal damage from surface fires, or, if crowns and trunks are killed, rapid vegetative regeneration from crowns, burls or roots occurs for many species. Following fire, recovery of shrubs may equal or exceed that of perennial grasses (Cornelius 1988). Prehistoric wildfires, which might typically occur at the end of the growing season when dry litter has accumulated and air temperatures are high, may have been more effective in causing woody plant mortality than prescribed fires implemented during cooler periods. However, reductions in grass production, basal area and survival following late-summer fires (Ewing and Engle 1988) may also generate windows of opportunity for invasive species.

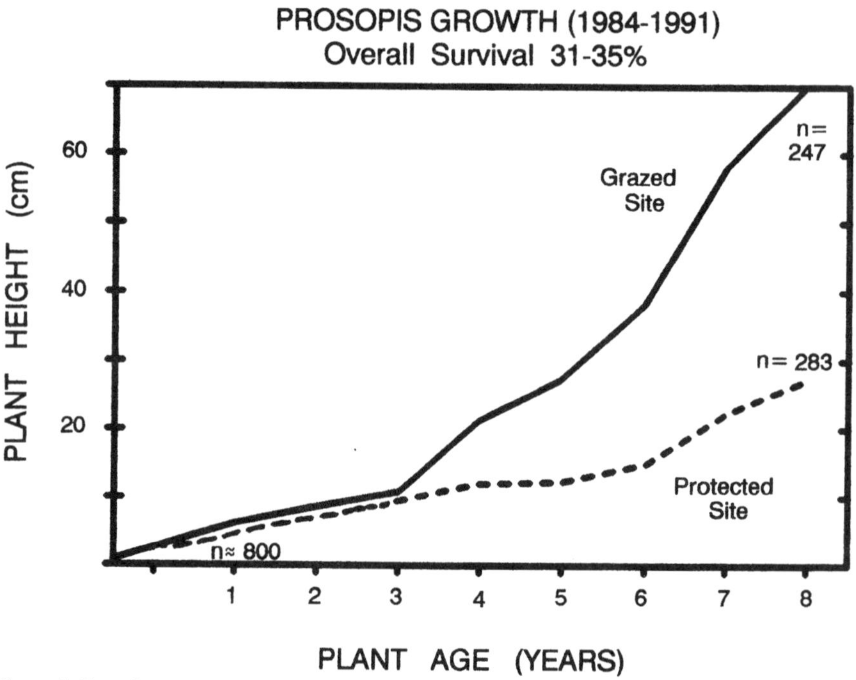

Figure 7. Recruitment and growth of *Prosopis glandulosa* seedlings on sites with contrasting long-term grazing histories and differing in herbaceous biomass production and composition (Archer, unpublished). See Brown and Archer (1989) for details on site characteristics and emergence and survival over first two years. About 800 seedlings emerged on each site; after 8 years, survival was highest on the site protected from grazing (283 vs. 247 plants) but growth rates were greatest on the grazed site. Note lack of shoot development for first 3-4 years on both sites.

Herbivore Determinants

Changes or fluctuation in broad-scale climatic regimes or atmospheric CO_2 enrichment in recent history cannot account for small-scale vegetation patterns and cannot explain why grasslands and savannas have persisted on some landscapes within a region, but not others. As spatial and temporal frames of observation are diminished and resolution increased, edaphic heterogeneity and disturbance assume greater importance in determining vegetation structure (Prentice 1986). The utilization of plants by animals is superimposed on a backdrop of topoedaphic heterogeneity and climatic variability to influence ecosystem processes. As a result, plant species whose adaptations to the prevailing climate and soils would make them competitive dominants of the community when herbivore populations are low may assume subordinate roles or even face local extinction as levels of herbivory increase.

Native Herbivores

Numerous examples exist to illustrate the role of vertebrate and invertebrate herbivores in regulating vegetation structure, composition and productivity, (see special issue of BioScience 38 (11) [1988] on "How Animals Shape Their Ecosystems"). Specific examples illustrate how browsing mammals can maintain grasslands, meadows and savannas and/or change closed woodland, thickets or heathland into open, grass-dominated systems (see Crisp and Lange 1976, Sinclair and Norton-Griffiths 1979, Hunter et. al. 1980, Belsky 1984, Berdowski 1987, Yeaton 1988, Cantor and Whitham 1989). Patterns of Holocene vegetation development subsequent to extinction of megafauna (Martin 1975, Lundelius 1976) were potentially much different from those which would have occurred had these herbivores persisted. The conversion of open, park-like woodlands and mosaic grasslands of the Pleistocene to the more uniform forests, shrublands and grasslands of today may, to some extent, reflect faunal extinctions and alteration of browsing and seed dispersal influences (Janzen 1986, Owen-Smith 1987).

Differential levels of grazing, browsing, and granivory on grasses and woody plants by various classes of herbivores can be a locally important control over vegetation composition. However, the influence of various classes of herbivores (e.g., root-feeding nematodes, grasshoppers, termites, rodents, lagomorphs, jackrabbits) on vegetation relative to that of more conspicuous ungulate herbivores is difficult to assess. When taken into account, the influence of native herbivores on vegetation can be pronounced. For example, in the 12 years following removal of kangaroo rats (*Dipodomys* spp.) from a Chihuahuan Desert site in southeastern Arizona, the cover of tall grasses increased (Fig. 8), bare ground decreased, litter accumulated and snow cover persisted longer relative to plots where kangaroo rats remained (Brown and Heske 1990). Exclusion of livestock only produced no significant or detectable change in vegetation during the same period. The herbaceous vegetation maintained by kangaroo rats (shortgrasses, annuals) may be less effective at competitively excluding woody vegetation. In addition, kangaroo rats may be an important agent of woody plant seed dispersal (Reynolds and Glendening 1949, Reynolds 1954). Changes in vegetation ostensibly attributed to livestock may thus be similarly effected by activities of less conspicuous rodent herbivores.

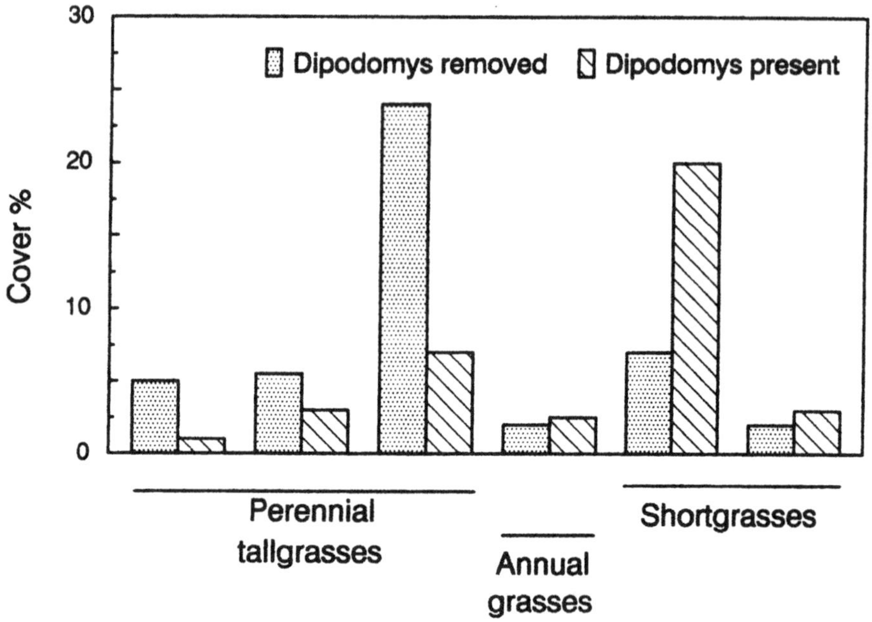

Figure 8. Differences in perennial tall and shortgrass cover on Chihuahuan Desert sites with kangaroo rats (*Dipodomys* spp.) and after 12 years of *Dipodomys* exclusion (from Brown and Heske 1990).

Prairie dogs (*Cynomys spp.*) represent another example of a once wide-spread native herbivore whose activities are known to influence grassland patch structure, nutrient cycling and feeding-site selection by other herbivores (Whicker and Detling 1988 a, b). However, their role in regulating the distribution and abundance of trees and shrubs in grasslands has not been widely considered. Available data (Weltzin 1990) suggests that on landscapes otherwise suited for woody plants, recruitment of trees and shrubs would be minimal when prairie dogs were present (Fig. 9). Spatial/temporal variation in prairie dog distribution may help explain inconsistencies in historical accounts of woody plant distribution and abundance. Elimination of prairie dogs from landscapes by natural (drought, famine, disease) or anthropogenic (poisoning) causes would remove a primary and locally pervasive mortality factor and either release suppressed populations of woody plants or make new habitats available for their colonization.

Livestock

Utilization of grassland vegetation by domesticated and feral livestock in southwestern North America dates back to the 1500s (Lehman 1969, Wagoner 1949, Hastings and Turner 1965, Jordan 1981, Hanselka and Kilgore 1987, Wagner 1989, Robinett 1990). By the early 1900s, uncontrolled grazing by horses, sheep, cattle and burrows had so degraded vegetation and soils that federal legislation (the Taylor Grazing Act of 1934) was enacted in an attempt to curtail further deterioration (Stoddart et al. 1975). Scenarios paralleling those of the North American Southwest occurred in Australia (Harrington et al. 1979) and South America (Schofield

Figure 9. (A) Landscape in the Rolling Plains of northern Texas illustrating a grassland maintained by prairie dogs (*Cynomys ludovicianus*) and a *Prosopis* shrubland off-colony. Seedlings and transplanted *Prosopis* saplings were eliminated on-colony, unless protected by exclosures (arrows denote transplanted saplings). Seedlings and sapling planted off-colony had high rates of survival both in and out of exclosures. (B) A prairie dog colony on this landscape was poisoned out in 1950. Aerial photographs from the late 1940s showed this site to be grassland with scattered *Prosopis* trees. By the time of this photograph (1989) a *Prosopis* woodland had developed. See Weltzin 1990 for additional details. (Photo by S. Archer).

and Bucher 1986, Bucher 1987). The widespread invasion of woody plants into grasslands, "thicketization" and increases in their stature and density in savannas (Table 1) typically coincide with the development of the livestock industry and intensification of grazing.

Grasslands of the Great Plains and southwest evolved with grazing animals. Yet, in contrast to native herbivores whose numbers or patterns of grazing may vary widely, concentrations of domestic livestock can be artificially maintained at consistently high levels by supplemental feeding and watering and by protection from natural predation and disease. Fences prevent migration to new areas when the abundance of preferred forages decreases, resulting in higher frequencies and intensities of defoliation and maintenance of grazing pressure over a greater portion of the year and over a higher frequency of years than might otherwise occur. The end results can be radical changes in species composition and significant soil erosion that are not reversible over time-frames relevant to management. The impacts of domestic herbivory on plants (see Briske 1991, Briske and Richards, this volume) and ecosystems (see Archer and Smeins 1991, Thurow 1991, Skarpe 1991, Pieper, this volume) have been recently reviewed. In this section, factors contributing to the replacement of grasses with woody plants are considered.

Preferential utilization of grasses variously tolerant of defoliation alters plant competitive interactions in a community and changes patterns of resource distribution and availability. Grazing by livestock can potentially increase the probability of woody plant recruitment in numerous, self-reinforcing ways:

- Livestock may effectively disperse woody plant seeds, particularly those of some leguminous shrubs and arborescents (see Brown and Archer 1987).
- Increased light levels at the soil surface could increase chances for germination and early establishment of woody seedlings.
- Concomitant reductions in transpirational leaf area, root biomass and root activity associated with grazing of grasses can
  - increase surficial soil moisture (Archer and Detling 1986) to enhance woody seedling establishment and growth of shallow-rooted woody species
  - increase the amount of water percolating to deeper depths and benefit established woody species with deep root systems
  - increase nutrient availability to woody plants (Fig. 10)
  - "release" suppressed populations of established tree or shrub seedlings (e.g., Merz and Boyce 1956, Harper 1977:634, Grubb 1977, Hara 1987) (Fig. 7)
- Grazing decreases plant basal area, increases mortality rates and decreases seed production and seedling establishment of palatable grasses (e.g., O'Connor 1991, O'Connor and Pickett 1992). Grazing may also increase susceptibility of grasses to other stresses such as drought (Nelson 1934, Paulsen and Ares 1962, Herbel et al. 1972, Clarkson and Lee 1988). These factors would combine to increase the rate of gap formation and available area for woody plant seedling establishment, especially in post-drought periods.

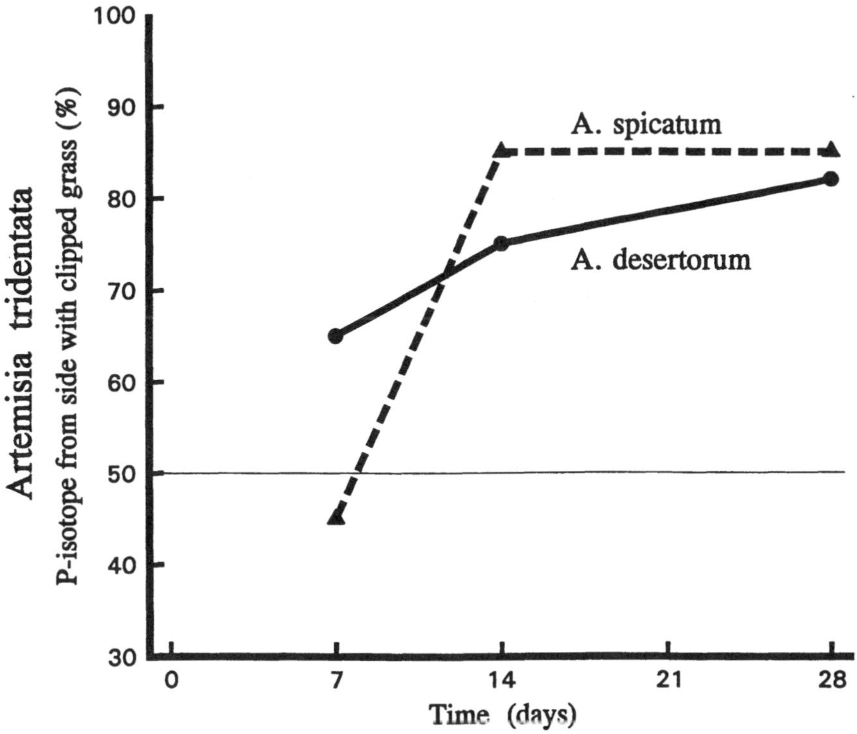

Figure 10. Phosphorus uptake by *Artemisia tridentata* plants with defoliated vs. non-defoliated *Agropyron spicatum* and *A. desertorum* neighbors (from Caldwell et al. 1987). P-uptake by *Artemisia* increased significantly from the side with defoliated grass plants within one week.

- Shifts in herbaceous composition accompanying grazing may be to assemblages less effective at sequestering resources and competitively excluding woody plants or limiting their growth and seed production.
- Reductions in fine fuel biomass and continuity would reduce fire frequency and intensity (Savage and Swetnam 1990).
- Invading woody species are often unpalatable relative to grasses and forbs and are thus not browsed with sufficient regularity or severity to limit establishment or growth.
- Reductions in soil fertility and alterations in physiochemical properties occur with loss of vegetative cover and erosion (Harrington et al. 1979, Denevan 1967, Thurow 1991). This would favor N_2-fixing woody plants (e.g., *Prosopis, Acacia*) and growth forms tolerant of low nutrient conditions (Cohn et al. 1989, Bush and Van Auken 1989, Van Auken and Bush 1989)(Fig. 6).
- Widespread eradication of prairie dogs, viewed as competitors with livestock for forage and whose burrows pose injury hazards, would have removed a significant barrier to woody plant recruitment on many landscapes (Weltzin 1990).

Modification of microclimate, plant competitive interactions, soil fertility and fire frequency associated with the defoliation and preferential utilization of grasses can thus result in increased likelihood of successful woody plant invasion, establishment and growth (Fig. 7)(Blackburn and Tueller 1970, Bush and Van Auken 1990, Van Auken and Bush 1987,1988, 1990, McPherson et al. 1988, Brown and Archer 1989, Owens and Norton 1989, McPherson and Wright 1990, Skarpe 1990), decreased time to reproductive maturity (McPherson and Wright 1987), increased frequency and magnitude of seed production, and extended woody plant longevity (West et al. 1979).

The rate, pattern and extent of livestock-induced change in community structure would be expected to vary with the type of animal and degree of grazing pressure (stocking rate, frequency, duration, season and intensity of plant utilization). Within a landscape or watershed, livestock grazing pressure is effectively controlled by distance from water and by topography (Andrew 1988, Stuth 1991). Vegetation change on grazed landscapes may be accentuated or mitigated, depending on climatic influences (see **Climatic Determinants** section) and may be non-linear and abrupt once critical disturbance thresholds are exceeded (Figs. 3 and 11). Once established on a site, woody plants may re-direct successional processes and persist, despite subsequent relaxation of grazing (Archer and Smeins 1991, Laycock, this volume).

The effects of livestock grazing may be most important in influencing establishment of woody plant seedlings. Most research on woody plant germination and establishment has focused on abiotic factors. Studies of woody plant seedling establishment in competition with grasses are often short-term or involve potted plants with restricted rooting volumes experiencing relatively low light intensities and (likely) artificially high CO_2 concentrations. Field investigations pertaining to biotic limitations, are scarce. Based on controlled-environment experiments and non-experimental field observations, it has been generally assumed that established, climax-dominant grassland species would exclude woody seedlings and that grazed areas are more prone to woody plant invasion, other factors held equal. There are few guidelines for relating changes in woody plant seedling establishment probabilities to grazing intensity. It is also well-known that herbaceous retrogression accompanies intensive grazing, yet there are few reliable guidelines for determining how grazing at different stages of retrogression alters susceptibility of sites to invasion by woody plants. In the case of honey mesquite, field experiments and diverse field observations indicate that level of grazing and grazing history may have little influence on *Prosopis* seedling emergence or survival in present-day systems (Brown and Archer 1989 and references therein; but see Bush and Van Auken 1991b). Thresholds of herbaceous biomass production or composition required to limit establishment of some woody seedlings may thus be exceeded even at low levels of grazing. However, site grazing history and herbaceous composition markedly influence *Prosopis* growth rate and time to reproductive maturity (Fig. 7).

Given the potential ability of *Prosopis* to successfully establish in stands of highly productive, late seral grasses, why has its abundance within its historic range increased since settlement? One hypothesis is that prior to settlement and the intro-

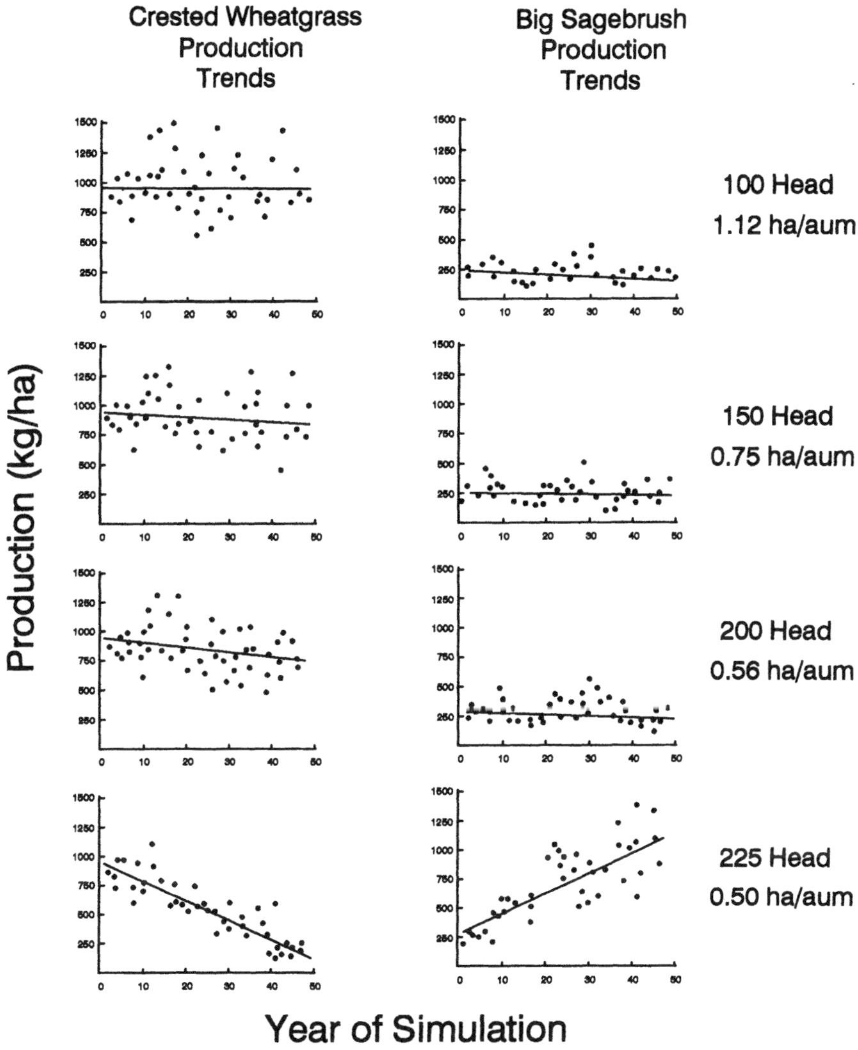

Figure 11. Predicted changes in crested wheatgrass and big sagebrush production under different stocking rates (from Torell 1984). Doubling the number of animals from 100 to 200 had little impact on production. However, increasing stocking from 200 to 225 head (12.5% increase) had a marked impact.

duction of livestock, *Prosopis* was absent from grass-dominated portions of the landscapes because of dispersal limitations (Brown and Archer 1987). Alternatively, *Prosopis* may have been present, but suppressed (not eliminated) by herbaceous competition and periodic fire (see Pyric Determinants section). With the advent of livestock grazing, seed dispersal would have been greatly increased, competition from grasses would have been relaxed and set-backs associated with periodic fire eliminated. *Prosopis* plants would then have rapidly increased their stature and dominance.

Evaluation of Proximate Causes of Woody Plant Invasion

The preceding sections highlight the potential independent role of climate, atmospheric CO_2, soils, fire and herbivory in regulating the balance between grasses and woody vegetation. Three important concepts in evaluating the relative contribution of these factors are: (1) interactions; (2) necessary and sufficient conditions; and (3) feedbacks. Case studies indicate that direct and indirect effects of livestock grazing (including reductions in fire frequency), augmented by climatic fluctuation, are reasonable explanations for replacement of grasslands and savannas by shrub and woodlands.

Interactions

In experimental manipulations, the significance of main effects and interactions are tested. In an ecosystem context, we are interested in identifying the primary factors governing ecosystem processes (i.e. the main effects) and the extent to which the influence of those primary factors might be mediated (lessened or magnified) when other factors are operational (i.e., interactions). Main effects, in the context of identifying factors controlling grass vs. woody plant composition, would include climate, soils, fire and grazing. Unfortunately, because of logistical constraints, most experimental research has focused on but one of these main effects. Results from such studies are thus out of context, since other potentially important factors and interaction terms are not taken into account. Simulation models are one tool for circumventing this problem. However, the degree to which such models can successfully represent a system will be directly related to our level of understanding and the quality and availability of information used in their construction. Conceptual models of woody plant encroachment into desert grassland (Fig. 12), sagebrush steppe (Fig. 13), *Prosopis* savanna (Fig. 14), and pinyon-juniper savanna (Fig. 15) illustrate the variety of factors that have interacted since settlement to produce the current vegetation. The driving force for the changes depicted in each of these models centers around livestock grazing.

Necessary and Sufficient Conditions

It is necessary for some conditions to change if ecosystems are to change. However, because of thresholds, inertia and negative feedbacks, a change in some conditions does not necessarily mean that a change in ecosystem status will occur. Thus, certain kinds of changes may be necessary, but by themselves are not sufficient to elicit a change in ecological systems. The influence of changing a certain variable(s) on ecosystem structure will be determined by the magnitude, direction and duration of the change and the extent to which the change is either augmented or negated by other variables.

Figure 3 shows changes in woody plant cover at a savanna parkland site in southern Texas. The site had experienced continuous, high levels of cattle grazing since the late 1800s, extending through the early 1980s. During the severe drought of the 1950s, woody cover decreased on all three sites. In the subsequent period characterized by normal to above-normal annual rainfall, woody cover increased 3- to 4-fold. Would an increase of this magnitude have been observed at pre-settlement

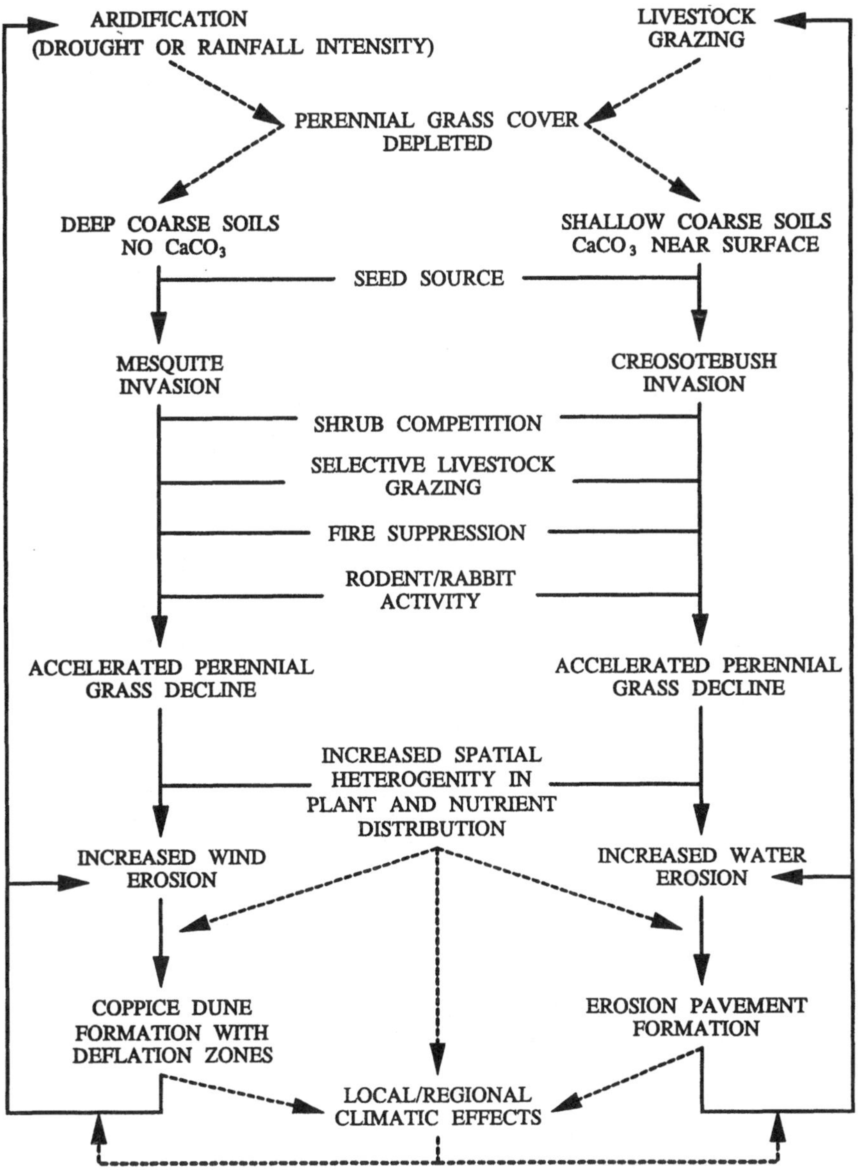

Figure 12. Conceptual model of *Prosopis* and *Larrea* invasion of desert grassland (from Grover and Musick 1990).

levels of atmospheric CO_2? Drought, with its adverse affect on grasses, may have been necessary to set the stage for woodland expansion, but would drought acting alone have been sufficient to produce the pattern and magnitude of change observed? Grazing by livestock during the drought period likely increased grass plant susceptibility to drought-related stresses. In addition, *Prosopis* pod production during the drought period was likely maintained or even stimulated (El Youssoufi 1991). Utilization of pods and subsequent dispersal of seed by livestock dur-

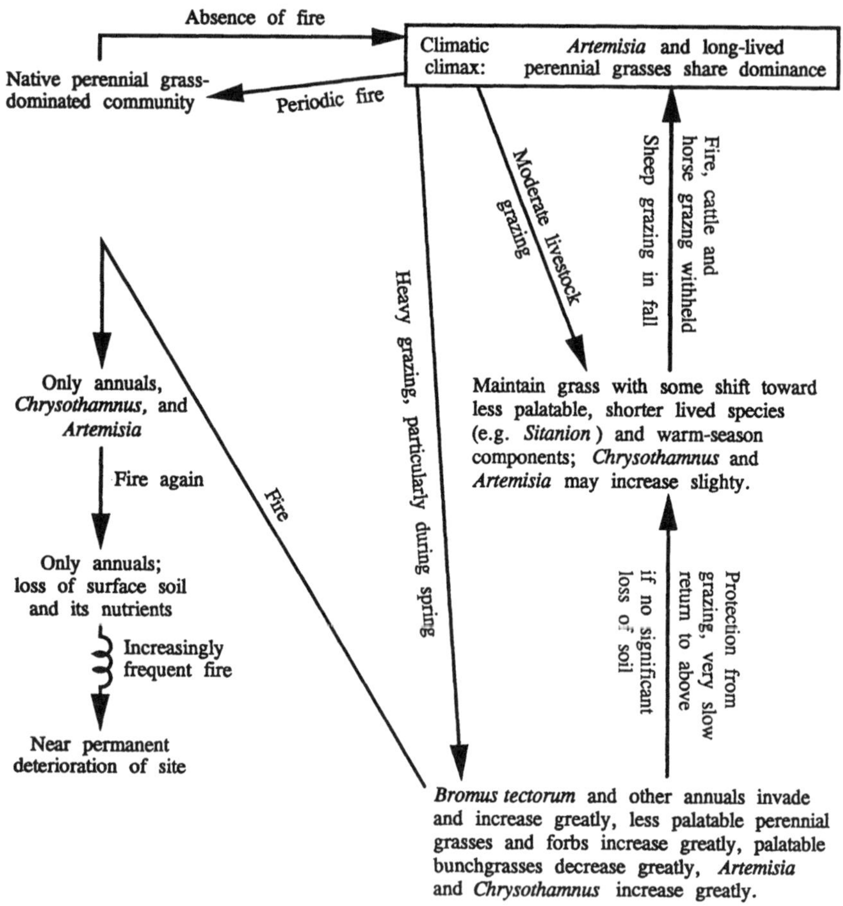

Figure 13. The role of grazing, fire and introduced annual weeds in changing grass-shrub balance in the Great Basin over the past 150 years (from West 1988).

ing the drought period would translate to increased opportunities for *Prosopis* recruitment in the post-drought period, especially where the density and productivity of grasses had been reduced. Grazing in the post-drought period would have slowed grass recovery and minimized the likelihood of fire. The interaction between drought, atmospheric CO_2 levels, and livestock grazing may therefore have generated a change in vegetation that could not have been produced by these factors operating independently.

Feedbacks

Interactions among ecosystem components are often regulated by positive or negative feedbacks which confer a degree of homeostasis. Positive feedbacks are those which reinforce deviations from a set point, whereas negative feedbacks are those which halt or reverse movement away from a steady-state condition. Positive feedbacks can thus accelerate change, whereas negative feedbacks may enable systems to resist change. Schlesinger et al. (1990) hypothesize that positive feed-

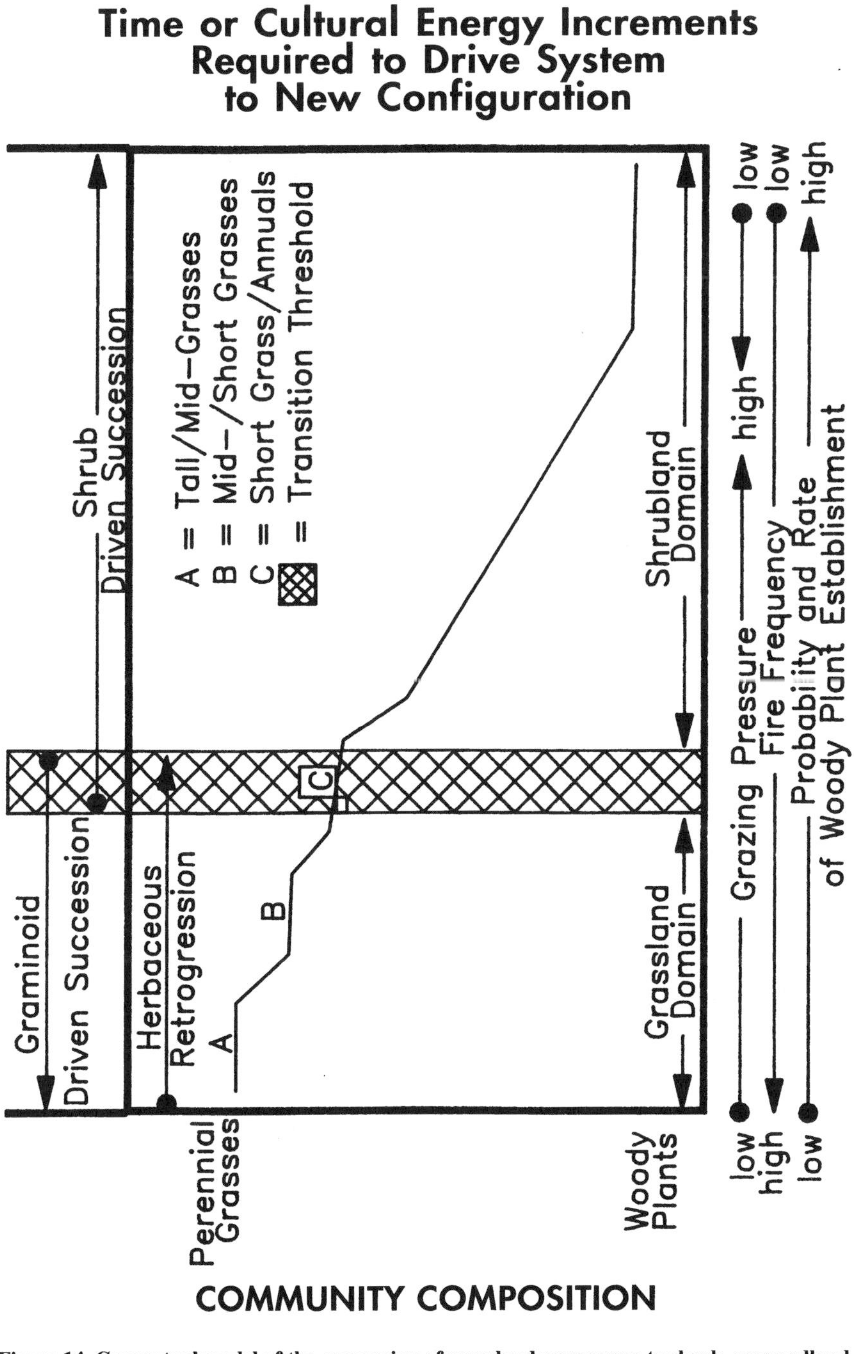

Figure 14. Conceptual model of the conversion of grassland or savanna to shrub- or woodland, postulating the existence of a threshold of grazing disturbance (e.g., Fig. 11) in triggering a successional transition between alternate steady states (from Archer 1989).

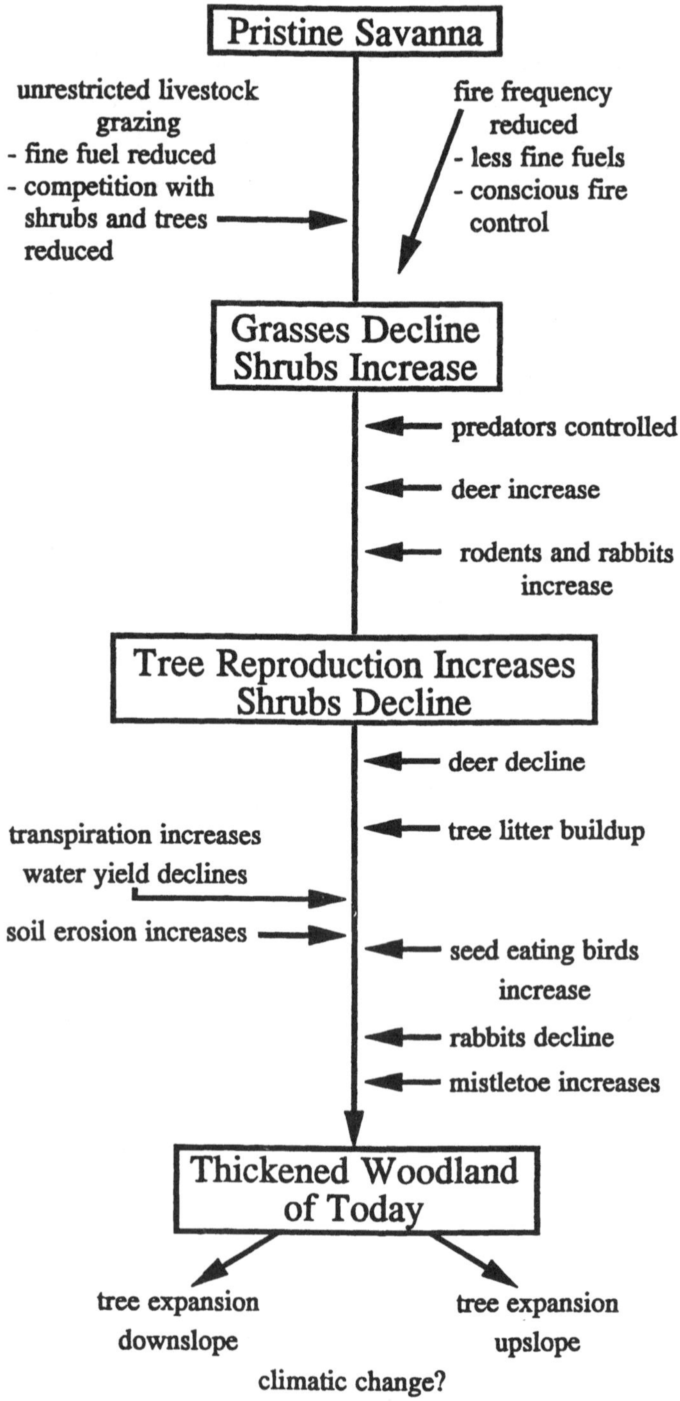

Figure 15. Tree expansion and conversion of Great Basin savanna to pinyon-juniper woodland (from West and Van Pelt 1987).

backs are operational in desertification. In their proposed scenario, long-term grazing of semi-arid grasslands by livestock produces spatial and temporal heterogeneity of soil resources such as nitrogen and water. Heterogeneity of soil resources subsequently promotes invasion by desert shrubs which leads to a further localization of soil resources under shrub canopies. In the grazed areas between shrubs, soil fertility is lost by erosion and gaseous emissions, further increasing susceptibility to shrub encroachment. As grassland is replaced by shrubland a greater percentage of the soil surface is exposed. Soil surface and air temperatures thus increase and produce a warmer, drier micro- and mesoclimate which favors drought-tolerant shrubs over grasses. This positive feedback process continues until a new steady-state is achieved (see Archer 1990, Grover and Musick 1990 and Laycock, this volume for other examples).

Lessons From Case Studies

As discussed in the Climatic Determinants section, the fact that succession from grassland and savanna to shrub- and woodland has occurred over similar time frames and over such broad geographic areas constitutes one line of evidence that broad-scale factors like climate change and atmospheric CO_2 enrichment may be operating. Local and regional studies also offer some evidence that seasonal patterns of rainfall and temperature may have changed in recent history to favor woody plants over grasses in portions of North America. In addition, grasslands and savannas which established under previous climatic regimes may be only marginally suited to the recent climate and were perhaps prone to woody plant invasion. However, changes or fluctuations in broad-scale climatic regimes cannot explain how grasslands and savannas have persisted on some sites within a climatic zone but not others. Most lines of evidence point to grazing effects as the proximate cause for recent vegetation shifts:

Case 1. In their analysis of primary production and carbon balance of the central Great Plains, Burke et al. (1991) concluded that land management decisions may be more important than climate change in affecting carbon balance. In many arid land ecosystems, these land management decisions center around livestock grazing.

Case 2. "Although the grasslands of southern New Mexico were extensive and dominated the area, they were on the xeric edge of the continental Grassland Formation. A single factor such as grazing was evidently enough to set in motion a series of relatively rapid events which culminated in a desert shrub vegetation" (York and Dick-Peddie 1969:165).

Case 3. The stability of grass-shrub mosaics in relation to climate, soils, fire and grazing is being investigated in an ongoing field project in southern Texas (Archer 1990). While all have interacted to influence rates and patterns of woody plant encroachment, grazing appears to have been the driving force. The potential natural vegetation of this region has been classified by plant geographers as *Prosopis-Acacia* savannas (Kuchler 1964), and historical accounts suggest the upland vegetation was grassland or savanna (Table 2). Field investigations indicate that upland soils are characterized by a laterally extensive argillic or clay pan horizon (fine-textured clays) which contain non-argillic inclusions (soil texture coarse throughout profile) (Loomis 1989). The historic grasslands or savannas may thus have been edaphically controlled, with mid- and shortgrasses dominating the upland, but interspersed with patches of tall grass species on the non-argillic inclusions (Fig. 4a). Alternatively, *Prosopis* and other shrubs may have dominated the non-argillic inclusions (Fig. 4b). In either case, woody vegetation would have been competitively excluded under the historic rainfall regimes where the argillic horizon was present. Such a system would have been prone to fire, which may have served as a secondary deterrent to woody plant establishment.

Present day landscapes in southern Texas contain high densities of discrete shrub clusters where the argillic horizon is present; groves of larger *Prosopis* trees occur where the argillic horizon is absent (Fig. 4c). Clusters dominated by C_3 shrubs occur on soils whose organic carbon reflects prior domination by C_4 grasses (Tieszen and Archer 1990). Models based on shrub growth rates (Archer 1989) and transition probabilities (Scanlan and Archer 1991) indicate this vegetation transformation occurred over the past 100-150 years. Three pathways of succession are hypothesized in Figure 4: (1) A→C (2) B→C or (3) A→B→C. Even though *Prosopis* trees growing on the non-argillic soils are much larger than the largest *Prosopis* plants now growing on argillic soils in terms of height, canopy area and trunk diameter, they are comparable in age (Table 5). The differences in size therefore reflect the fact that the growth rate of *Prosopis* plants on the non-argillic soils is greater than that of plants on the argillic soils. This suggests that the prior state of the vegetation would have been as shown in Figure 4a, with *Prosopis* invasion occurring across the landscape, but growing and developing faster on the non-argillic microsites.

A climatic explanation for the observed shift in vegetation structure would have to center around increased rainfall or a shift to increased winter rainfall (see Climatic and Edaphic Determinants Sections; Scanlan and Archer 1991). However, there is little evidence to suggest this has occurred (Norwine 1978). An alternative explanation centered around livestock grazing seems plausible. Reductions in transpiring leaf area, concomitant with reductions in initiation, extension, and biomass of grass roots, would increase soil moisture in the fine-textured upper horizons for establishing shrub seedlings. Shifts in herbaceous species composition to assemblages potentially less effective at sequestering soil resources would also occur. At the same time, livestock would have greatly increased *Prosopis* seed dispersal from populations restricted to intermittent drainages into surrounding grassland (Brown and Archer 1987) while fire frequency decreased. Periodic drought may have magnified grazing-induced stresses on the herbaceous vegetation and set the stage for subsequent pulses of woody plant recruitment (Fig. 3). The result has been a rapid, non-linear shift from grassland to woodland over a short (150-200 y) period, coincident with the intensification of livestock grazing.

Case 4. On a local scale, fire, grazing and soil properties interact within a variable climate to determine the balance between grasses and woody plants. How might the relative contributions of these interacting factors be determined? In a unique field study in Utah, Madany and West (1983) have documented a case in which a savanna protected from cattle grazing was maintained, whereas nearby edaphically-similar sites subjected to cattle grazing changed from savanna to dense woodland soon after the introduction of livestock in the late 1800s. Low frequencies of fire were documented on the savanna site protected from grazing, indicating that frequent fire was not required to maintain the grass/tree balance in this system. Both sites experienced the same climate and atmospheric conditions (e.g., CO_2 concentrations), yet one changed dramatically over the past 100 years, whereas the other did not. Such data indicate that livestock grazing has been the proximate cause of vegetation change on this site, not changes in fire or climatic regimes. Although climate change or fluctuation in recent history may have been necessary to have caused the change on the grazed site, it was not sufficient to cause change on the ungrazed site.

Case 5. The photograph in Figure 16 depicts a fenceline contrast in the Mitchell grass plains of central Australia. The properties on both sides of the fence have similar soils, have experienced the same climate and have the same general history of livestock grazing. However, prickly acacia (*Acacia nilotica*) plants from Africa were introduced along water courses on the property to the left of the fence in the 1940s (J.O. Carter, pers. comm.). At the time of this photograph (1988), prickly acacia had spread throughout the pasture and formed dense

Figure 16. Fenceline contrast in the Mitchell Grasslands, Queensland, Australia showing stand of *Acacia nilotica* which has developed since the 1940s on one property, but not the other. Local contrasts in physiognomy suggest differences in land management practices as the cause for vegetation change and not broad-scale factors such as changes in climatic or atmospheric conditions. (Photo by S. Archer).

stands. The plant was not introduced to the property on the right. The fact that the spread of the plant has been limited by the fence suggests that livestock rather than native fauna were the principal agents of seed dispersal. Field observations indicate that seed production and seedling establishment of prickly acacia away from watercourses is episodic, occurring only during periods of sufficient rainfall which are infrequent and widely spaced (Burrows et al. 1986, 1990). Thus, while climatic variables are a key component in dictating the dynamics of prickly acacia seedling establishment and stand development, they cannot account for the differences in vegetation structure on these adjacent properties.

Case 6. Woody plant densities in an arid Botswana savanna fluctuated over a 5-year period in areas with no and moderate grazing by cattle (Skarpe 1990). Over the same period, the density and growth of shrubs increased on areas receiving heavy grazing .

Case 7. Redberry juniper (*Juniperus pinchotii)* densities on sites inaccessible to livestock in western Texas were an order of magnitude lower than those observed on edaphically similar sites with a history of livestock grazing (288 vs. 2123 plants/ha)(McPherson et al. 1988).

Case 8. With the implementation of the Taylor Grazing Act of 1934, land management practices in the southwestern United States were modified. Such was not the case in Mexico. Differences in grazing management practices since that time have produced sharp discontinuities in vegetative composition and cover along the international border in the Sonoran Desert (Bahre and Bradbury 1978). There are manifested in remotely sensed surface temperatures, greenness, soil moisture and convective cloud formation (Bryant et al. 1990). Border-line and fence-line contrasts (e.g., Case 7) reflect the importance of grazing and overall land use practices in shaping ecosystem structure relative to that of climate (see also Knight 1991).

Case 9. "About cause, then, the best answer seems to be that the new vegetation . . . has not arisen from climatic variation alone, but in response to the unique combination of climatic and cultural stress imposed by the events of the past eighty years; that climate and cattle have united to produce it" (Hastings and Turner 1965:289).

Case 10. Gehlbach (1981) examined vegetation in repeat photography along the international boundary between the Rio Grande and the Colorado River and concluded "man's landscape tinkering has been more influential than climate in creating unidirectional trends . . ." and that ". . . short-term climatic cycles not climatic change, have exacerbated man-made vegetative changes" (pp. 239-241).

Case 11. Bahre (1991) used repeat photography to demonstrate changes in woody plant distribution and abundance in southeastern Arizona since the 1870s. After a detailed review of existing literature and historic land-use practices, he concluded: (1) "probably more time has been spent on massaging the climatic change hypothesis than on any other factor of vegetation change, and yet it remains the least convincing" (p.105); and (2) "the increase of woody plants in rangelands . . . is most likely the result of grazing and fire exclusion" (p.187).

Summary

Replacement of grasslands and savannas with shrub- and woodlands dominated by unpalatable species appears to have been widespread since European settlement. Available data suggest these changes have been (1) rapid, with substantial changes occurring over 50- to 100-year time spans, (2) non-linear, (3) accentuated by climatic fluctuation, (4) locally influenced by topoedaphic factors, and (5) non-reversible over time frames relevant to management.

Post-industrial atmospheric CO_2 enrichment and climate change may have facilitated shifts from grass to woody plant domination. However, case studies documenting differences in the rate and pattern of woody plant encroachment on nearby landscapes with similar topoedaphic properties suggest these were not the proximate factors driving vegetation change. In contrast, numerous case studies have established a strong link between livestock grazing and the encroachment of unpalatable woody plants. Direct and indirect effects of livestock grazing (preferential utilization of grasses, alteration of soil structure and chemistry, seed dispersal, reduction of fire) generally favor woody plant establishment and stand development. In addition, differences in livestock species and levels of grazing pressure across landscapes over time would explain why rates and patterns of vegetation change have varied substantially for similar habitats within climatic zones.

Grasslands and savannas are globally significant for their production of forage, food and fiber. An understanding of factors controlling the balance between grasses and woody plants is fundamental to developing natural resource management plans for sustained utilization in these ecosystems. Shifts in grass and woody plant abundance have broad implications for biodiversity, primary and secondary productivity, soil development and stability, livestock and wildlife composition and carrying capacity, recreational opportunities, water quality and water distribution. Changes from herbaceous- to wood plant-domination also constitute a potentially important global climate feedback affecting carbon sequestering, non-methane hydrocarbon emissions and biophysical land surface-atmosphere interactions (albedo, evapotranspiration, surface roughness, boundary layer dynamics). Additional information is required if we are to understand mechanisms and predict the rate, pattern and extent of change in grass and woody plant distribution in the face of expanding human populations and human-induced climate change.

Acknowledgments

Thanks to P.W. Barnes, A.J. Belsky, R.C. Flinn, G.R. McPherson, M.K. Owens, and O.W. Van Auken for reading earlier drafts and making constructive suggestions and to Teddi Walker and Carol McDonald for assistance with word processing.

This work was supported in part by Hatch Project 6717, NSF Grant BSR-9109240 and NASA grant NAGW-2662 and is published with the permission of the Director, Texas Agricultural Experiment Station as TA-31056.

Literature Cited

Acocks, J.P.H. 1964. Karroo vegetation in relation to the development of deserts, pp. 100-112. *In*: D.H.S. Davis (ed), Ecological studies of southern Africa. Dr. Junk Publishers, The Hague.

Adamoli, J., E. Sennhauser, J.M. Acero, and A. Rescia. 1990. Stress and disturbance: vegetation dynamics in the dry Chaco region of Argentina. J. Biogeography 17:491-500.

Aerts, R. 1989. Above ground biomass and nutrient dynamics of *Calluna vulgaris* and *Molina caerulea* in dry heathland. Oikos 56:31-38.

Amos, B.B. and F.R. Gehlbach. (eds). 1988. Edwards plateau vegetation. Baylor Univ. Press, Waco, TX. USA.

Andrew, M.H. 1988. Grazing impact in relation to livestock watering points. Trends Ecol. Evol. 3:336-339.

Archer, S. 1989. Have Southern Texas savannas been converted to woodlands in recent history? Am. Nat. 134:545-561.

Archer, S. 1990. Development and stability of grass/woody mosaics in a subtropical savanna parkland, Texas, U.S.A. J. Biogeography 17:453-462.

Archer, S. J.K. and Detling. 1986. The potential role of herbivores in mediating plant water status. Oikos 47:287-291.

Archer, S. and R.C. Flinn. 1991. Longevity of *Prosopis glandulosa var. glandulosa* ramets and genets in a subtropical thorn woodland. Bul. Ecological Soc. America 72:57.

Archer, S., C.J. Scifres, C.R. Bassham, and R. Maggio. 1988. Autogenic succession in a subtropical savanna: conversion of grassland to thorn woodland. Ecol. Monogr. 58:111-127.

Archer, S. and F.E. Smeins. 1991. Ecosystem-level processes, pp. 109-139. *In*: R.K. Heitschmidt and J.W. Stuth (eds), Grazing management: an ecological perspective. Timberline Press, Portland, OR.

Arno, S.F. and G.E. Gruell. 1986. Douglas fir encroachment into mountain grasslands in southwestern Montana. J. Range Manage. 39:272-276.

Bahre, C.J. and D.E. Bradbury. 1978. Vegetation change along the Arizona-Sonora boundary. Ann. Assoc. Amer. Geog. 68:145-165.

Bahre, C.J. 1991. A legacy of change: historic human impact on vegetation in the Arizona borderlands. Univ. Arizona Press, Tucson. 231 p.

Baisan, C.H. and T.W. Swetnam. 1990. Fire history on a desert mountain range: Rincon Mountain Wilderness, Arizona, USA. Can. J. For. Res. 20:1559-1569.

Bazzaz, F.A. 1990. The response of natural ecosystems to the rising global CO_2 levels. Ann. Rev. Ecol. Syst. 21:167-196.

Beatley, J.C. 1974. Phenological events and their environmental triggers in Mojave Desert ecosystems. Ecology 55:856-863.

Belsky, A.J. 1984. Role of small browsing mammals in preventing woodland regeneration in the Serengeti National Park, Tanzania. Afr. J. Ecol. 22:271-279.

Belsky, A.J. 1991. Tree/grass ratios in East African savannas: a comparison of existing models, pp. 139-145. *In*: P.A. Werner (ed), Savanna ecology and management: Australian perspectives and intercontinental comparisons. Blackwell Scientific Publications, London.

Berdowski, J.J.M. 1987. Transition from heathland to grassland initiated by the heather beetle. Vegetatio 72:167-173.

Betancourt, J.L., T.R. Van Devender, and P.S. Martin. 1990. Synthesis and prospectus, pp. 435-447. *In*: J.L. Betancourt, T.R. Van Devender and P.S. Martin, (eds), Packrat middens: the last 40,000 years of biotic change. University of Arizona Press, Tucson.

Blackburn, W.H. and P.T. Tueller. 1970. Pinyon and juniper invasion in black sagebrush communities in east-central Nevada. Ecology 51: 841-848.

Bogusch, E.R. 1952. Brush invasion of the Rio Grande Plains of Texas. Texas J. Sci. 4:85-91.

Booth, C.A. and P.J. Barker. 1981. Shrub invasion on sandplain country west of Wanaaring, New South Wales. Soil Conservation Service of New South Wales. 37:65-70

Box, E.O. 1981. Microclimate and plant forms: an introduction to predictive modeling in phytogeography. Dr. W. Junk Publishers, The Hague, Netherlands.

Bragg, T.B. and L.C. Hulbert. 1976. Woody plant invasion of unburned Kansas bluestem prairie. J. Range Manage. 29:19-23.

Branson, F.A. 1985. Vegetation changes on western rangelands. Range Monograph No. 2, Soc. Range Management, Denver, CO.

Briske, D.D. 1991. Developmental physiology and morphology of grasses, pp. 85-108. *In*: R.K. Heitschmidt and J.W. Stuth. (eds), Grazing management: an ecological perspective. Timber Press, Inc. Portland, OR.

Brown, A.L. 1950. Shrub invasions of southern Arizona desert grasslands. J. Range Manage. 3:172-177.

Brown, J.R. and S. Archer. 1987. Woody plant seed dispersal and gap formation in a North American subtropical savanna woodland: the role of domestic herbivores. Vegetatio 73:73-80.

Brown, J.R. and S. Archer. 1989. Woody plant invasion of grasslands: establishment of honey mesquite (*Prosopis glandulosa var. glandulosa*) on sites differing in herbaceous biomass and grazing history. Oecologia (Berl.) 80:19-26.

Brown, J.R. and S. Archer. 1990. Water relations of a perennial grass and seedling versus adult woody plants in a subtropical savanna, Texas. Oikos 57:366-374.

Brown, J.H. and E.J. Heske. 1990. Control of a desert-grassland transition by a keystone rodent guild. Science 250:1705-1707.

Bryant, N.A., L.F. Johnson, A.J. Brazel, R.C. Balling, C.F. Hutchinson and L.R. Beck. 1990. Measuring the effect of overgrazing in the Sonoran Desert. Climate Change 17:243-264.

Bryson, R.A. 1974. A perspective on climatic change. Science 184:753-760.

Bucher, E.H. 1987. Herbivory in arid and semi-arid regions of Argentina. Revista Chilena de Historia Natural 60:265-273.

Buffington, L.D. and C.H. Herbel. 1965. Vegetational changes on a semidesert grassland range from 1958 to 1963. Ecol. Monogr. 35:139-164.

Burke, I.C., T.G.F. Kittel, W.K. Lauenroth, P. Snook, C.M. Yonker, and W.J. Parton. 1991. Regional analysis of the central Great Plains: sensitivity to climate variability. BioScience 41:685-692.

Burkhardt, J.W. and E.W. Tisdale. 1969. Natural and successional status of western juniper vegetation in Idaho. J. Range Manage. 22:264-270.

Burrows, W.H., I.F. Beale, R.G. Silcock, and A.J. Pressland. 1985. Prediction of tree and shrub population changes in a semi-arid woodland, pp. 207-211. *In*: J.C. Tothill and J.J. Mott (eds), Ecology and management of the world's savannas. Australian Academy of Science, Canberra.

Burrows, W.H., J.O. Carter, E.R. Anderson, and M.P. Bolton. 1986. Prickly acacia (*Acacia nilotica*) invasion of Mitchell grass (*Astrebla* spp.) plains in central and northern Queensland. Proc. 4th Bienn. Conf. Australian Rangeland Soc., Armidale, pp. 104-106.

Burrows, W.H., J.O. Carter, J.C. Scanlan, and E.R. Anderson. 1990. Management of savannas for livestock production in north-east Australia: contrasts across the tree-grass continuum. J. Biogeography 17:503-512.

Bush, J.K. and O.W. Van Auken. 1989. Soil resource levels and competition between a woody and herbaceous species. Bul. Tor. Bot. Club 116:22-30.

Bush, J.K. and O.W. Van Auken. 1990. Growth and survival of *Prosopis glandulosa* seedlings associated with shade and herbaceous competition. Bot. Gaz. 151:234-239.

Bush, J.K. and O.W. Van Auken. 1991a. Importance of time of germination and soil depth on growth of *Prosopis glandulosa* (Leguminosae) seedlings in the presence of a C_4 grass. Am. J. Bot. 78:1732-1739.

Bush, J.K. and O.W. Van Auken. 1991b. Growth and survival of *Prosopis glandulosa* seedlings associated with shade and herbaceous competition. Bot. Gaz. 151:234-239.

Caldwell, M.M., D.M. Eissenstat, J.H. Richards, and M.F. Allen. 1985. Competition for phosphorus: differential uptake from dual isotope-labeled soil interspaces between shrub and grass. Science 229:384-386.

Caldwell, M.M., J.H. Richards, J.H. Manwaring, and D.M. Eissenstat. 1987. Rapid shifts in phosphate acquisition show direct competition between neighboring plants. Nature 327:615-616.

Canfield, R.H. 1957. Reproduction and life span of some perennial grasses of southern Arizona. J. Range Manage. 10:199-203.

Cantor, L.F. and T.G. Whitham. 1989. Importance of below ground herbivory: pocket gophers may limit aspen to rock outcrop refugia. Ecology 70:962-970.

Carter, M.G. 1964. Effects of drouth on mesquite. J. Range Manage. 17:275-276.

Chapin, F.S., III. 1980. The mineral nutrition of wild plants. Annu. Rev. Ecol. Syst. 11:233-260.

Chapin, F.S., III and G.R. Shaver. 1989. Differences in growth and nutrient use among arctic plant growth forms. Functional Ecology 3:73-80.

Chew, R.M. 1982. Changes in herbaceous and suffrutescent perennials in grazed and ungrazed desertified grassland in southeastern Arizona, 1958-1978. Am. Midl. Naturalist 108:159-169.

Clark, J.S. 1988. Effect of climate change of fire regimes in northwestern Minnesota. Nature 334:233-235.

Clarkson, N.M. and G.R. Lee. 1988. Effects of grazing and severe drought on a native pasture in the Traprock region of southern Queensland. Tropical Grasslands 22:176-183.

Coffin, D.P. and W.K. Lauenroth. 1990. A gap dynamics simulation model of succession in a semiarid grassland. Ecological Modelling 49:229-266.

Cohn, E.J., O.W. Van Auken, and J.K. Bush. 1989. Competitive interactions between *Cynodon dactylon* and *Acacia smallii* seedlings at different nutrient levels. Am. Midl. Nat. 121:265-272.

Cole, K. 1985. Past rates of change, species richness and a model of vegetation inertia in the Grand Canyon, Arizona. Am. Nat. 125:289-303.

Collins, S.L. 1987. Interaction of disturbances in tallgrass prairie: a field experiment. Ecology 68:1243-1250.

Cornelius, J.M. 1988. Vegetation response to fire in a northern Chihuahuan desert grassland. Ph.D. Thesis, New Mexico State Univ., Las Cruces.

Crisp, M.D. and R.T. Lange. 1976. Age structure, distribution and survival under grazing of the arid-zone shrub *Acacia burkittii*. Oikos 27:86-92.

Cunningham, G.M. and P.J. Walker. 1973. Growth and survival of mulga *Acacia aneura* F. Muell, ex Benth.) in western New South Wales. Trop. Grassland. 7: 69-77.

Davis, M.B. 1982. Quaternary history and the stability of forest communities, pp. 132-153. *In*: D.C. West, H.H. Shugart and D.D. Botkin (eds), Forest succession: concepts and application. Springer-Verlag, New York.

Davis, S.D. and H.A. Mooney. 1985. Comparative water relations of adjacent California shrub and grassland communities. Oecologia (Berl.) 66:522-529.

de Faria S.M., G.P. Lewis, J.I. Sprent, and J.M. Sutherland. 1989. Occurrence of nodulation in the Leguminosae. New Phytol. 607-619.

Denevan, W.M. 1967. Livestock numbers in nineteenth century New Mexico, and the problem of gullying in the southwest. Ann. Assn. American Geographers 57:691-703.

Eissenstat, D.M. and Caldwell, M.M. 1988. Competitive ability is linked to rates of water extraction: a field study of two arid land tussock grasses. Oecologia (Berl.) 75:1-7.

El Youssoufi, M. 1991. Factors influencing recruitment of honey mesquite (*Prosopis glandulosa* var. *glandulosa*) in a savanna woodland, Texas. Ph.D. Dissertation.

Ellis, D. and J.L. Schuster. 1968. Juniper age and distribution on an isolated butte in Garza County. Southwestern Naturalist 13:343-348.

Emanuel, W.R., H.H. Shugart and M. Stevenson. 1985a. Climatic change and the broad-scale distribution of terrestrial ecosystem complexes. Climatic Change 7:29-43.

Emanuel, W.R., H.H. Shugart and M. Stevenson. 1985b. Response to comment: climatic change and the broad-scale distribution of terrestrial ecosystem complexes. Climatic Change 7:457-460.

Ewing, A.L. and Engle, D.M. 1988. Effects of late summer fire on tallgrass prairie microclimate and community composition. Am. Midl. Nat. 120:212-223.

Eyre, S.R. 1963. Vegetation and soils. Aldine, Chicago.

Flinn, R.C., S. Archer, and T.W. Boutton. 1991. Tree-ring analysis of *Prosopis glandulosa var. glandulosa* from stands of known age. Bul. Ecological Soc. America 72:114.

Foran, B.D. 1986. The impact of rabbits and cattle on an arid calcareous shrubby grassland in central Australia. Vegetatio 66:49-59.

Forman, R.T.T. and E.W.B. Russell. 1983. Evaluation of historical data in ecology. Ecol. Soc. Amer. Bul. 64:5-7.

Fritts, H.C. Tree rings and climate. Academic Press, New York.

Gardner, J.L. 1951. Vegetation of the creosote bush area of the Rio Grande Valley in New Mexico. Ecological Monographs 21:379-403.

Gehlbach, F. R. 1981. Mountain islands and desert seas. Texas A&M Univ. Press, College Station.

Gleason, H.A. 1913. The relation of forest distribution and prairie fires in the Middle West. Torreya 13:173-181.

Glendening, G.E. 1952. Some quantitative data on the increase of mesquite and cactus on a desert grassland range in southern Arizona. Ecology 33:319-328.

Goldberg, D.E. 1982. The distribution of evergreen and deciduous trees relative to soil type: an example from the Sierra Madre, Mexico, and a general model. Ecology 63:942-951.

Goldberg, D.E. and R.L. Turner. 1986. Vegetation change and plant demography in permanent plots in the Sonoran Desert. Ecology 67:695-712.

Gornitz, V. and NASA. 1985. A survey of anthropogenic vegetation changes in West Africa during the last century - climatic implications. Climatic Change 7:885-325.

Graumlich, L.J. 1991. Subalpine tree growth, climate and increasing CO_2: an assessment of recent growth trends. Ecology 72:1-11.

Graumlich, L.J., L.B. Brubaker, and C.C. Grier. 1989. Long-term trends in forest net primary productivity: Cascade Mountains, Washington. Ecology 70:405-410.

Gray, J.T. 1983. Nutrient use by evergreen and deciduous shrubs in California. I. Community nutrient cycling and nutrient-use efficiency. J. Ecology 71:21-41.

Gray, J.T. and W.H. Schlesinger. 1983. Nutrient use by evergreen and deciduous shrubs in California. II. Experimental investigations of the relationship between growth, nitrogen uptake and nitrogen availability. J. Ecology 71:43-56.

Griffin, G.F. and M.H. Friedel. 1985. Discontinuous change in central Australia: some implications of major ecological events for land management. J. Arid Environ. 9:63-80.

Grime, J.P. 1977. Evidence for the existence of three primary strategies in plants and its relevance to ecological and evolutionary theory. Am. Nat. 111:1169-1194.

Grime, P.J. 1979. Plant strategies and vegetation processes. John Wiley and Sons, New York.

Gross, F.A. and W.A. Dick-Peddie. 1979. A map of primeval vegetation in New Mexico. Southwestern Naturalist 24:115-122.

Grover, H.D. and H.B. Musick. 1990. Shrub land encroachment in southern New Mexico, USA: an analysis of desertification processes in the American Southwest. Climate Change 17:305-330.

Grubb, P.J. 1977. The maintenance of species richness in plant communities: the importance of the regeneration niche. Biol. Rev. 52:107-145.

Hallmark, C.T. and B.L. Allen. 1975. The distribution of creosotebush in west Texas and eastern New Mexico as affected by selected soil properties. Soil Sci. Soc. Amer. Proc. 39:120-124.

Hanselka, C.W. and Kilgore, D.E. 1987. The Nueces River valley: the cradle of the western livestock industry. Rangelands 9:195-198.

Hara, M. 1987. Analysis of seedling banks of a climax beech forest: ecological importance of seedling sprouts. Vegetatio 71:67-74.

Harper, J.L. 1977. Population biology of plants. Academic Press, New York.

Harrington, G.N. 1979. The effects of feral goats an sheep on the shrub populations in a semi-arid woodland. Australian Rangeland Journal 1:334-345.

Harrington, G.N. 1991. Effects of soil moisture on shrub seedling survival in semi-arid grassland. Ecology 72:1138-1149.

Harrington, G.N. and K.C. Hodgkinson. 1986. Shrub-grass dynamics in mulga communities of eastern Australia, pp. 26-28. *In*: Rangelands: a resource under siege. P.J. Joss, P.W. Lynch, and O.B. Williams (eds), Australian Academy of Science, Canberra.

Harrington, G.N., R.E. Oxley, and D.J. Tongway. 1979. The effects of European settlement and domestic livestock on the biological system in poplar box (*Eucalyptus populnea*) lands. Australian Rangeland Journal 1:271-279.

Hastings, J.R. and R.L. Turner. 1965. The changing mile: an ecological study of vegetation change with time in the lower mile of an arid and semi-arid region. University of Arizona Press, Tucson.

Haynes, C.V. Jr. 1982. Great Sand Sea and Selima Sand Sheet, eastern Sahara: geochronology of desertification. Science 217:629-633.

Heddle, E.M. and R.L. Sprecht. 1975. Dark Island heath (Ninety-Mile Plain, South Australia). VIII. The effects of fertilizers on composition and growth, 1950-1972. Aust. J. Bot. 23:151-164.

Heitschmidt, R.K., D.L. Price, R.A. Gordon and J.R. Frasure. 1982. Short duration grazing at the Texas experimental ranch: effects on above-ground net primary production and seasonal growth dynamics. J. Range Manage. 35:367-372.

Hennessy, J.T., R.P. Gibbens, J.M. Tromble and M. Cardenas. 1983. Vegetation changes from 1935 to 1980 in mesquite dunelands and former grasslands of southern New Mexico. J. Range Manage. 36:370-374.

Henry, G.H.R., B. Freedman and J. Svoboda. 1986. Effects of fertilization on three tundra plant communities of a polar desert oasis. Can. J. Bot. 64: 2502-2507.

Herbel, C.H. 1979. Utilization of grass and shrublands of the southwestern United States, pp. 161-203. *In*: B.H. Walker (ed), Management of semi-arid ecosystems. Elsevier, New York.

Herbel, C.H., F.N. Ares, and R.A. Wright. 1972. Drought effects on a semi-desert grassland. Ecology 53:1084-1093.

Hobbs, R.J. and H.A. Mooney. 1986. Community changes following shrub invasion of grassland. Oecologia (Berl.) 70:508-513.

Hogberg, P. 1989. Root symbiosis of trees in savannas, pp. 121-136. *In*: J. Proctor, (ed), Mineral nutrients in tropical forest and savanna ecosystems. Blackwell Scientific Publ., London.

Holdridge, L.R. 1947. Determination of world plant formations from simple climatic data. Science 105:367-368.

Holdridge, L.R. 1964. Life zone ecology. Tropical Science Center, San José, Costa Rica.

Humphrey, R.R. 1987. 90 years and 535 miles: vegetation changes along the Mexican border. Univ. New Mexico Press, Albuquerque.

Humphrey, R.R. and Mehrhoff, L.A. 1958. Vegetation change on a southern Arizona grassland range. Ecology 39:720-726.

Humphrey, R.R. 1958. The desert grasslands: a history of vegetational changes and an analysis of causes. Bot. Rev. 24:193-252.

Hunt, H.W., M.J. Trlica, E.F. Redente, J.C. Moore, J.K. Detling, T.G.G. Kittel, D. E. Walter, M.C. Fowler, D.A. Klein, and E.T. Elliott. 1991. Simulation model for the effects of climate change on temperate grassland ecosystems. Ecological Modelling 53:205-246.

Hunter, R.B., E.M. Romney and A. Wallace. 1980. Rodent-denuded areas of the northern Mojave Desert. Great Basin Naturalist Memoirs 4:208-211.

Huston, M. and T. Smith. 1987. Plant succession: life history and competition. American Naturalist 130:168-198.

Idso, S.B. 1992. Shrubland expansion in the American southwest. Climate Change 22:85-86

Inglis, J.M. 1964. A history of vegetation on the Rio Grande Plains. Texas Parks and Wildlife Dept. Bull. 45, Austin, 122 p.

Jameson, D.A. 1987. Climax or alternative steady states in woodland ecology, pp. 9-13. *In*: R.L. Everett (ed), Proceedings — pinyon-juniper conference. USDA Forest Service Gen. Tech. Rep. INT-215, Ogden, Utah. 581 pp.

Janzen, D.H. 1986. Chihuahuan Desert nopaleras: defaunated big mammal vegetation. Ann. Rev. Ecol. Syst. 17:595-636.

Johnson, H.B. and H.S. Mayeux. 1990. *Prosopis glandulosa* and the nitrogen balance of rangelands: extent and occurrence of nodulation. Oecologia (Berl.) 84:176-185.

Johnson, H.B., H.W. Polley and H.S. Mayeux. 1993. Increasing CO_2 and plant-plant interactions: effects on natural vegetation. Vegetatio 104-105:157-170.

Johnson, K.L. 1987. Sagebrush over time: a photographic study of rangeland change, pp. 223-252. *In*: E.D. McArthur and B.L. Welch (eds), Biology of *Artemisia* and *Chrysothamnus*. USDA For. Serv. Gen. Tech. Rep. INT-200. Ogden, Utah.

Johnson, R.W. and J.C. Tothill. 1985. Definition and broad geographic outline of savanna lands, pp. 1-13. *In*: J.C. Tothill and J.J. Mott (eds), Ecology and management of the world's savannas. Australian Academy of Science.

Johnston, M.C. 1963. Past and present grasslands of southern Texas and northeastern Mexico. Ecology 44:456-466.

Jordan, T.G. 1981. Trails to Texas: southern routes of western cattle ranching. Univ. Nebraska Press, Lincoln. 220 p.

Katz, R.W. and B.G. Brown. 1992. Extreme events in a changing climate: variability is more important than averages. Climatic Change 21:289-302.

Klemmedson, J.O. 1979. Ecological importance of actinomycete-nodulated plants in the western United States. Bot. Gaz. 140 (Suppl.):591-596.

Knight, D.H. 1991. Congressional incentives for landscape research. Ecol. Soc. Amer. Bul. 72:195-203.

Knoop, W.T. and B.H. Walker. 1985. Interactions of woody and herbaceous vegetation in a southern African savanna. J. Ecol. 73:235-253.

Kuchler, A.W. 1964. The potential natural vegetation of the conterminous United States. Amer. Geographical Soc., New York, NY.

Kuchler, A.W. 1978. Natural vegetation, pp. 16-17. *In*: E.B. Espenshade, Jr. and J.L. Morrison (eds), Goode's World Atlas, 15th ed., Rand-McNally, Chicago.

Kummerrow, J., D. Krause and W. Jow. 1977. Root systems of some chaparral shrubs. Oecologia (Berl.) 29:163-177.

La Marche, V.C., Jr., D.A. Graybill, H.C. Fritts and M.R. Rose. 1984. Increasing atmospheric carbon dioxide: tree ring evidence for growth enhancement in natural vegetation. Science 225:1019-1021.

Lechowicz, J.J. and G.R. Shaver. 1982. A multivariate approach to the analysis of factorial fertilizer experiments in Alaskan arctic tundra. Ecology 63:1029-1038.

Lehman, V.W. 1969. Forgotten legions: sheep in the Rio Grande Plains of Texas. Texas Western Univ. Press, Univ. Texas, El Paso.

Leps, J., J. Osbornova-Kosinova and M. Rejmanek. 1982. Community stability, complexity and species life history strategies. Vegetatio 50:53-63.

Lewin, R. 1985. Plant communities resist climatic change. Science 228:165-166.

Liang, Y.M., D.L. Hazlett and W.K. Lauenroth. 1989. Biomass dynamics and water use efficiencies of five plant communities in the shortgrass steppe. Oecologia (Berl.) 80:148-153.

Lieth, H. 1972. Modelling the primary productivity of the world. Nature and Resources 8:5-10.

Lonard, R.I. and F.W. Judd. 1985. Effects of a severe freeze on native woody plants in the Lower Rio Grande Valley, Texas. Southwestern Naturalist 30:397-403.

Lonsdale, M. and R. Braithwaite. 1988. The shrub that conquered the bush. New Scientist 15:52-55.

Loomis, L.E. 1989. Plant-soil relationships in grassland-to-woodland succession. PhD Dissertation, Texas A&M University, College Station.

Lowe, C.H. and D.E. Brown. 1982. Introduction, pp. 8-16. *In*: D.E. Brown (ed), Biotic communities of the American Southwest-United States and Mexico. University of Arizona Press, Tucson (Special Issue of Desert Plants, vol. 4).

Lundelius, E.L. 1976. Vertebrate paleontology of the Pleistocene: an overview. Geoscience and Man 13:45-59.

MacMahon, J.A. 1980. Ecosystems over time: succession and other types of change, pp. 27-58. *In*: R. Waring (ed), Forests: fresh perspectives from ecosystem analyses. Oregon State Univ. Press, Corvallis.

Madany, M.H. and N.E. West. 1983. Livestock grazing-fire regime interactions within montane forests of Zion National Park, Utah. Ecology 64:661-667.

Malin, J.C. 1953. Soil, animal, and plant relations of the grassland, historically recorded. Scientific Monthly 76:207-220.

Martin, P.S. and P.J. Mehringer Jr. 1965. Pleistocene pollen analysis and biogeography of the Southwest, pp. 433-451. *In*: H.E. Wright, Jr. (ed), The late Quaternary of the United States. Yale University Press, New Haven.

Martin, S.C. 1975. Ecology and management of southwestern semidesert grass-shrub ranges: the status of our knowledge. USDA/For. Serv. Res. Paper RM-156.

Martin, S.C. and R.M. Turner. 1977. Vegetation change in the Sonoran Desert region, Arizona and Sonora. Arizona Acad. Sci. 12:59-69.

Mayeux, H.S., Johnson, H.B. and H.W. Polley. 1991. Global change and vegetation dynamics, pp. 62-74. *In*: L.F. James, J.O. Evans, M.H. Ralphs, and B.J. Sigler (eds), Noxious range weeds. Westview Press, Boulder, CO.

McArthur, E.D., E.M. Romney, S.D. Smith and P.T. Tueller. 1990. Cheatgrass invasion, shrub die-off and other aspects of shrub biology and management. USDA/Forest Service Gen. Tech. Rep. INT-276, Ogden, Utah. 351 pp.

McAuliffe, J.R. 1988. Markovian dynamics of simple and complex desert plant communities. Am. Nat. 131:459-490.

McBride, J.R. and H.F. Heady. 1968. Invasion of grassland by *Baccharis pilularis* DC. J. Range Manage. 21:106-108.

McCraw, D.J. 1985. Phytogeographic history of *Larrea* in southwestern New Mexico illustrating the historical expansion of the Chihuahuan Desert. MA Thesis, University of New Mexico, Albuquerque, USA.

McGraw, J.B. and F.S. Chapin III. 1989. Competitive ability and adaptation to fertile and infertile soils in two *Eriophorum* species. Ecology 70:736-749.

McLaughlin, S.P. and J.E. Bowers. 1982. Effects of wildfire on a Sonoran Desert plant community. Ecology 63:246-248.

McNaughton, S.J. 1983. Serengeti grassland ecology: the role of composite environmental factors and contingency in community organization. Ecol. Monogr. 53:291-320.

McMaster, G.S., W.M. Jow and J. Kummerow. 1982. Response of *Adenostoma fasciculatum* and *Ceanothus greggii* chaparral to nutrient additions. J. Ecology 70:745-756.

McPherson, G.R. and H.A. Wright. 1987. Factors affecting reproductive maturity of redberry juniper (*Juniperus pinchotii*). Forest Ecol. & Manage. 21:191-196.

McPherson, G.R. and H.A. Wright. 1990. Effects of cattle grazing and *Juniperus pinchotii* canopy cover on herb cover and production in western Texas. Am. Midl. Nat. 123:144-151.

McPherson, G.R., H.A. Wright and D.B. Wester. 1988. Patterns of shrub invasion in semiarid Texas grasslands. The American Midland Naturalist 120:391-397.

Merrill, L.B. and V.A. Young. 1959. Effect of drouth on woody plants. Texas Agricultural Progress 5:9-10.

Merino, J., C. Field and Mooney, H.A. 1984. Construction and maintenance costs of mediterranean-climate evergreen and deciduous leaves. Oecologia Plantarum 5:211-229.

Merz, R.W. and S.G. Boyce. 1956. Age of oak "seedlings". J. Forestry 54:774-775.

Mitchell, J.F.B., and D.A. Warrilow. 1987. Summer dryness in northern mid-latitudes due to increased CO_2. Nature 330:238-240.

Mitchell, J.M. 1980. History and mechanisms of climate, p. 31-42. *In*: H. Oeschger, B. Messerli and M. Svilar (eds), Das klima-analysen und modelle, geschichte und zukunft, Springer-Verlag, Berlin.

Mooney, H.A., Drake, B.G., Luxmore, R.J., Oechel, W.C. and Pitelka, L.F. 1991. Predicting ecosystem responses to elevated CO_2 concentrations. BioScience 41:96-104.

Morison, C.G.T., A.C. Hoyle and J.F. Hope-Simpson. 1948. Tropical soil-vegetation catenas and mosaics. J. Ecology 36:1-84.

Neftel, A., E. Moor, H. Oeschger, and B. Stauffer. 1985. Evidence from polar ice cores for the increase in atmospheric CO_2 in the past two centuries. Nature 315:45-47.

Neilson, R.P. 1986. High resolution climatic analysis and southwest biogeography. Science 232:27-34.

Neilson, R.P. and Wullstein, L.H. 1985. Comparative drought physiology and biogeography of *Quercus gambelii* and *Quercus turbinella*. Am. Midl. Nat. 114:259-271.

Nelson, E.W. 1934. The influence of precipitation and grazing upon black grama grass range. USDA Tech. Bul. 409, 32 p.

Nelson, J.T. and P.L. Beres. 1987. Was it grassland? A look at vegetation in Brewster County, Texas through the eyes of a photographer in 1899. Texas J. of Agric. and Nat. Res. 1:34-37.

Norby, R.J., C.A. Gunderson, S.D. Wullschleger, E.G. O'Neill and M.K. McCracken. 1992. Productivity and compensatory responses of yellow-poplar trees in elevated CO_2. Nature 357:322-324.

Norwine, J. 1978. Twentieth-century semiarid climates and climatic fluctuations in Texas and northeastern Mexico. J. Arid Environments 1:313-325.

O'Connor, T.G. 1991. Local extinction in perennial grasslands: a life history approach. Amer. Nat. 137:753-773.

O'Connor, T.G. and G.A. Pickett. 1992. The influence of grazing on seed production and seed banks of some African savanna grasses. J. Applied Ecol. 29:247-260.

Owen, D.F. 1979. Drought and desertification in Africa: lessons from the Nairobi Conference. Oikos 33:139-151.

Owen-Smith, N. 1987. Pleistocene extinctions: the pivotal role of megaherbivores. Paleobiology 13:351-362.

Owens, M.K. and B.E. Norton. 1989. The impact of 'available area' on *Artemisia tridentata* seedling dynamics. Vegetatio 82:155-162.

Owensby, C.E., P.I. Coyne, J.M. Hamm, L.M. Auen and A.K. Knapp. 1993. Biomass production i a tallgrass prairie ecosystem exposed to ambient and elevated levels in CO^2. Ecological Applications. (In Press).

Parsons, A.J., A.D. Abrahams and J.R. Simanton. 1992. Microtopography and soil-surface materials on semi-arid piedmont hillslopes, southern Arizona. J. Arid Environ. 22:107-115.

Paulsen, H.A., Jr. and F.N. Ares. 1962. Grazing values and management of black grama and tobosa grasslands and associated shrub ranges of the Southwest. USDA/Forest Serv. Tech. Bull. 1270, 56p.

Petranka, J.W. and J.K. McPherson. 1979. The role of *Rhus copallina* in the dynamics of the forest-prairie ecotone in north-central Oklahoma. Ecology 60:956-965.

Post, W.M., W.R. Emanuel, P.J. Zinke and A.G. Stangenberger. 1982. Soil carbon pools and world life zones. Nature 298:156-159.

Potter, L.D. and D.L. Green. 1964. Ecology of ponderosa pine in western North Dakota. Ecology 45:10-23.

Prentice, I.C. 1986. Vegetation responses to past climatic variation. Vegetatio 67:131-141.

Pressland, A.J. 1973. Rainfall partitioning by an arid woodland (*Acacia aneura* F. Muell.) in southwestern Queensland. Aust. J. Bot. 21:235-245.

Progulske, D.R. 1974. Yellow ore, yellow hair and yellow pine: a photographic study of a century of forest ecology. Bulletin 616, South Dakota State Univ. Ext. Serv., Brookings, South Dakota.

Progulske, D.R. and F.J. Schideler. 1983. Following Custer. Bulletin 674, Agricultural Experiment Station, South Dakota State University. Brookings, South Dakota.

Pyke, D.A. and J.P. Dobrowolski. 1989. Shrub die-back in the Great Basin. Utah Science 50:66-71.

Reynolds, H.G. 1954. Some interrelations of the Merriam kangaroo rat to velvet mesquite. J. Range Manage. 7:176-180.

Reynolds, H.G. and G.E. Glendening. 1949. Merriam Kangaroo Rat as a factor in mesquite propagation on southern Arizona rangelands. J. Range Manage. 2:193-197.

Riechers, G.D. and B.R. Strain. 1988. Growth of blue grama (*Bouteloua gracilis*) in response to atmospheric carbon dioxide enrichment. Canadian Journal of Botany 66:1570-1573.

Robertson, G. 1988. Effect of rainfall on biomass, growth and dieback of pastures in arid grazing systems. Australian J. Ecology 13:519-528.

Robinett, D. 1990. Tohono O'Odham range history. Rangelands 12:296-300.

Robinson, T.W. 1965. Introduction, spread and areal extent of saltcedar (*Tamarix*) in the western States. US Geol. Survey Prof. Paper 491-A, 12 p.

Rosen, E. 1988. Shrub expansion in Alvar grasslands on Oland. Acta Phytogeography Sue. 76:87-100.

Rosenzweig, M.L. 1968. Net primary productivity of terrestrial communities: prediction from climatological data. Am. Naturalist 102:67-74.

Sala, O.E., R.A. Golluscio, W.K. Lauenroth and A. Soriano. 1989. Resource partitioning between shrubs and grasses. Oecologia (Berl.) 81:505-505.

Sala, O.E. and W.K. Lauenroth. 1982. Small rainfall events: an ecological role in semi-arid regions. Oecologia (Berl.) 53:301-304.

Sala, O.E., W.J. Parton, L.A. Joyce and W.K. Lauenroth. 1988. Primary production of the central grassland region of the United States: spatial pattern and major controls. Ecology 69:40-45.

San Jose, J.J. and M.R. Farinas. 1983. Changes in tree density and species composition in a protected *Trachypogon* savanna, Venezuela. Ecology 64(3):447-453.

Savage, M. and T.W. Swetnam. 1990. Early 19th-century fire decline following sheep pasturing in a Navajo ponderosa pine forest. Ecology 71:2374-2378.

Scanlan, J.C. and S. Archer. 1991. Simulated dynamics of succession in a North American subtropical *Prosopis* savanna. J. Vegetation Science 2:625-634.

Schlesinger, W.H., J.F. Reynolds, G.L. Cunningham, L.F. Huenneke, W.M. Jarrell, R.A. Virginia and W.G. Whitford. 1990. Biological feedbacks in global desertification. Science 247:1043-1048.

Schofield, C.J. and E.H. Bucher. 1986. Industrial contributions to desertification in South America. Trends in Ecology and Evolution 1:78-80.

Scifres, C.J. and J.L. Mutz. 1975. Secondary succession following extended inundation of Texas coastal rangeland. J. Range Manage. 279-282.

Sharifi, M.R., E.T. Nilsen and P.W. Rundel. 1982. Productivity and biomass of *Prosopis glandulosa* (mesquite) in the Sonoran Desert of California. Amer. J. Bot. 69: 760-767.

Shreve, F. 1942. The desert vegetation of North America. Bot. Rev. 8:195-246.

Shreve, F. and A.L. Hinckley. 1937. Thirty years of change in desert vegetation. Ecology 18:463-478.

Sinclair, A.R.E. and M. Norton-Griffiths. (eds). 1979. Serengeti: dynamics of an ecosystem. Univ. Chicago Press. 389 pp.

Skarpe, C. 1990. Shrub layer dynamics under different herbivore densities in an arid savanna, Botswana. J. Applied Ecology 27:873-885.

Skarpe, C. 1991. Impact of grazing in savanna ecosystems. Ambio 20:351-356.

Smeins, F.E. 1984. Origin of the brush problem-a geographical and ecological perspective of contemporary distributions, pp. 5-16. *In*: K.W. McDaniel (ed.), Proc. Brush Management Symposium. Texas Tech Press, Lubbock.

Smeins, F.E. and L.B. Merrill. 1988. Long-term change in semi-arid grassland, pp. 101-114. *In*: B.B. Amos and F.R. Gehlbach (eds), Edwards plateau vegetation. Baylor Univ. Press, Waco, TX. USA.

Smith, D.A. and E.M. Schmutz. 1975. Vegetative changes on protected versus grazed desert grassland ranges in Arizona. J. Range Manage. 28:453-457.

Soriano, A. and O. Sala. 1983. Ecological strategies in a Patagonian arid steppe. Vegetatio 56:9-15.

Snook, E.C. 1985. Distribution of eastern redcedar on Oklahoma rangelands, pp. 45-52. *In*: R.F. Wittwer and D.M. Engle (eds), Eastern redcedar in Oklahoma. Conference Proceedings, Coop. Extension Serv., Div. Agriculture, Oklahoma State Univ., Stillwater.

Sparks, S.R., N.E. West and E.B. Allen. 1990. Changes in vegetation and land use at two townships in skull valley, western Utah, pp. 27-36. *In*: E.D. McArthur, E.M. Romney, S.D. Smith and P.T. Tueller (eds), Cheatgrass invasion, shrub die-off and other aspects of shrub biology and management. USDA/Forest Service Gen. Tech. Rep. INT-276, Ogden, Utah. 351 pp.

Specht, R.L., D.J. Connor and H.T. Clifford. 1977. The heath-savanna problem: The effects of fertilizers on sand-heath vegetation of North Stradbroke Island, Queensland. Aust. J. Ecology 2:179-186.

Stahle, D.W. and M.K. Cleavland. 1988. Texas drought history reconstructed and analyzed from 1698 to 1980. J. Climate 1:59-74.

Stein, R. and J.A. Ludwig. 1979. Vegetation and soil patterns on a Chihuahuan Desert bajada. American Naturalist 102:28-37.

Steinauer, E.M. and T.B. Bragg. 1987. Ponderosa pine (*Pinus ponderosa*) invasion of Nebraska sandhills prairie. American Midland Naturalist 118:358-365.

Stockton, C.W. and D.M. Meko. 1975. A long-term history of drought occurrence in western United States as inferred from tree rings. Weatherwise 28:245-249.

Stoddart, L.A., A.D. Smith and T.W. Box. 1975. Range management. McGraw Hill, New York.

Streng, D.R. and P.A. Harcombe. 1982. Why don't east Texas savannahs grow up to forest? Amer. Midl. Nat. 108:278-294.

Stuart-Hill, G.C. and N.M. Tainton. 1989. The competitive interaction between *Acacia karroo* and the herbaceous layer and how this is influenced by defoliation. J. Applied Ecology 26:285-298.

Stuth, J.W. 1991. Foraging behavior, pp. 65-83. *In*: R.K. Heitschmidt and J.W. Stuth (eds), Grazing management: an ecological perspective. Timber Press, Portland, OR.

Swetnam, T.W. and J.L. Betancourt. 1990. Fire-southern oscillation relations in the southwestern United States. Science 249:1010-1020.

Tharp, B.C. 1926. Structure of Texas vegetation east of the 98th Meridian. Univ. Texas Bulletin 2606.

Thurow, T.L. 1991. Hydrology and erosion, pp. 141-160. *In*: R.K. Heitschmidt and J.W. Stuth. (eds), Grazing management: an ecological perspective. Timber Press, Portland, OR.

Tieszen, L.L. and S. Archer. 1990. Isotopic assessment of vegetation changes in grassland and woodland systems, pp. 293-321. *In*: C.B. Osmond, L.F. Pitelka and G.M. Hidy. (eds). Plant biology of the basin and range. Ecological Studies 80, Springer-Verlag, New York.

Tisdale, E.W. and M. Hironaka. 1981. The sagebrush-grass region: a review of the ecological literature. Idaho For. Wildl. Range Exp. Sta. Bul. 33, Moscow, ID.

Torell, L.A. 1984. Economic optimum stocking rates and retreatment schedules for crested wheatgrass stands. Ph.D. Dissertation, Utah State Univ., Logan.

Trewartha, G.T. 1968. An introduction to climate. McGraw-Hill, New York.

Trollope, W.S.W. 1982. Ecological effects of fire in South African savannas, pp. 292-306. *In*: B.J. Huntley and B.H. Walker (eds), Ecology of tropical savannas. Ecological Studies 42, Springer-Verlag, Berlin.

Turner, R.M. 1990. Long-term vegetation change at a fully protected Sonoran Desert site. Ecology 71:464-477.

Van Auken, O.W. and J.K. Bush. 1987. Influence of plant density on growth of *Prosopis glandulosa var. glandulosa* and *Buchloe dactyloides*. Bul. Tor. Bot. Club 114:393-401.

Van Auken, O.W. and J.K. Bush. 1988. Competition between *Schizachyrium scoparium* and *Prosopis glandulosa*. Amer. J. Bot. 75:782-789.

Van Auken, O.W. and J.K. Bush. 1989. *Prosopis glandulosa* growth: influence of nutrients and simulated grazing of *Bouteloua curtipendula*. Ecology 70:512-516.

Van Auken, O.W. and J.K. Bush. 1990. Influence of light levels, soil nutrients and competition on seedling growth of *Baccharis neglecta* (Asteraceae). Bul. Tor. Bot. Club 117:438-444.

Van Devender, T.R. 1980. Holocene plant remains from Rocky Arroyo and Last Chance Canyon, Eddy County, New Mexico. Southwestern Naturalist 25:361-372.

Van Devender, T.R. and W.G. Spaulding. 1979. Development of vegetation and climate in the southwestern United States. Science 204:701-710.

Van Vegten, J.A. 1983. Thornbush invasion in a savanna ecosystem in eastern Botswana. Vegetatio 56:3-7.

Vasek, F.C. 1980. Creosote bush: long-lived clones in the Mojave Desert. American J. Botany 67:246-255.

Verstraete, M.M. 1986. Defining desertification: a review. Climatic Change 9:5-18.

Wagner, F.H. 1989. Grazers, past and present, pp. 151-162. *In*: L.F. Huenneke and H.A. Mooney (eds), Grassland structure and function: California annual grassland. Kluwer Academic Publishers, Boston.

Wagoner, J.J. 1949. History of the cattle industry in southern Arizona, 1540-1940. Social Sciences Bul. 20, Univ. Arizona, Tucson.

Walter, H. 1979. Vegetation of the Earth and ecological systems of the geobiosphere. Springer-Verlag, New York.

Weltzin, J.F. 1990. The role of prairie dogs (*Cynomys ludovicianus*) in regulating the population dynamics of the woody legume *Prosopis glandulosa*. MS Thesis, Texas A&M University, College Station.

West, N.E. 1988. Inter-mountain deserts, shrubsteppes and woodlands, pp. 209-230. *In*: M.G. Barbour and W.D. Billings (eds), North American terrestrial vegetation. Cambridge Univ. Press, New York.

West, N.E., K.H. Rea and R.O. Harniss. 1979. Plant demographic studies in sagebrush-grass communities in southeasten Idaho. Ecology 60:376-388.

West, N.E. and N.S. Van Pelt. 1987. Successional patterns in pinyon-juniper woodlands, pp. 43-52. *In*: R.L. Everett (ed), Proceedings—Pinyon-juniper conference. USDA/For. Serv. INT-215, Ogden, Utah.

Western, D. and C. Van Praet. 1973. Cyclical changes in the habitat and climate of an East African ecosystem. Nature 241:104-106.

Whicker, A.D. and J.K. Detling. 1988a. Ecological consequences of prairie dog disturbances. BioScience 38:778-785.

Whicker, A.D. and J.K. Detling. 1988b. Modification of vegetation structure and ecosystem processes by North American grassland mammals, pp 301-316. *In*: M.J.A. Werger, P.J.M. van der Aart, H.J. During and J.T.A. Verhoeven (eds). Plant form and vegetation structure. SPB Academic Publishing, The Hague, The Netherlands.

Whitowski, E.T.F. 1989. Response to nutrient additions by the plant growth forms of sand-plain lowland fynbos, South Africa. Vegetatio 79:89-97.

Whittaker, R.H. 1975. Communities and ecosystems. (2nd ed). MacMillan, New York.

Whittaker, R.H., L.E. Gilbert and J.H. Connell. 1979. Analysis of a two-phase pattern in a mesquite grassland, Texas. J. Ecology 67:935-952.

Williams, K., R.J. Hobbs and S.P. Hamburg. 1987. Invasion of an annual grassland in Northern California by *Baccharis pulularis ssp. consanguinea*. Oecologia (Berl.) 72: 461-465.

Wondzell, S.M. 1984. Recovery of desert grassland in Big Bend National Park following 36 years of protection from grazing by domestic livestock. MS Thesis, New Mexico State University, Las Cruces.

Woodward, R.A., K.T. Harper and A.R. Tiedemann. 1984. An ecological consideration of the significance of cation-exchange capacity of roots of some Utah range plants. Plant and Soil 79:169-180.

Wright, H.A. and A.W. Bailey. 1982. Fire ecology. John Wiley, New York.

Wright, H.A., S.C. Bunting and L.F. Neunschwander. 1976. Effect of fire on honey mesquite. J. Range Manage. 29:467-471.

Wright, R.G. and G.M. Van Dyne. 1976. Environmental factors influencing semidesert grassland perennial grass demography. Southwestern Naturalist 21:259-274.

Yeaton, R.I. 1988. Porcupines, fires and the dynamics of the tree layer of the *Burkea africana* savanna. J. Ecology 76:1017-1029.

York, J.C. and W.A. Dick-Peddie. 1969. Vegetation changes in southern New Mexico during the past one hundred years, pp. 155-166. *In*: W.G. McGinnes and B.J. Goldman (eds), Arid lands in perspective. University of Arizona Press, Tucson.

Young, J.A. and R.A. Evans. 1981. Demography and fire history of a western juniper stand. J. Range Manage. 34:501-506.

Effects of Grazing on Ecosystems of the Great Plains

W. K. Lauenroth, D. G. Milchunas, J. L. Dodd, R. H. Hart, R. K. Heitschmidt, and L. R. Rittenhouse

Abstract

Grazing by large herbivores has been part of the evolutionary and developmental environment of the Great Plains for the past 10,000 years. There is evidence that past grazing has had an important positive effect on the ability of the ecosystems of the Great Plains to withstand grazing by domestic livestock. Settlement of the region by European agriculturalists in the past 150 years has had major impacts but there is no conclusive evidence that the change from wild ungulates to properly managed domestic livestock has had detrimental ecological impacts.

The major west-east and south-north climatic gradients provide the best explanations for differences in the responses of ecosystems to grazing. Experiments conducted in the western, driest portion of the region suggests that plant communities in these areas are the least susceptible to detrimental impacts of improper grazing management. These areas also have the lowest productivity both of plants and animals. By contrast, plant communities in the eastern, wettest part of the plains have the capacity to support the largest number of animals per unit area but they are most susceptible to changes in community composition, diversity, and physiognomy under well managed light to moderate grazing. These communities are also potentially the most vulnerable to the detrimental impacts of poor management. Aboveground net primary production generally declines with grazing, although increases in the wettest areas have been attributed to opening dense canopies and reducing litter layers.

Interactions between livestock grazing and wildlife is a complex issue in the Great Plains as it is everywhere else. The same grazing management system can have positive impacts for one wildlife species and negative for others. Furthermore, it is clear in may cases that the effects of one management system on a single species can be a mixture of positive and negative impacts.

Despite the fact that grazing by domestic livestock does not have simple relationships to the other elements of the structure and function of Great Plains ecosystems, management of grazing animals using current technology appears to be a sustainable resource use. The conversion of rangeland to cropland has many implications for both sustainable resource use and global environmental issues.

Introduction

The Great Plains of North America is the vast treeless area that occupies the majority of the center of the continent. At the time of settlement of North America by Europeans the Great Plains was the home of largely nomadic tribes of native Americans and large numbers of bison, prairie dogs, elk, pronghorn, bighorn sheep, and deer. These herbivores have been present on the plains since the retreat of the Pleistocene glaciers and have had large influences on the ecosystems of the region. This long history of herbivory in the Great Plains has resulted in ecosystems that appear to be very well adapted to withstand grazing by domestic livestock (Mack and Thompson 1982).

In the 1800s settlers began systematically reducing the herds of bison across the plains and replacing them with domestic livestock. The plains livestock enterprise began in Texas before the Civil War and by 1876 had spread over the entire region (Webb 1931). In 1880 the Great Plains including Texas supported 11 million head of cattle (Webb 1931). No sooner had the range livestock industry expanded to include the entire plains area than it began to decline. In 1862 congress passed the Homestead Act. In 1874 the first barbed-wire was sold in the United States (Webb 1931). By the end of the century the forces of climate, economics, and social pressure had changed the plains cattle industry from an open range exclusive-use enterprise to a ranch based industry that coexisted with cropland.

The early 1900s was a time of extensive cropland development throughout the plains. High prices and favorable weather resulted in homesteaders occupying large areas of the plains including many that were marginal or unsuitable for sustained crop production. The depression of the early 1930s accompanied by a major drought resulted in widespread abandonment of farmland in the plains and the creation of large federal landholdings that would become the National Grasslands in the 1950s. Landuse patterns established after the 1930s have largely remained stable to the present with the exceptions of increases in the area of irrigated cropland on uplands associated with the development of the Ogalla aquifer and fluctuations in the amount of cropland caused by weather and economic conditions.

The livestock industry and the use of native vegetation by domestic grazing animals is well established in the Great Plains and is an important economic and social force in the region. Current interest and concern about the use of public lands in the United States is resulting in widespread questioning of the wisdom of use of public lands by domestic livestock. The objective of this paper is to examine the currently available information about impacts of domestic livestock grazing on ecosystems of the Great Plains and make projections about the future.

Description Of The Great Plains

Geography and Physiography

The Great Plains has several geographic definitions. First it is a physiographic region defined by landform extending north-south from southern Canada to central and southwestern Texas and east-west from approximately the 98th meridian to the base of the Rocky Mountains. This is a precise definition that is independent of political boundaries. Secondly in the United States there is an agricultural region that is referred to as the Great Plains that is based upon agency and organizational considerations that correspond to state boundaries (Fig. 1). This Great Plains is a 10-state region that includes the physiographic Great Plains. Finally there is a popular non-technical definition of the Great Plains that corresponds to the entire central portion of the North American continent that was largely grassland at the time of settlement.

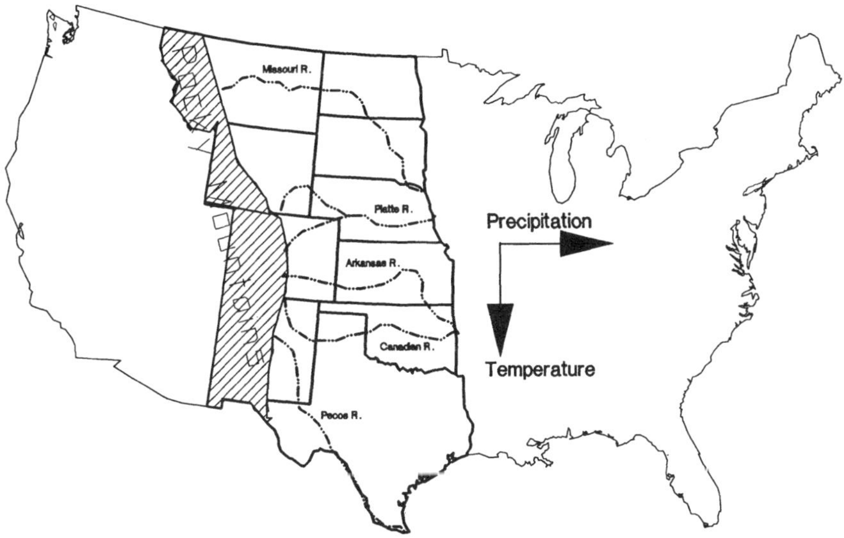

Figure 1. Geographic location of the Great Plains of the United States showing major rivers and climatic gradients.

We will use the term Great Plains to refer to the 10-state agricultural region that includes North and South Dakota, Nebraska, Kansas, Oklahoma, Texas and the plains region of Montana, Wyoming, Colorado, and New Mexico (Fig. 1). We will refer to the remainder of the grassland area in the center of the continent as the Central Grassland Region.

The major surface features of the Great Plains are the result of complex interactions between wind and water driven processes as well as glaciation in the northeastern portion. Material eroded from the Rocky Mountains and carried eastward by major river systems has been deposited on the plains over the past several million years. After deposition this material has been reworked by both wind and water erosion to produce the current surface. As it exists today, the Great Plains is a relatively flat surface that slopes from the base of the Rocky Mountains to approximately the 98th meridian. Elevation at the base of the mountains is approximately 1500 meters, and 150 meters where it meets the Mississippi River. The flatness of the surface is only relative and local topography results in elevation differences of up to several hundred meters over short distances.

Five important river systems cross the region from west to east. In the north the Missouri River flows through Montana, North and South Dakota, Nebraska, and

Kansas. In the center of the region the Platte and Arkansas Rivers drain the east slope of the central Rockies and flow across Wyoming, Colorado, Nebraska, Kansas, and Oklahoma. In the south the Canadian River flows across New Mexico, Texas, and Oklahoma. Because water is such a critical resource in the region, these river systems had a major impact on the patterns of settlement and currently have a large impact on the geography of regional development and landuse.

Except in the northeast, glaciation during the Pleistocene had a relatively minor direct impact on the physical features of the Great Plains . The southern extent of continental glaciers during the Pleistocene was the Missouri River. South of the boundary the region was ice-free. A major indirect influence of glaciation was the transport of billions of tons of sediment onto the plains by the melt water from mountain glaciers.

Climate

The distinctive features of the Great Plains climate are the result of two important geographic characteristics; its location in the center of the North American continent and the presence of the Rocky Mountains as its western boundary. The general characteristics of the climate are a north to south temperature gradient and a west to east precipitation gradient (Fig. 2). Annual average temperatures are near 0°C at the northern extension of the Great Plains and near 20 degrees in the south (Fig. 2). The climate is classified as temperate in the north and subtropical in the south. Annual precipitation is 300 mm in the semiarid western portion and 900 mm in the subhumid east (Fig. 2). These gradients in precipitation and temperature produce important gradients in the structure and dynamics of ecosystems across the region.

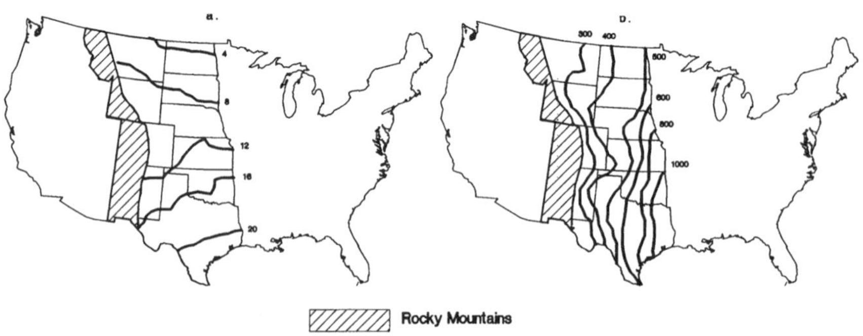

Figure 2. Isotherms for mean annual temperature (°C) and isohyets for mean annual precipitation (mm) for the Great Plains of the United States.

The climate of the Great Plains is influenced by three air masses, each with unique properties. Winters are dry and dominated by strong westerly flow and the influence of a temperate maritime air mass that has it origins in the northern Pacific Ocean. To reach the Great Plains this air must cross a number of mountain ranges and consequently it is typically a source of cool relatively dry air. The water content of this air and the amount of winter precipitation a particular location receives is dependent upon where it crossed the mountains. There is a decreasing winter precipitation gradient in the Great Plains from north to south. The position of the jet stream determines how often and how much of the region is subjected to the influence of a very cold dry polar air mass. The frequency of exposure to these conditions also decreases from north to south.

Spring and summer are the wet seasons across the Great Plains. Peak precipitation tends to occur in the spring (April and May) in the north grading to summer peaks (July and August) in the south. The strong westerly flow of winter begins weakening in the spring and the influence of a warm, wet subtropical air mass with origins over the Gulf of Mexico begins to be felt in the region. In addition to an increase in precipitation in the spring, there is also a shift from snow or a mixture of snow and rain to rain.

Ecosystems and Landuse

Grasslands

Prior to settlement of the Great Plains by Europeans, grasslands were the dominant ecosystem on uplands. Ecosystems dominated by woody plants were restricted to gallery forests along the major river channels and to sites with particularly favorable characteristics. The current distribution of ecosystems on uplands is a mosaic of native grasslands and croplands. Much of the area along the major river channels now supports irrigated crops. Ecosystems dominated by woody plants still survive very close to river channels and on favorable sites not suitable for cropland.

The natural or presettlement vegetation has been described by many authors (Shantz 1923, Weaver and Clements 1938, Weaver 1954, Weaver and Albertson 1956, Küchler 1964). Previous to the Küchler map most authors divided the vegetation into either shortgrass and tallgrass types or shortgrass, mixedgrass, and tallgrass types. Küchler (1964) departed from that tradition by dividing the region into areas dominated by particular species or groups of species. Thus the Küchler map was more detailed and necessarily more complicated than previous maps. Dodd (1979) produced a useful grassland type map of the region by using the Küchler map as a starting point and surveying all previous classifications to arrive at a new synthesis of the information about the distributions of grassland types (Fig. 3). The Dodd map is a classification of the vegetation into grassland types each with specific ecological and management characteristics. An additional value of the Dodd map is that the grassland types are in most cases an aggregation of Küchler's vegetation types and therefore each type can be broken down into its constituent vegetation units if necessary.

The Great Plains comprises four grassland types (Fig. 3). Three of the types are contained entirely in the Great Plains; approximately one-half of the fourth type, the tallgrass prairie, lies outside the region. The distribution of the grassland types is almost entirely determined by the climatic gradients across the region. The tallgrass prairie occupies the wettest, eastern-most portion of the Great Plains. The dominant species in the tallgrass type are all C_4 grasses in the central and southern parts and C_4 and C_3 grasses in the north. The most common C_4 species are *Andropogon gerardii, Schizachyrium scoparium, Panicum virgatum*, and *Sorgas-*

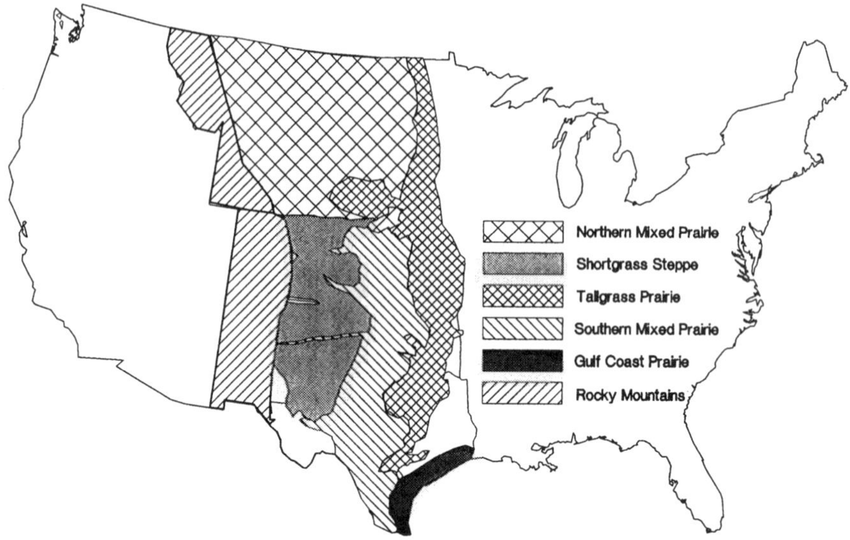

Figure 3. Geographic distribution of grassland types in the Great Plains of the United States (Adapted from Dodd 1979).

trum nutans. The most common C_3 species is *Stipa spartea*. In addition to occupying the wettest portion of the Great Plains, the tallgrass prairie is the most productive of the grassland types (Fig. 4)(Sala et al. 1988). By contrast the shortgrass steppe occupies the driest and one of the least productive parts of the region. The dominant species *Bouteloua gracilis* and *Buchloe dactyloides* are both C_4 shortgrasses. The northern mixed prairie occupies the area north of the shortgrass steppe and west of the tallgrass prairie. The boundary between the northern mixed prairie and the tallgrass prairie occurs at approximately the 500 mm/yr isohyet (Figs. 2 and 4). Annual aboveground net primary production is less than 300 g/m^2 throughout the northern mixed prairie. The reason for describing it as a 'mixed' prairie is that the dominants are a mixture of short and mid-height grasses and a mixture of C_3 and C_4 species. The dominant species are the C_3 grasses *Agropyron smithii* and *Stipa comata* as well as the C_4 shortgrasses from the shortgrass steppe. The southern mixed prairie is located between and south of the shortgrass steppe and tallgrass prairies. This is a 'mixed' prairie because the dominant species are a mixture of mid-height and short C_4 grasses. The mid-height grasses are *Bouteloua curtipendula* and *Schizachyrium scoparius*. The shortgrasses are dominants from the shortgrass steppe. Aboveground production ranges from 200 to 400 g/m^2 per year (Fig. 4).

Croplands

Current patterns of landcover in the region include a substantial portion of cropland in addition to native grassland (Fig. 5). The proportion of the grassland types that have been converted to cropland depends upon a combination of soil and climatic factors. The wettest areas have been exploited most heavily. The tallgrass prairie has been largely converted to cropland except in those areas of very shal-

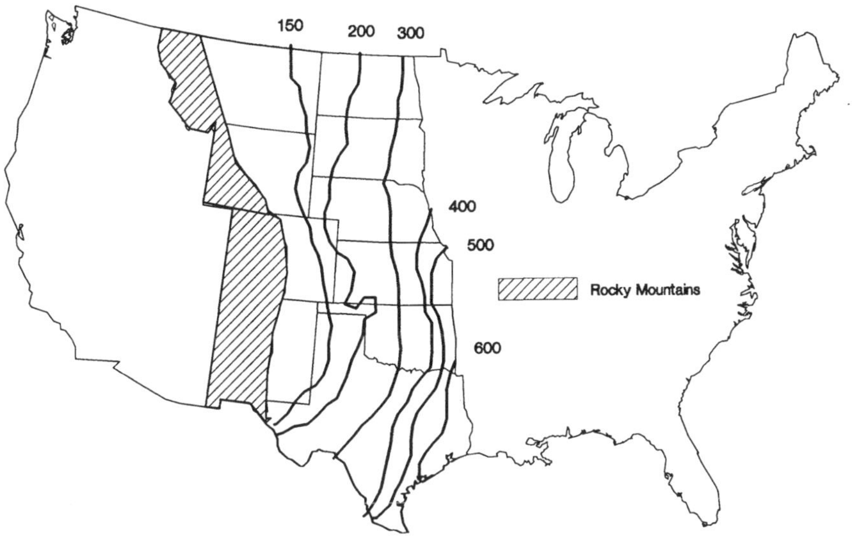

Figure 4. Isopleths of average annual aboveground net primary production (g/m²) for the Great Plains of the United States (Adapted from Sala et al. 1988).

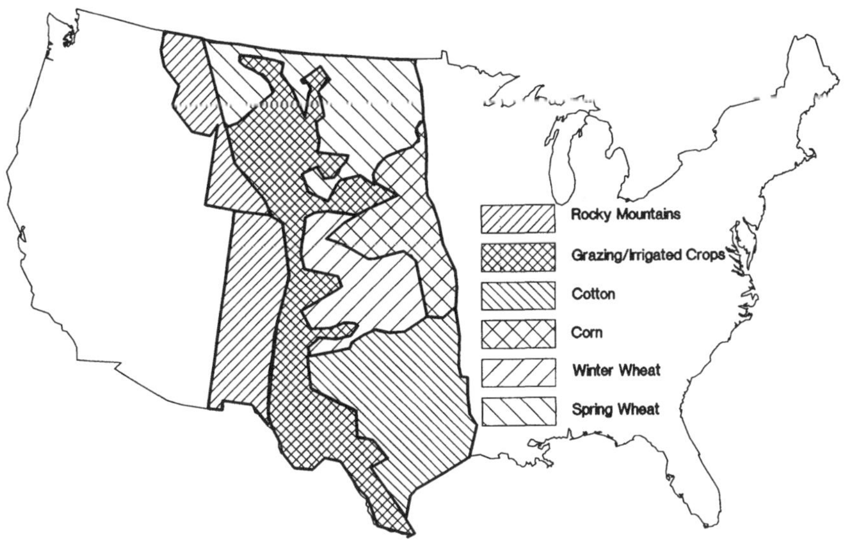

Figure 5. Geographic distribution of crop regions in the Great Plains of the United States. (Adapted from Dregne and Willis 1983)

low soils such as the Flint and Osage Hills of Kansas and Oklahoma, or very sandy soils such as the Sand Hills of Nebraska. The shortgrass steppe and the northern mixed prairie have the smallest proportion of cropland of the four grassland types.

The region can be divided into four crop zones and a zone that is a combination of grazingland and irrigated cropland. The principal crops in the Great Plains are cotton in the southeast, winter wheat in the center, corn in the east central, and spring wheat in the north. The western edge of the region is largely grazingland except for local areas in which irrigation water is available. The west to east projections in this zone are associated with large rivers and reflect irrigated cropland.

Approximately 58% of the area of the Great Plains is rangeland supporting mostly native vegetation (Table 1). The states with the largest proportions of rangeland are those in the western portion of the Great Plains adjacent to the eastern edge of the Rocky Mountains. Thirty-one percent of the region is in crops and the eastern tier of states have the highest proportions of croplands corresponding to highest precipitation.

Table 1. Agricultural landuse (hectares / %) for the ten Great Plains states (adapted from Newman 1988).

	Pasture	Range	Forest	Crop	Total
Colorado	5.26×10^5 3.2	9.80×10^6 60.3	1.62×10^6 10.0	4.29×10^6 26.4	1.62×10^7 100.0
Kansas	8.91×10^5 4.5	6.84×10^6 34.6	2.43×10^5 1.2	1.18×10^7 59.6	1.98×10^7 100.0
Montana	1.21×10^6 4.7	1.53×10^7 59.8	2.11×10^6 8.2	6.96×10^6 27.2	2.56×10^7 100.0
Nebraska	8.50×10^5 4.5	9.35×10^6 50.0	2.83×10^5 1.5	8.22×10^6 43.9	1.87×10^7 100.0
New Mexico	8.10×10^4 0.4	1.66×10^7 84.9	1.90×10^6 9.7	9.72×10^5 5.0	1.96×10^7 100.0
North Dakota	5.26×10^5 3.3	4.41×10^6 27.5	1.62×10^5 1.0	1.09×10^7 68.2	1.60×10^7 100.0
Oklahoma	2.87×10^6 17.6	6.11×10^6 37.5	2.63×10^6 16.1	4.70×10^6 28.8	1.63×10^7 100.0
South Dakota	1.09×10^6 6.3	9.23×10^6 53.0	2.43×10^5 1.4	6.84×10^6 39.3	1.74×10^7 100.0
Texas	2.83×10^6 4.8	3.86×10^7 65.8	3.77×10^6 6.4	1.35×10^7 23.0	5.87×10^7 100.0
Wyoming	2.83×10^5 2.2	1.09×10^7 86.2	4.05×10^5 3.2	1.05×10^6 8.3	1.26×10^7 100.0
Great Plains	1.12×10^7 5.1	1.27×10^8 57.6	1.34×10^7 6.0	6.92×10^7 31.3	2.21×10^8 100.0

Response of Great Plains Ecosystems to Grazing

Large grazing animals have a wide range of effects on ecosystems. Some are the direct result of grazing and occur quite rapidly; others are indirect and may take decades to occur (Ellison 1960, Belsky 1986). Within a particular ecosystem, these effects are a function of the intensity, frequency, seasonal timing, and duration of grazing. Among ecosystems, a particular grazing regime can result in very different responses depending on climatic conditions and structural, functional, and historical attributes. The above general statements represent potential effects of grazing, potentials for management of grazing, and different outcomes depending on the specific location. While the ecological effects of grazing on a particular parcel of land can be variable, public opinion of grazing tends to assume only negative effects are possible. A recent symposium at the annual meeting of the Society for Range Management was entitled "Can livestock be used as a tool to enhance wildlife habitat?" (Severson 1990). Does livestock grazing have only negative impacts on ecosystems? Are there situations where livestock grazing is not a negative, destabilizing force? In this section, we will attempt to answer these questions by examining some examples of the effects of livestock grazing on plant communities, wildlife and other animal populations, and soil properties across the Great Plains.

From the standpoint of public opinion, judgments concerning livestock grazing are based primarily upon impacts on vegetation and wildlife. Stories from early explorers of grass-up-to-the-stirrups and news reports of shooting of raptors and poisoning of prairie dogs form a basis for generalizations. From a traditional ranchers standpoint, judgments concerning livestock grazing are primarily based upon animal weight gain (economics) and changes in species composition of the rangeland. Rangeland condition and trend analysis ranks sites from poor to excellent condition based upon species composition using concepts developed for secondary succession, and such ranking is the primary method of assessment on federal and private grazinglands. An ecologist may question grass-up-to-stirrups in several of the Great Plains grassland types, the degree of competition or interference between livestock and wildlife, and succession based models, while acknowledging the concerns of the public, the necessity of a meat industry, and the necessity of simple means of assessing vast expanses of rangeland. An ecologist would also consider aspects of grazing such as long-term soil nutrient status and the ability to sustain a productive system, the magnitude of ammonia volatilization from urine, ruminal methane production, and carbon budgets of rangelands in reference to global climate change. The issue of grazing effects is sociologically as well as ecologically complex. Different publics view grazing from different perspectives. While the effects of grazing can vary from one location to another, measuring any one variable within a particular site may not provide an indication of the response of the ecosystem to grazing; we examine this in the following sections.

Plant community composition

The most widely used model for rangeland condition and trend in the U.S. is based upon changes in plant community composition as a result of grazing (Lauenroth and Laycock 1989). This model is based upon the concept of secondary succession following a disturbance and conceptualizes grazing as a force which moves a plant community backwards along the pathway that would be followed to produce the community. While this model has some applicability to the Great Plains, there are enough concerns about it (Lauenroth and Laycock 1989) that we have chosen to use a general model of the response of grasslands to grazing (Milchunas et al. 1988a). This model uses evolutionary grazing history and environmental moisture to explain differences among grasslands in their responses to grazing by large herbivores. We will use the Milchunas et al. (1988a) model as a framework for understanding the effects of grazing on plant community composition across Great Plains grasslands, and make contrasts with the traditional condition and trend model.

Grasslands in the Great Plains generally have a long evolutionary history of grazing (Mack and Thompson 1982, Milchunas et al. 1988a); therefore we will primarily focus on the response of plant communities to grazing in relation to the moisture/productivity gradient (Fig. 2 and 4). We will group studies with the four grassland types: shortgrass steppe, northern mixed prairie, southern mixed prairie, and tallgrass prairie.

Lauenroth and Milchunas (1991) calculated that total above-and belowground herbivory at the Central Plains Experimental Range (CPER) in the northern shortgrass steppe is equally divided among ruminants, arthropods, and nematodes. However, cattle account for 84% of the aboveground herbivory and consume 40% of the aboveground production in moderately grazed rangeland. Levels of herbivory on heavily grazed shortgrass steppe are similar to those reported for heavily grazed Serengeti grasslands, and these levels are among the highest of any communities in the world (see McNaughton 1976, Milchunas et al. 1989).

Light, moderate, and heavy grazing treatments and grazing exclosures were established at the CPER in 1939 (Klipple and Costello 1960), and have been sampled at various times since their inception. Based upon data from 1940-42, 1946-48, and 1952-53, Klipple and Costello concluded: 1) changes in plant cover as a result of grazing were slow, and climate affected changes far more rapidly, 2) light grazing (20% removal of plant production) increased cover, 3) moderate grazing (40% removal) maintained range condition, and 4) heavy grazing (60% removal) caused some declines in palatable species. Sampling in 1962-63, Hyder et al. (1966) concluded that grazing at the different intensities for 23 years had "not affected the frequency percentages of species to a great extent". Sampling in 1970, Grant (1971) concluded that "grazing by large herbivores appears to have no significant effect on the proportional plant species composition at Pawnee or Pantex" (CPER and a shortgrass site in Texas, respectively). Sampling in 1984 and 1986, Milchunas et al. (1989) concluded that shortgrass plant communities were relatively invariant with respect to grazing, and that grazing is not a disturbance to plant communities at the CPER.

We calculated dissimilarity indices (Whittaker (1952) index of community association) from the above data sets, where a value of 0 indicates no difference between grazing treatments and a value of 1 indicates completely different species composition. Dissimilarity values for ungrazed versus respective light, moderate, and heavy grazing treatments were 0.161, 0.148, and 0.068 after 8 years of treatment (data from Klipple and Costello 1960), 0.133, 0.133, and 0.101 after 13 years of treatment, 0.319, 0.211 and 0.218 after 31 years of treatment (data from Hyder et al. 1966), and 0.100 (ungrazed vs heavy grazed) after 47 years of treatment (data from Milchunas et al. 1989). The relatively high values for the 31-year-data are probably because of sampling methods, i.e., biomass versus density or frequency for other studies.

Additional studies of the structure of shortgrass steppe vegetation at the CPER in response to grazing found that long-term heavy grazing by cattle: 1) increased total plant basal cover and basal cover of the dominant species (Milchunas et al. 1989) 2) decreased abundance of native and exotic opportunistic species (Milchunas et al. 1989, 1990), 3) decreased differences in plant communities along topographic gradients and lowered within-site diversity (Milchunas et al. 1989), and produced a more uniform horizontal distribution of aboveground, crown, and root biomass, with only minor effects on the quantity of root biomass (Milchunas and Lauenroth 1989). Comparisons of the grazing treatments to other large-scale disturbances indicated that ungrazed plant communities were more similar to disturbed plant communities than were heavily-grazed communities (Milchunas et al. 1990). A demographic study using seed additions and plant removals indicated that the ungrazed plant communities were more susceptible to invasion by opportunistic species (weeds) than were heavily-grazed communities, primarily due to the indirect effects of cattle on microenvironments and plant competition than to direct effects of injury and mortality from defoliation (Milchunas et al. 1992). Management of wildlife exclosures by the U. S. Forest Service in the Pawnee National Grasslands includes periodic grazing by cattle to reduce populations of introduced invader species such as *Cirsium undulatum* (Godbolt, pers. commun.).

Changes in species composition with grazing on northern mixed prairie are generally of a greater magnitude than on shortgrass steppe, although change may be slower than in more productive grassland types. Mixed prairie has a component of shortgrass steppe species, and it is this component which usually increases with grazing and decreases with exclosure to grazing. Sampling four-year-old ungrazed exclosures and grazed areas with 35% utilization, Vogel and Van Dyne (1966) found essentially the same basal cover of grasses, sedges, forbs, litter, and bare soil on grazed and protected sites in Montana. They concluded that four years was too short a time for dramatic changes in cover to occur. Hofmann and Ries (1989) observed similar results on a four-year exclosure in North Dakota; no change in basal cover classes and only one significant change for an individual species. Dissimilarity indices for grazed versus ungrazed treatments calculated from the data of these two studies, 0.148 and 0.190, indicated small changes. Even after 41 years of exclosure, changes in species composition can be relatively small when the site is in the dry, low production portion of northern mixed prairie. Smoliak et al. (1972) reported that *Bouteloua gracilis* (the dominant species of shortgrass range) increased with increasing intensity of grazing and *Stipa comata* and *Agropyron smithii* (both mid-height species) decreased. However, dissimilarity indices for this dry (310 mm/yr precipitation) grassland are only 0.150, 0.188, 0.259 for the ungrazed versus the light, moderate, and heavy grazing treatments, respectively, indicating relatively small changes.

After 11 and 12 years of protection, dissimilarity indices for data from Pace and Thilenius (1968) were 0.125 and 0.280, respectively, for northern mixed prairie in a 432 mm/yr zone in the Black Hills of South Dakota. Dissimilarity indices after 27 years of protection were 0.594, 0.689, and 0.706 for three other grazed-ungrazed comparisons. Working in a 356 mm/year precipitation zone in North Dakota, Brand and Goetz (1986) found increases in the shortgrass *B. gracilis* with grazing, but the midgrass and tallgrass category was more dominant in only one out of four exclosures. These four 40-year-old exclosures and 45% removal grazing treatments had dissimilarity indices averaging 0.418. Rauzi (1963) observed increases in *B. gracilis* composition from 1.7% in 21-year-old ungrazed to 53.5 and 74.7% in moderate and heavy grazed treatments, respectively, in a 415 mm/yr North Dakota mixed prairie. Dissimilarity indices calculated from Rauzi's data were 0.573 and 0.740 for ungrazed versus moderate and heavy grazed treatments, respectively.

On northern mixed prairie in Wyoming, botanical composition did not differ between light and heavy grazing after 6 years (Hart et al. 1988). However, basal cover of *Agropyron smithii* decreased and that of *B. gracilis* increased after 10 years of heavy grazing (Hart unpublished). The most profitable stocking rate was calculated to be 58-64 steer-days/ha, 60-80% above that recommended by the Soil Conservation Service to maintain range condition. Torell et al. (1989) re-examined gains of cattle grazing Colorado sandhills range and the response of vegetation (Sims et al. 1976), and concluded that long-term changes in vegetation would be minimal at the stocking rate returning the highest short-term profits.

Southern mixed prairies generally display responses to grazing similar to those of northern mixed prairies; increases in shortgrasses under heavy grazing at the expense of the mid-height grasses. Wood and Blackburn (1984) observed that heavy grazed pastures in the Texas Rolling Plains were generally dominated by shortgrass communities, moderately grazed by midgrass communities with only 30% of the area in shortgrass communities, and ungrazed 20 year-old exclosures had no shortgrass communities. Dissimilarity indices for this 624 mm/yr precipitation site averaged 0.369 and 0.690 for ungrazed versus moderate and heavy grazing treatments, respectively. Sampling the same treatments two years later, Heitschmidt et al. (1985) found similar responses in mixtures of short- and midgrasses to grazing. Average dissimilarity indices based upon Heitschmidt et al.'s data were 0.535 and 0.602 for the ungrazed versus moderate and heavy grazed treatments, respectively.

Grazed/ungrazed comparisons for three different southern mixed prairies in Kansas were provided by Tomanek and Albertson (1953, 1957). Sites near Hays, Kansas (590 mm annual precipitation), that were ungrazed, lightly grazed since settlement, grazed at moderate rates, and very heavy rates for "many" years, displayed the following responses: 1) basal cover was greatest on the heavily grazed treatment and least on the ungrazed, 2) tallgrasses were dominant in the ungrazed areas, mid-grasses in moderately grazed, and shortgrasses in the very heavily grazed treatments, and 3) dissimilarity indices calculated from their data averaged 0.561 for ungrazed versus moderate grazed and 0.701 for the ungrazed versus the very heavily grazed comparisons. Very similar results were observed at the other two Kansas grasslands that were in different soil types but similar levels of precipitation. Short buffalo grass (*Buchloe dactyloides*) averaged 10% of basal cover composition on ungrazed areas and increased to 90% with heavy use by cattle. Dissimilarity indices calculated from these data averaged 0.328 and 0.684 for ungrazed versus moderate and heavy grazing, respectively, at the sites receiving 567 mm annual precipitation, and 0.470 and 0.840, respectively, for the site receiving 594 mm of precipitation.

Grazed/ungrazed comparisons for the tallgrass prairie are limited from the standpoint that levels of removal by the livestock are not well documented and years of exclosure and grazing treatment are generally short. When grazing intensities are provided or can be calculated from stocking rates, what is considered moderate levels of grazing in tallgrass prairie is less intensive than moderate levels in the other grassland types. Productive grasslands with the potential for canopy closure are most susceptible to changes in species composition because tall species adapted to competing for light are not generally tolerant of grazing (Milchunas et al. 1988a).

In a two-year-study of grazing and fire treatments in tallgrass prairie, Collins (1987) reported that grazing reduced the cover of large, dense-patch-forming perennial grasses by 20 and 26% on burned and unburned treatments, respectively, and increased the number and density of annuals and ruderal forbs. After 10 to 13 years of exclosure to grazing, Penfound (1964) observed invasions of forest shrubs and trees into unburned tallgrass prairie in central Oklahoma. He predicted that the prairie would be eliminated in less than 30 years if left ungrazed. We calculated grazing intensities of only 30 to 34% removal, with dissimilarity indices averaging 0.492 for the Penfound site with 783 mm annual precipitation. Kelting (1954), working at the same grassland in different exclosures, concluded that large amounts of plant litter in unburned virgin tallgrass prairie prevented seedling establishment and probably accounted for the low cover of living vegetation compared with grazed treatments. However, grazing at only 30% removal for 25 years resulted in "weedy species which are characteristic of disturbed areas". Further, only 39% of the total species found in virgin prairie were encountered in the grazed pasture. We calculated a dissimilarity index of 0.470 from the Kelting data.

Drew (1947) and Kucera (1956) studied the same prairie in north-central Missouri after at least 70 and 83 years of exclosure to grazing, respectively. This relic prairie was burned frequently and represents the oldest studied exclosure we know of in tallgrass prairie, but the intensity of grazing on the adjacent land is not known. Drew observed greater total plant density and greater density and frequency of "cultivated weeds" in the grazed than the relic area. Grazing caused a decrease in woody species, but invasion into the ungrazed relic area was not great. Kucera found a 41% reduction in the number of species with grazing and additional increases in "invader" species. Dissimilarity indices calculated from the Drew and Kucera studies were 0.408 and 0.514, respectively, for this site which receives 939 mm annual precipitation. Dix (1959) sampled 24 pairs of grazed/ungrazed sites in Wisconsin. He found that tall, erect and tussock-forming species decreased with grazing and short, rhizomatous species increased. Native "weeds" confined to small mammal disturbances in ungrazed prairie occupied much larger areas of grazed pastures and introduced *Poa spp.* spread rapidly. We calculated an average dissimilarity index of 0.580 for these 24 comparisons.

Traversing the west-to-east gradient of increasing precipitation, from shortgrass steppe, to mixed prairie, to tallgrass prairie, dissimilarity in species composition with grazing generally increase (Fig. 6). Within precipitation zones, the degree of change is a function of years of grazing, intensity of grazing, the production potential of the site in terms of soil nutrient status, the particular assemblage of plant species, and precipitation during the particular years in which studies are conducted. Deciphering the complex set of factors that regulate the response of a plant community to grazing is further complicated by the different methods researchers employed. Gradients such as that presented in figure 6 may help us explain and predict differences between rangeland that is grazed versus ungrazed, but does not indicate the direction of the change. This requires an understanding of background conditions. Plant communities that have co-evolved with large herbivores for thou-

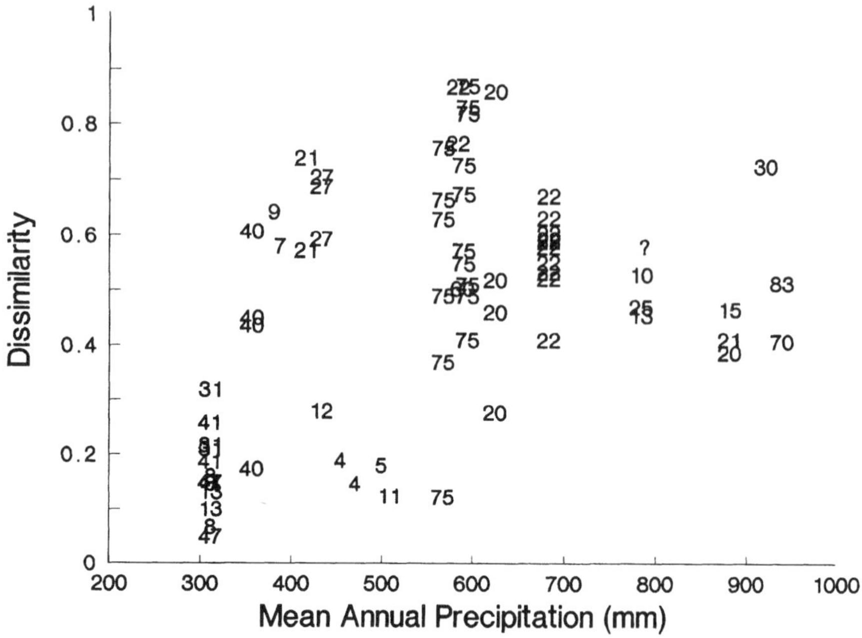

Figure 6. Calculated dissimilarity in species composition between grazed and ungrazed grasslands in the Great Plains of the United States. Numbers are the duration in years of the individual studies from which the data were taken (this is usually the age of the ungrazed exclosure).

sands of years before domestic grazers were introduced will more likely have a negative response to the removal of grazers rather than to grazing by domestic livestock. This has been demonstrated for the shortgrass steppe (Milchunas et al. 1990). Plant communities with a short evolutionary history of grazing are more likely to change when grazed by domestic animals than are those with a long evolutionary history of grazing (Mack and Thompson 1982, Milchunas et al. 1988a), and the change is more likely to be away from background conditions (Milchunas et al. 1988a).

In terms of species diversity, we will compare the two extremes in the precipitation/productivity gradient across the Great Plains. In the semiarid shortgrass steppe, combining data of Milchunas et al. (1989) and Gill (1985) resulted in a diversity index (Exp H') gradient of ungrazed > light, moderate grazing > heavy grazing for preferentially grazed lowland communities. Ungrazed treatments had greater richness (total number of species encountered) than heavily grazed treatments (Milchunas et al. 1989). However, part of the relatively high diversity and richness of the ungrazed treatment was due to greater densities of native and introduced opportunistic species. In contrast, Collins (1987) found diversity was lowest on ungrazed-burned tallgrass prairie and highest on grazed-burned treatments, and species richness increased with increasing disturbance intensity. Grazing in tallgrass prairie increases species diversity due to dynamics within grazed and ungrazed patches (Collins and Barber 1985). However, early studies of relatively virgin tallgrass prairie by Kelting (1954), Drew (1947), and Kucera (1956) suggest large reductions in the numbers of native prairie species with grazing.

Several conclusions can be reached concerning the effects of grazing on plant community structure in the Great Plains. Grazing and/or burning levels in tallgrass prairie can in many cases cause shifts in the community from forest to tallgrass prairie to mixed prairie. Ungrazed relatively mesic mixed prairie can shift towards tallgrass prairie and heavily grazed mixed prairie can resemble shortgrass steppe. Grazed and ungrazed shortgrass steppe are very similar and large shifts in community composition are not evident. Invasion by exotic species is enhanced by grazing in tallgrass prairie and decreased by grazing in shortgrass steppe. However, it is important to note that any grassland can be overgrazed, and heavily utilized areas in the immediate vicinity of watering tanks on shortgrass range, and improperly managed riparian zones, are often dominated by exotic and native opportunistic species. Ungrazed shortgrass steppe is more diverse than is grazed, but the opposite has been observed for tallgrass prairie. Diversity is not an indicator of the impact domestic grazers may have on rare or endangered species, and early studies have shown that less abundant species can be eliminated when virgin prairie is grazed. These responses are in agreement with Milchunas et al. (1988a) model, but are sometimes contradictory to the condition and trend model. Ungrazed shortgrass steppe communities are more characteristic of early successional stages than are heavily grazed communities.

Plant community productivity

Forage production of light, moderate, and heavily grazed treatments at the CPER shortgrass steppe site were first measured 11-15 years after the treatments were initiated, with five year averages of 74, 69, and 54 g/m^2 (Klipple and Costello 1960). Moderate grazing had not affected productivity compared with light grazing, but removal of 60% of the aboveground biomass (heavy grazing) had significantly reduced production. Cattle weight gains showed the same relationship with grazing treatment. Sampling the same areas after 31-33 years of treatment, Sims and Singh (1978) found 172 and 103 g/m^2 ANPP for the ungrazed and heavily grazed treatments, respectively. The same authors' reported 257 and 225 g/m^2 for ungrazed and heavily grazed shortgrass steppe in Texas after 5-7 years of treatment. Cattle grazing at the CPER is three times more intense on lowland communities than on upland communities for pastures stocked at either moderate (Senft 1983) or heavy (Milchunas et al. 1989) rates. Milchunas et al. (1992) observed ANPP reductions averaging 15 and 37% for uplands and lowlands, respectively, on heavily stocked versus ungrazed treatments after 47 and 49 years of treatment. Based upon 8 years of data, Lauenroth and Milchunas (unpubl. data) estimated average ANPP in a 21 year-old upland exclosure to be 91 g/m^2 compared with 60 g/m^2 for a moderately grazed upland at the CPER. Even though grazing reduced ANPP on shortgrass steppe sites, Milchunas et al. (1992) observed equal-compensation to current-year removals on long-term heavily grazed uplands and lowlands, and Williamson et al. (1989) observed some cases of over-compensation to grasshopper grazing of ungrazed shortgrass steppe. Equal-compensation refers to the situation where grazed and temporarily caged/ungrazed plots have the same standing biomass at the end of the growing season, whereas overcompensation indicates an even greater stimulation in production due to current-year herbivory.

It is possible to obtain equal- or over-compensation even when long-term grazing treatments have less production than ungrazed treatments.

Change in ANPP of northern mixed prairies in response to grazing are generally negative or not significant. After four years of treatment, Hofmann and Ries (1989) observed 31%, and Vogel and Van Dyne (1966) observed 12% greater production in protected areas. Six to eight years of heavy grazing at 65% removal reduced production by 37% in South Dakota, and nine years of heavy grazing in North Dakota by 13%, compared with ungrazed sites (Sims and Singh 1978). Grazing in the Black Hills of South Dakota reduced production an average of 34 and 36% after 11-12 and 27 years of treatment, respectively. Brand and Goetz (1986) reported greater production in protected areas in southwestern North Dakota in only one of four different sites after 40 years of removal of 45% of the ANPP.

Grazed-ungrazed comparisons of ANPP in southern mixed prairie are similar to those for northern mixed prairie. Three different moderately grazed (avg. 23% removal) and heavily grazed (avg. 32% removal) plant communities in Texas displayed no significant differences in ANPP compared with that in 22 year-old exclosures (Heitschmidt et al. 1985). A 60 year-old exclosure and 50% removal grazing treatment in Kansas displayed no significant difference in ANPP. The following three comparisons assessed moderate grazing at 50% removal and\or heavy grazing at 75% removal with virgin prairie. Tomanek and Albertson (1957) observed average changes in ANPP of -24 and -53% compared with ungrazed for moderate and heavy grazing in northwestern Kansas, respectively, and +9 and -28% in southwestern Kansas, respectively. Some moderately grazed pastures produced more than ungrazed sites. In central Kansas, Tomanek and Albertson (1953) found a 42% reduction in ANPP with heavy grazing and litter seven-times greater in the ungrazed.

Two grazed-ungrazed comparisons in tallgrass prairie both show a 28% increase in ANPP with grazing (Kelting 1954, Sims and Singh 1978). The exclosures were 15 and 25 years old, and levels of removal by the animals were 30 and 65%. Greater accumulations of litter on ungrazed compared with grazed sites is commonly reported for all grassland types. However, the very thick layer of litter on ungrazed tallgrass sites is often cited as the reason for lower production with protection from grazing when burning is not a treatment. Penfound (1964) stated "apparently the effects of undergrazing, and of no grazing, are due to prodigious mulch . . . much of this mulch lies flat on the ground and is of sufficient thickness to prevent the emergence of seedlings and culms of the dominant grasses". Kelting (1954) concluded "the actual amount of dead material disposed horizontally over the soil appears to be the controlling factor". However, Beebe and Hoffman (1968) found decreases of 43 and 66% with 15 and 49% removals, respectively, in comparisons with a 22 year-old ungrazed area that had been mown for hay once a year.

Most comparisons of grazing in this paper have used exclosures as reference sites or controls because factors such as litter accumulation in ungrazed communities and compensatory growth after defoliation may otherwise distort the gradient. Several other studies reporting aboveground production for tallgrass prairie not having exclosures include: 1) Hazel (1967), Oklahoma, light grazing = 414 g/m^2 and heavy grazing = 348 g/m^2, 2) Herbel and Anderson (1959), Kansas, 233,

196, and 148 g/m^2 for light, moderate, and heavy grazing, respectively, and 3) Weaver and Tomanek (1951), Nebraska, 344, 432, and 168 g/m^2 for light, moderate, and heavy grazing, respectively.

On the west to east precipitation gradient in the Great Plains, changes in ANPP in response to grazing were largely negative (Fig. 7). Twenty percent of the 61 comparisons were positive and most of these were for sites receiving greater than 400 mm/yr annual precipitation. Protection from grazing in shortgrass steppe resulted in communities more similar to those of disturbed early seral stages, but of greater productivity. Grazing tallgrass prairie often prevented forest from developing and was more likely to result in greater abundance of opportunistic invader species, but in many cases increased aboveground productivity. This result is not consistent with the condition and trend model which views grazing as a force which always moves a community towards early seral stages.

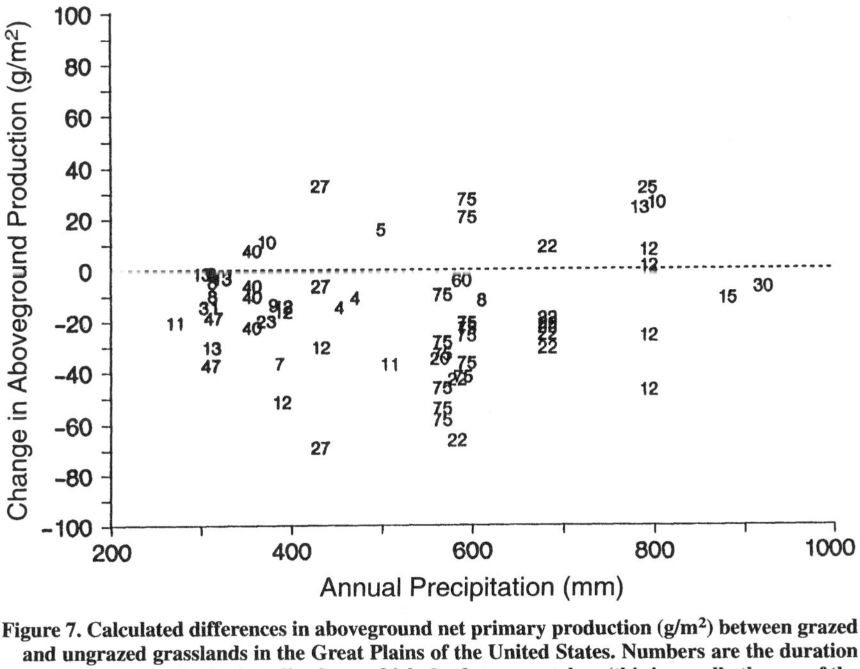

Figure 7. Calculated differences in aboveground net primary production (g/m^2) between grazed and ungrazed grasslands in the Great Plains of the United States. Numbers are the duration in years of the individual studies from which the data were taken (this is usually the age of the ungrazed exclosure).

Changes in aboveground productivity with grazing do not necessarily indicate how grazing may be affecting belowground (BNPP) or total plant productivity (NPP). However, traditional methods of estimating root production have many serious problems (Singh et al. 1984). We will review studies of the effect of grazing on belowground plant organs (roots and crowns) using biomass instead of BNPP when possible. While changes in root or crown biomass with grazing may not be an accurate indication of changes in BNPP, we believe this approach is less biased than one using estimates of BNPP.

In the shortgrass steppe, Sims et al. (1978) observed average crown biomass of 254 and 299 g/m^2 for ungrazed and heavy grazed treatments at the CPER in Colorado, and 299 and 315 g/m^2 at a site in Texas. Three-year average root biomass for the above treatment-sites was 1149 and 1329, and 620 and 725 g/m^2. Grazing tended to result in increased crown and root biomass in these studies. Leetham and Milchunas (1985) found no difference between root biomass of ungrazed and heavy grazed treatments at the CPER after 33 years of treatment. Milchunas and Lauenroth (1989) found 12% less biomass in 0-10 cm roots, and no difference in crown biomass, on heavy grazed compared with ungrazed treatments at the CPER after 47 years of treatment. The greatest effect of grazing on roots and crowns was in the more uniform horizontal distribution of the biomass than that on ungrazed treatments. Differences between these studies may be due to years of treatment or weather conditions during the years of study, but none of the results parallel the large decrease reported for aboveground response to grazing.

Sims et al. (1978) observed greater root biomass on grazed compared with ungrazed sites in all three years of sampling a northern mixed prairie site, with averages of 2177 and 1421 g/m^2, respectively. Two years of data for crowns averaged 300 and 235 g/m^2 for grazed and ungrazed, respectively. Sims et al. (1978) found no difference in root biomass between grazed and ungrazed sites at a southern mixed prairie site. Other studies of the response of root biomass to grazing in mixed prairie include +51, +22, -12, and +4% after 40 years of 45% removal treatment (Brand and Goetz 1986), and +5, +12, and +40% for light, moderate, and heavy grazing, respectively, compared with 41 years of exclosure (Smoliak et al. 1972). One study in tallgrass prairie gave three-year averages of 247 and 175 g/m^2 for crowns, and 964 and 924 g/m^2 for roots, for ungrazed and grazed treatments, respectively (Sims et al. 1978). Root or crown responses to grazing do not parallel the aboveground responses for the same study. Based on data from grazing studies across the globe, Milchunas and Lauenroth (1993) found that opposite responses to grazing often occur in ANPP and root biomass, and there was no simple relationship between change in species composition and ANPP or root mass.

Nutrient cycling and soil properties

Grazing by large herbivores generally increases nutrient cycling rates, thereby introducing the potential for greater losses. Nutrients consumed, digested, and deposited in feces and urine return to the soil more rapidly than through senescence-decomposition pathways. Eructation of gases produced by rumen microbial populations, volatilization of ammonia from urea, and animal harvest represent additional nitrogen loss pathways. Grazing can influence soil organic matter or carbon pools through its effect on plant productivity, erosion from runoff after soil compaction and/or by wind if percent cover of bare ground increases, and by altering decomposition rates through effects on soil moisture and temperature.

Lauenroth and Milchunas (1991) calculated a nitrogen budget for shortgrass range which included estimates of the role of cattle. Nitrogen losses due to animal harvest, gases via eructation, and volatilization of ammonia from urea were estimated to be 871, 550, and 110 g/ha/yr, respectively. Schimel et al. (1986) and Milchunas et al. (1988b) found that nitrogen losses due to urine deposition were relatively small for shortgrass steppe compared with those reported for many other grasslands. Harvest, eructation, and volatilization represented a total loss of 376 g/ha/yr compared with other non-cattle related losses of 2500 g/ha/yr (Lauenroth and Milchunas 1991). Furthermore, atmospheric inputs were estimated to be 2665 and 1500 g/ha/yr for wet and dry deposition, respectively. These estimates for moderately grazed shortgrass steppe do not include the possible input or output effects of cattle through alterations of production-decomposition-eolian related pathways. Further, Milchunas et al. (unpublished) observed soil nitrogen values of 0.20 and 0.14%, and soil carbon values of 2.1 and 1.5%, for lowlands of 47-year-old ungrazed and heavily grazed treatments, respectively, with no effect of grazing on the less utilized uplands. However, it is not known whether the ungrazed lowlands are aggrading or the heavily grazed areas are degrading in terms of these soil nutrients.

Water infiltration rates at the CPER were not significantly different on light and moderate grazing treatments, but both were greater than on heavy grazed treatments in two out of three soil types tested (Rauzi and Smith 1973). Leetham and Milchunas (1985) found average soil water contents on heavy grazing treatments that were 19% higher than on light grazing treatments.

Hanson et al. (1970) reported increased runoff under 20 years of heavy compared to light grazing in the northern mixed prairie. Both Naeth et al. (1990) in Alberta, and Rauzi (1963) in North Dakota, reported decreasing rates of water infiltration with increasing grazing intensities. Increased runoff was associated with decreased standing crop of vegetation. Working in 19-year-old treatments in northern mixed prairie, Smoliak et al. (1972) observed highest spring soil moistures in ungrazed, intermediate in light and moderate grazed, and lowest in heavy grazed treatments, with no differences between grazing treatments in soil bulk densities of the Ah horizon. While there was no effect of grazing treatments on soil nitrogen and total or available phosphorus, soil carbon increased with increasing grazing intensity. In this particular study, the differences were attributed to a shift from deep to shallow rooted species with grazing. Wood and Blackburn (1984) found little difference in soil organic matter of different grazing treatments of 4 to 20 years in Texas. In contrast, Smith (1940) observed that soil organic matter declined steadily as 'overgrazing continues' and was reduced to 34% below that of undisturbed prairie in Oklahoma.

For a tallgrass site in South Dakota, Beebe and Hoffman (1968) found large decreases in soil nitrogen, phosphorus, and moisture with both light and heavy grazing compared to an ungrazed treatment after 22 years. This site displayed large negative impacts of grazing on ANPP and large changes in species composition. Kelting (1954) observed that 25 years of moderate grazing, which had large effects on species composition and increased ANPP 28%, increased soil carbon by 29% and decreased soil pH and soil water contents compared with exclosures. Drew (1947) found no difference in soil organic matter of virgin prairie (burned often and mown intermittently since mid-1800's) and grassland grazed for approximately 125 years by cattle, horses, and sheep.

A conclusion that can be reached from this limited amount of research on soil properties is that no consistent patterns emerge either within or between grassland types. A possible exception is that grazing usually tends to increase soil compaction, but 1) this effect may often disappear or decrease after seasonal wetting and drying or freezing and thawing (Risser et al.. 1981, Abdel-Magid et al. 1987), and 2) although related to infiltration rates, increased compaction does not mean lower soil water because the effects of grazing on evapotranspiration may be greater. Further, a common perception is that soil nitrogen and organic matter parallel long-term trends in root biomass in grasslands where the majority of biomass is belowground. For studies in Colorado (Milchunas et al. unpublished), Alberta (Smoliak et al. 1972), and Khakase, USSR (Afanasyev and Rotowa 1986, pers. commun.; Titlyanova pers. commun.) which measured all three, there was no consistent relationship between root biomass and soil carbon and/or nitrogen (Milchunas and Lauenroth 1993). Reliable methods for measuring BNPP and decomposition have only recently been developed, and their application to grazing studies will help to interpret these responses.

Wildlife

Livestock can influence or interact with wildlife via food resources, modification of habitat structure, and altering nutrient cycling rates and pathways. Each of these three means of influence can be negative, neutral, or positive and the interaction can be competitive, nonexistent, or mutualistic depending on the particular species involved. Combinations of negative, neutral, and positive influences can occur simultaneously in any one situation.

Although the long evolutionary history of grazing by native ungulates in the shortgrass steppe probably played a large role in the ability of plant species to tolerate defoliation (Larson 1940, Mack and Thompson 1982, Milchunas et al. 1988a), replacement of bison by cattle may have had an influence on plant and wildlife communities due to differences between bison and cattle in diet selection and grazing behavior. A study of bison, pronghorn antelope, cattle, and sheep grazing the same range indicated that dietary overlap with pronghorn is greatest with sheep, intermediate with cattle, and least with bison. Cattle exert relatively greater pressure on lowland than upland plant communities, and consume relatively greater amounts of cool-season grasses, compared with the other ungulates, and bison are the least selective feeder of the four (Schwartz and Ellis 1981). Hornaday [1889, quoted by Larson (1940)] reported that bison "at times so completely consumed the herbage of the plains that detachments of the United States Army found it difficult to find sufficient grass for their mules and horses." Cattle defoliation events are probably of a greater frequency but a lighter intensity. Based on these and other studies reviewed in Lauenroth and Milchunas (1991), we suggest that cattle versus bison grazing resulted in 1) greater competition with pronghorn and a greater negative effect on winter-stressed wildlife obtaining spring forage, and 2) greater community heterogeneity at the square meter scale but greater homogeneity at the catena and landscape scales.

Although cattle reduce the quantities of available forage, pronghorn maintained similar diet qualities on lightly and heavily grazed pastures at the CPER (Schwartz et al. 1977). For example, winter (January) forage availability on light and heavy grazing treatment of 52 and 20 g/m^2 resulted in pronghorn dietary crude protein contents of 7.3% on both grazing treatments. Dietary phosphorus and digestibility were greater in August on the light than the heavy grazing treatment, but dietary fiber varied little between grazing treatments. Uresk and Sims (1975) found no differences among light, moderate, or heavy grazing treatments in the crude protein content of *B. gracilis* at any time during the year.

Lagomorphs are the largest component of native mammalian biomass at the CPER (Lauenroth and Milchunas 1991). Cover is an important aspect of habitat for many wildlife species, especially in the shortgrass steppe where grass canopy height does not conceal even a small lagomorph, raptor numbers are large, and shade in the treeless, hot environment is sought in terms of burrows and shrubs. Densities of lagomorphs at the CPER in long-term light, moderate, and heavy summer grazed and moderate winter grazed treatments illustrate how the influence of cattle on vegetation structure can affect closely related species in different ways. Although mean annual dietary overlap between the black-tailed and white-tailed jackrabbit is approximately 89% (Harris 1971), black-tailed jackrabbits were most abundant in light and moderate grazed treatments and white-tailed jackrabbits showed no preference for any grazing treatment (Flinders and Hansen 1975). Desert cottontail rabbits were more abundant on the moderately grazed treatment than on either light or heavy grazed treatments. Black-tailed jackrabbits preferred lowland plant communities where cattle graze more frequently, white-tailed jackrabbits preferred upland plant communities, and cottontails showed no preference.

The percent of land area covered by pocket gopher mounds at the CPER was reported at 8, 2.5, and 6.5% on heavily, lightly, and ungrazed treatments, respectively (Grant et al. 1980); suggesting no relationship of pocket gopher density with decreasing grazing intensity. The effects of grazing on other rodent populations at the CPER has not been directly studied. However, other small rodents (mice, rats, and ground squirrels) consume seeds and arthropods (French et al. 1976, Lauenroth and Milchunas 1991). Seed production was much greater on ungrazed versus grazed sites with coarse-textured soil, with only small differences on fine textured soil (Coffin and Lauenroth 1992). Aboveground macroarthropods were more abundant in lightly than in heavily grazed treatments (Andrews 1979). The effects of grazing on rodent food resources would suggest decreased carrying capacity of rodents in areas grazed by cattle. Rodent species probably respond differentially to grazing modifications of cover similar to that described above for lagomorphs. Prairie dogs are viewed as highly competitive with cattle, and this view is reinforced in standard range management texts such as Stoddard and Smith (1955). Hanson and Gold (1977) found prairie dog versus cattle diet similarity at the CPER was 64%, but cottontail vs cattle diet similarity was also high (61%). Consumption of vegetation in prairie-dog-towns was proportioned as 13% prairie dogs, 10% cottontails, and 77% cattle (Hanson and Gold 1977, calculations based on several assumptions). Although diet overlap between prairie dogs and cattle is moderate, cattle are heavy grazers on prairie-dog-towns.

Pawnee National Grasslands and the contiguous CPER is an important breeding ground for mountain plovers. Graul (1973, 1975) found mountain plovers were most abundant on heavily grazed shortgrass steppe. Ryder (1980), in an intensive 5-year study, found mountain plover densities of 1.3 nests/8 ha in long-term heavily grazed treatments and zero in light grazed treatments. Management for mountain plovers by the U.S. Forest Service in the Pawnee National Grasslands entails grazing certain areas more heavily than is allowed on other allotments. These areas appear overgrazed, even more so than the heavy 60+% removal treatments at the CPER (Milchunas pers. observation). The heavy grazed treatments at the CPER are stocked with a herd consisting of equal numbers of animals from each supplying rancher so that the burden of low animal weight gain is shared (Forwood pers. commun.). This is one unusual situation where management for wildlife entails grazing at a rate higher than what the ranchers want. However, managing for the benefit of one species is almost always detrimental to others. In addition to mountain plovers, horned larks and McCown's longspurs were abundant nesters on the heavily grazed treatment, but chestnut-collared longspurs, western meadowlarks, and lark buntings were more abundant on lighter grazed areas (Ryder 1980). The long-billed curlew also prefers grazed areas on the Comanche National Grassland in southern Colorado (King 1978).

Raptors are a conspicuous component of wildlife at the Pawnee National Grasslands and the CPER. The ratio of passerine to raptor biomass for Pawnee Grasslands averages 65 (Olendorff 1973). The influence of grazing on raptors is probably mediated by rodent and lagomorph responses to grazing, since these groups constitute a large proportion of raptor diets (Olendorff 1973). An indirect effect of ranches, the use of artificial nesting sites (wind-breaks, old buildings, etc.) ranged from 4% for golden eagles (*Aquila chrysaetos*) to 52% for ferruginous hawks (*Buteo regalis*) (Olendorff 1972).

Long-term grazing had relatively greater effects on macro-than microarthropod populations at the CPER. Andrews (1979) found biomass and numbers of macroarthropods were highest on lightly grazed and ungrazed and lowest on heavily grazed treatments. Grazing decreased aboveground and increased belowground macroarthropods and increased scavengers and slightly decreased other trophic groups. Western harvester ants (*Pogonomyrmex occidentalis*) (Rogers et al. 1972) and grasshoppers (Bhatnagar and Pfadt 1973, Capinera and Sechrist 1982) were lower on heavy compared to light grazing treatments. Leetham and Milchunas (1985) found very few differences in microarthropod populations of light and heavy grazing treatments and no significant differences in trophic structures. This may have been due to the unusual characteristic of the shortgrass steppe in which approximately half of the microarthropod community was observed below a depth of 30 cm. The vertical distribution of microarthropods did not parallel root distributions, and no differences in root biomass were observed between the grazing treatments in this study.

Herbivory by nematodes is approximately equal to that by cattle on moderately grazed shortgrass steppe (Lauenroth and Milchunas 1991). Nematodes have not been studied in relation to grazing treatments at the CPER, but their interactions with cattle are potentially significant. Fungi make up 60% of the microbial biomass at the CPER (Lauenroth and Milchunas 1991), and their numbers have been reported to increase in response to cattle grazing (Christansen and Scarborough 1969).

Facilitation of grazing of a small herbivore by a large herbivore is commonly reported in various parts of the world (Rasmussen 1941, Vesy-Fitzgerald 1960,

Hamilton and Bath 1970, Bell 1971, McNaughton 1976, Gordon 1988). The mechanism attributed to the facilitation is usually an increased availability of young or green versus old or dead tissue or increased leaf to stem ratios. This mechanism is less likely to occur in habitats such as the shortgrass steppe where canopies are low and sparse, and more likely to occur as ANPP and canopy development increase. Facilitation or no effects of livestock grazing on other large herbivores probably turns to detrimental effects as the grazing intensity reaches some critical level.

Detling and co-workers (see Whicker and Detling 1988 for an overview), working in the mixed prairie of South Dakota, report on an unusual case of grazing facilitation in which grazing by a small herbivore facilitated grazing of a large herbivore. Prairie dog towns were selectively grazed by bison, nearly 90% of all habitat use in 39% of the area (Coppock et al. 1983b). Vegetation in the very heavily grazed dog towns was more digestible, had a higher protein content, and lower dead:live biomass ratio (Coppock et al. 1983a). Peak live biomass on uncolonized prairie was twice that on areas colonized 3 to 8 years, but was not different from that on areas colonized for >26 years as forbs and shrubs replaced grasses. Although it is not known whether cattle display similar behavior in response to prairie dog towns, it is interesting to note that an animal considered a major pest facilitated grazing by an animal so similar to a cow.

As in the shortgrass steppe, the effects of grazing on passerines in the mixed prairie is favorable in some cases and detrimental in others. Maher (1979) found greater numbers of Baird's sparrows, Sprague pipits, and savannah sparrows in ungrazed grassland and greater numbers of horned larks and chestnut-collared and McCown's longspurs in grazed areas.

Grazing can have the most detrimental or the most beneficial impacts on wetlands and riparian zones. Heavy grazing or overgrazing of these areas not only has large influences on the vegetation, but increases soil erosion and water turbidity, and alters shoreline structure. However, because these areas are so productive, light to moderate grazing can often improve wildlife habitat. Whereas Braun (1978) reported that 55 studies showed grazing detrimental to waterfowl production and only one showed higher nesting success, Kantrud's (1990) review summarized studies showing 15 positive, 8 negative, 2 neutral, and 6 part positive-part negative effects of grazing on waterfowl. The negative effects were primarily examples of overgrazing or heavy grazing, and the positive effects were the result of opening dense stands of vegetation by light to moderate grazing. As with passerines, some studies reported differences in the amounts of cover or 'openness' preferred by various species, and therefore positive and negative effects of various degrees of light to moderate grazing depending on the species of waterfowl. Trampling of nests can be a negative aspect of grazing, whereas reduced predation of nests can be a positive aspect.

We might expect different and/or greater effects of grazing on wildlife populations at the other end of the precipitation gradient in the tallgrass prairie. Risser (1981) reported on a small mammal study conducted at the IBP Osage site where the grazing treatments included a 20-year-old ungrazed pasture, a lightly grazed pasture, and an exclosure ungrazed during the sampling period. Both the mean and variance of density of small mammals increased as grazing decreased. Small mammal biomass increased and diversity and equitability changed, within a single growing season of exclosure, to equal that of ungrazed prairie. Grazing reduced plant cover, and shifted species composition towards those common in the southern mixed prairie or shortgrass steppe. Some species were favored by grazing and others adversely affected. Risser concluded that grazing had a "profound effect" on small mammals and the "lag time appears to be relatively short in the True Prairie, probably because of the high cover provided by tallgrasses, even before a litter layer has developed".

Risser (1981) reviewed the literature on domestic livestock grazing and arthropod, avian, and microorganism populations in the tallgrass prairie. Grasshopper abundance increased and Coleoptera decreased under overgrazing. Based upon a limited number of studies, species diversity, density, and biomass of breeding birds appeared to increase at moderate grazing intensities compared with ungrazed conditions. The reduction in plant cover and litter depth and the increase in vegetation heterogeneity and forb densities increased the suitability of the habitat for dickcissels and grasshopper sparrows. Grazing tended to "lead to larger populations of some of the more representative True Prairie species". Microbial activity of the soil increased under moderate grazing. This was attributed to higher soil temperatures of grazed versus ungrazed grassland, although mechanical breakdown of the vegetation and/or addition of extra nitrogen via feces may have 'primed' the decomposition process.

We started this section by suggesting that the interactions between wildlife and domestic livestock could be complicated. It is apparent from this review that our understanding of the effects of grazing are limited because individual studies usually examined one group of organisms or closely related processes. However, several generalizations may be gleaned from the sum of studies. First, any grassland can be overgrazed by livestock and produce detrimental effects on wildlife. However, management of natural communities for wildlife often entails manipulating vegetation structure, and grazing can be one means of doing this. The potential for grazing to alter vegetation structure increases along the west to east productivity gradient of the Great Plains. The boundaries between shortgrass steppe, mixed prairie, tallgrass prairie, and forest can be shifted by grazing. Grazing can be used to manage for specific species, but other species will undoubtedly be adversely affected. Some plant communities have evolved with frequent fires, and fire is now a part of the management of these systems since the natural occurrence of the events is either suppressed or not possible due to changes in landuse. The same can be said of grazing, although the wild ungulates have been replaced by domestic ones. Defining what the historical levels of grazing were in these systems is difficult and decisions about management objectives will always be controversial.

The Future of Grazing in the Great Plains

The future of grazing in the Great Plains cannot be evaluated without also considering two other related topics. These are the dynamics of landuse and global climate change. While the major decisions about landuse that affected the entire region were made following the drought of the 1930s, important fluctuations in the proportion of cropland compared to grassland have occurred in the past 50 years. Unrelenting increases in world population will eventually begin placing increased pressure on all areas with crop production potential. Interacting with pressures to increase world food production will be the effects of climate change on the productive potential of regions and continents.

The Great Plains have been continuously grazed since the close of the Pleistocene epoch. Previous to the settlement of the area by Europeans, huge herds of wildlife, principally bison, were the dominant grazers. Settlement replaced the bison with domestic cattle and for the past 150 years the grazing regime has been one of intensively managed herds that remain on ranches the entire year. Despite the change in the seasonality and perhaps the frequency with which any individual unit of rangeland is grazed, domestic livestock grazing appears to be a sustainable use of Great Plains grasslands. At the present time approximately 63% of the area of the Great Plains is grazed by domestic livestock (Table 1). The majority of the remaining area is in crop production. A portion of the cropland is very likely grazed at some time of the year. As examples, fields of corn harvested for grain are frequently grazed following harvest while winter wheat is often grazed during the winter. Livestock grazing is an important component of the current economic and social environment of the Great Plains and will very likely remain so into the foreseeable future.

An important reason why the grasslands of Great Plains are well suited to support livestock grazing is related to their evolutionary history. The large herds of bison that grazed the plains over the past 10,000 years provided selection pressures for plants and other animals that resulted in an ecosystem that appears to be sustainable with domestic livestock as the dominant large herbivore. The same situation does not hold for crop production. Plowing the soil for row-crop production is different from anything in the developmental history of Great Plains ecosystems. Both the intensity of the disruption and the spatial scale are novel.

Cultivation of grassland soils results in depletion of soil organic matter (SOM) and reduced site fertility (Russel 1929, Tieszen et al. 1982, Aguilar et al. 1988). Tillage increases both erosion and decomposition, leading to decreased SOM and a decrease in the ability of soil to retain mineral nutrients. Reductions of up to 50% in SOM have been reported for long-term cultivation (Haas et al. 1957, Burke et al. 1989). The magnitude of losses depends upon interrelationships among intensity and duration of cultivation, fertilization, susceptibility to wind and water erosion, soil texture, and parent material. Losses of mineral nutrients are proportional to losses of soil C, but with perhaps greater relative losses of C than N and P to the extent that oxidation is important. Increased decomposition and mineralization, combined with decreasing SOM, induce conditions that increase N availability for an unknown length of time (Schimel 1986).

Recent investigations of the potential response of the central Great Plains to climate change (Burke et al. 1991) and to long-term cultivation (Burke et al. 1989) provides information for speculating about the response of the entire region. Burke et al. (1991) evaluated the response of soil organic carbon stores to a climate change scenario. Soil organic carbon was chosen as a response variable because it is the single best indicator of stability and sustainability of semiarid grassland ecosystems and it has important dynamics at decadal but not annual time scale. The climate change scenario consisted of a 5°C increase in mean annual temperature for the entire central Great Plains and a variable increase or decrease in precipitation of up to 50 mm depending upon location. The scenario was generated by the Goddard Institute for Space Science (GISS)-Global Circulation Model (Hansen et al. 1984) using doubled CO_2.

Simulation for 50 years, assuming the entire region was moderately grazed grassland, suggested that soil organic carbon would decrease as a result of climate change. Losses ranged from 50 to 350 g C/m^2; areas with the largest initial values suffered the greatest losses. Summing losses over the central Great Plains resulted in a predicted loss over 50 years of 7.0 x 10^{13} gC or approximately 3% of the total (Fig. 8). The conclusion from this result is that the central Great Plains will become a source of CO_2 to the atmosphere if the GISS climate change scenario comes to pass.

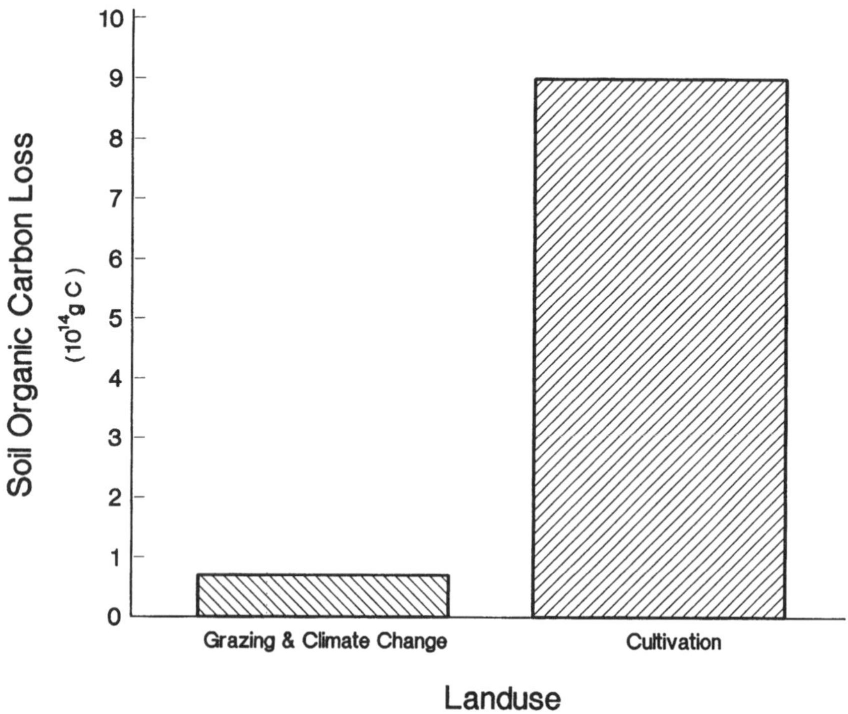

Figure 8. Estimated differences between soil carbon losses as a result of climate change compared to losses as a result of converting grassland to cropland. Data from Burke et al. (1991).

There are two ways to ask if such an afflux of CO_2 to the atmosphere is environmentally significant. First we can compare the simulated values to estimates of total increases in atmospheric CO_2. Tans et al. (1990) estimated that between 1981 and 1987 the average rate of increase in atmospheric carbon was 3.0 x 10^{15} g C/yr, 22% of which was contributed by landuse modification and 78% by fossil fuel burning. By comparison the predictions of annual losses from the central Great

Plains represent 0.09% of the annual landuse increment. The area of the central Great Plains represents approximately 0.4% of the terrestrial surface of the earth, suggesting that under a climate change condition the region may contribute less carbon to the atmosphere than might be predicted from knowing its area. The bulk of the increment is the result of destruction of tropical forest. The second way is to compare the simulated climate change contribution to the effects of other regional scale phenomena. Burke et al. (1989) used data from cultivated and uncultivated lands throughout the U.S. grassland region to evaluate carbon losses as a result of cultivation. Using these results for the central Great Plains suggests a 50 year loss of 9.0 x 10^{14} g C (Fig. 8). Calculating the per year rate and comparing to the results from Tans et al. (1990) indicates that conversion of grassland to cropland in the central Great Plains could contribute as much as 1.2% of the annual increment in CO_2 from landuse modification.

Sustainability of resource production has to be the primary goal of any management plan for the Great Plains. As a portion of the efforts associated with such a plan we must continue to pay close attention to how we manage the 63% of the area in grazinglands. Future research and development must explore the long-term consequences of current management methods as well as to seek new and innovative ways to ensure sustainability and meet multiple objectives. *We conclude from analyses of effects of climate change and cultivation on soil organic carbon that decisions to convert grasslands to cropland may be among the most critical management decisions in terms of long-term effects on Great Plains ecosystems as well as in affecting the relationship of the Great Plains to the rest of the biosphere (Fig. 8.* This is an area in which those with a concern for the entire region must be particularly vigilant. The current trend of converting cropland to grassland associated with the Conservation Reserve Program (Mitchell 1988) is positive but unfortunately with no guarantees that it will be long-term. Any decision to return those lands to crop production must not be based only on economic and social pressures; the ecological consequences must also be considered. If sustainability of resource production is an important goal for the Great Plains, grazing will be a predominant landuse for the foreseeable future.

Literature Cited

Abdel-Magid, A. H., G. E. Schuman, and R. H. Hart. 1987. Soil bulk density and water infiltration as affected by grazing systems. J. Range Manage. 40:307-309.

Afanasyev, N. A. and N. P. Rotova. 1986. The effect of grazing intensity on steppe ecosystems. pp. 59-61. *In:* Productivity of hayfields and pastures (A. A. Titlyanova, ed.). Science, Siberian Division, Novosibirsk.

Aguilar, R., E. F. Kelly, and R. D. Heil. 1988. Effects of cultivation on soils in Northern Great Plains rangelands. Soil Sci. Soc. Am. J. 52:1081-1085.

Andrews, R. M. 1979. A shortgrass prairie macroarthropod community: organization, energetics and effects of grazing by cattle. Tech. Rep. 208, US IBP Grassland Biome, Natural Resource Ecology Lab., Colorado St. Univ., Ft Collins, Colorado. 48 pp.

Beebe, J. D. and G. R. Hoffman. 1968. Effects of grazing on vegetation and soils in Southeastern South Dakota. Am. Midl. Nat. 80:96-110.

Bell, R. H. V. 1971. A grazing ecosystem in the Serengeti. Sci. Am. 225:2-9.

Belsky, A. J. 1986. Does herbivory benefit plants? A review of the evidence. Am. Nat. 127:870-892.

Bhatnagar, K. N. and R. E. Pfadt. 1973. Growth, density, and biomass of grasshoppers in the shortgrass and mixed-grass associations. Tech. Rep. 225, US IBP Grassland Biome, Natural Resource Ecology Lab., Colorado St. Univ., Ft Collins, Colorado. 120 pp.

Brand, M. D. and H. Goetz. 1986. Vegetation of exclosures in Southwestern North Dakota. J. Range Manage. 39:434-437.

Braun, C. E. 1978. Management of National Wildlife Refuges in the United States: its impact on birds. Conservation Comm. Rep. Wilson Bull. 90:309-321.

Burke, I.C., C. M. Yonker, W. J. Parton, C. V. Cole, K. Flach, and D. S. Schimel. 1989. Texture, climate, and cultivation effects on soil organic matter content in U.S. grassland soils. Soil Sci. Soc. Am. J. 53:800-805.

Burke, I.C., T. G. F. Kittel, W. K. Lauenroth, P. Snook, and C. M. Yonker. 1991. Regional analysis of the central Great Plains: Sensitivity to climate variability. BioScience 41:685-692.

Capinera, J. L. and T. S. Sechrist. 1982. Grasshopper (acrididae)-Host plant associations: Response of grasshopper populations to cattle grazing intensity. Can. Ent. 114:1055-1062.

Christansen, M. and A. M. Scarborough. 1969. Soil microfungi investigations, Pawnee Site. Tech. Rep. 23, US IBP Grassland Biome, Natural Resource Ecology Lab., Colorado St. Univ., Ft Collins, Colorado. 14 pp.

Coffin D. P. and W. K. Lauenroth. 1992. Spatial variability in seed production of the perennial grass *Bouteloua gracilis* (H.B.K.) LAG. ex GRIFFITHS. American Journal of Botany 79:347-353.

Collins, S. L. 1987. Interaction of disturbances in tallgrass prairie: A field experiment. Ecology 68:1243-1250.

Collins, S. L. and S. C. Barber. 1985. Effects of disturbance on diversity in mixed-grass prairie. Vegetatio 64:87-94.

Coppock, D. L., J. K. Detling, J. E. Ellis, and M. I. Dyer. 1983a. Plant-Herbivore interactions in a North American mixed-grass prairie. I. Effects of black-tailed prairie dogs on intraseasonal aboveground plant biomass and nutrient dynamics and plant species diversity. Oecologia 56:1-9.

Coppock, D. L., J. K. Detling, J. E. Ellis, and M. I. Dyer. 1983b. Plant-Herbivore interactions in a North American mixed-grass prairie. II. Responses of bison to modification of vegetation by prairie dogs. Oecologia 56:10-15.

Dix, R.L. 1959. The influence of grazing on the thin-soil prairies of Wisconsin. Ecology 40:36-49.

Dodd, J. L. 1979. North American grassland map. Fronticepiece *In:* N. R. French (ed.) Perspectives in grassland ecology. Springer-Verlag, New York, NY. 204 pp.

Dregne, H. E. and W. O. Willis. 1983. Dryland agriculture. American Society of Agronomy, Madison, Wisconsin. 622 pp.

Drew, W. B. 1947. Floristic composition of grazed and ungrazed prairie vegetation in North-Central Missouri. Ecology 28:26-41.

Ellison, L. 1960. Influence of grazing on plant succession of rangelands. Bot. Rev. 26:1-78.

Flinders, J.T., and R.M. Hansen. 1975. Spring population responses of cottontails and jackrabbits to cattle grazing shortgrass prairie. J. Range Manage. 28:290-293.

French N. R., W. E. Grant, W. Grodzinski, and D. M. Swift. 1976. Small mammal energetics in grassland ecosystems. Ecol. Monogr. 46:201-220.

Gill, J. L. 1985. The effects of grazing intensity on plant cover and diversity on a shortgrass steppe. M.S. Thesis, Colorado State University, Fort Collins, CO

Gordon, I. J. 1988. Facilitation of red deer grazing by cattle and its impact on red deer performance. J. Appl. Ecol. 25:1-10.

Grant, W. E. 1971. Comparisons of aboveground plant biomass on ungrazed pastures vs. pastures grazed by large herbivores, 1970 season. Tech. Rep. 131, US IBP Grassland Biome, Natural Resource Ecology Lab., Colorado St. Univ., Ft Collins, Colorado. 19 pp.

Grant, W. E., N. R. French and L. J. Folse Jr. 1980. Effects of pocket gopher mounds on plant production in shortgrass prairie ecosystems. Southwestern Nat. 25:215-224.

Graul, W. D. 1973. Adaptive aspects of the mountain plover social system. Living Bird 12:69-94.

Graul, W. D. 1975. Breeding biology of the mountain plover. Wilson Bull. 87:6-31.

Hamilton, D. and J. G. Bath. 1970. Performance of sheep and cattle grazed separately or together. Aust. J. Exp. Agric. Animal Husb. 10:19-26.

Hansen, J., A. Lacis, D. Rind, G. Russell, P. Stone, I. Fung, R. Ruedy, and J. Lerner. 1984. Climate sensitivity: analysis of feedback mechanisms. pp 130-163. *In*: Hansen, J. E. and T. Takahashi (eds). Climate processes and climate sensitivity. American Geophysical Union, Washington, DC.

Haas, H.J., C.E. Evans, and E.R. Miles. 1957. Nitrogen and carbon changes in soils as influenced by cropping and soil treatments. USDA Technical Bulletin 1164, US Government Printing Office, Washington, DC.

Hansen, R. M. and I. K. Gold. 1977. Blacktail prairie dogs, desert cottontails and cattle trophic relations on shortgrass range. J. Range Manage. 30:210-214.

Hanson, C. L., A. R. Kuhlman, C. J. Erickson, and J. K. Lewis. 1970. Grazing effects on runoff and vegetation on a western South Dakota rangeland. J. Range Manage. 23:418-420.

Harris, L. D. 1971. A precis of small mammal studies and results in the Grassland Biome. pp 213-240. *In:* Preliminary analysis of structure and function in grasslands. N. R. French, ed. Science Series 10. Range Science Dept., Colorado State Univ., Ft Collins, Colorado.

Hart, R. H. 1991. Sustaining livestock production and profit on range: managing for risk. pp 157-164. *In*: J. D. Hanson, M. J. Shaffer, D. A. Ball, and C. V. Cole. (eds.) Sustainable Agriculture for the Great Plains. Symposium Proceedings. USDA, ARS-89.

Hart, R. H., M. J. Samuel, P. S. Test, and M. A. Smith. 1988. Cattle, vegetation, and economic responses to grazing systems and grazing pressure. J. Range Manage. 41:282-286.

Hazel, D. B. 1967. Effect of grazing intensity on plant composition, vigor, and production. J. Range Manage. 20:249-253.

Heitschmidt, R. K., S. L. Dowhower, R. A. Gordon, and D. L. Price. 1985. Response of vegetation to livestock grazing at the Texas Experimental Ranch. Texas Agric. Exp. St. Bull. B-1515, College Station, Texas.

Herbel, C. H. and K. L. Anderson. 1959. Response of true prairie vegetation on major Flint Hills range sites to grazing treatment. Ecol. Monogr. 29:171-186.

Hofmann, L. and R. E. Ries. 1989. Animal performance and plant production from continuously grazed cool-season reclaimed and native pastures. J. Range Mange. 42:248-251.

Hyder, D. N., R. E. Bement, E. E. Remmenga, and C. Terwilliger, Jr. 1966. Vegetation-soils and vegetation-grazing relations from frequency data. J. Range Manage. 19:11-17.

Kantrud, H. A. 1990. Effects of vegetation manipulation on breeding waterfowl in prairie wetlands-A literature review. pp. 93-123. *In*: Can livestock be used as a tool to enhance wildlife habitat? USDA Forest Service Gen. Tech. Rep. RM-194. 123 pp.

Kelting, R. W. 1954. Effects of moderate grazing on the composition and plant production of a native tall-grass prairie in Central Oklahoma. Ecology 35:200-207.

King, R. 1978. Habitat use and relative behaviors of breeding long-billed curlews. M.S. Thesis, Colorado State Univ., Ft Collins, Colorado. 69 pp.

Klipple, G. E. and D. F. Costello. 1960. Vegetation and cattle responses to different intensities of grazing on short-grass ranges on the Central Great Plains. USDA Tech. Bull. 1216. Washington D. C. 82 pp.

Kucera, C. L. 1956. Grazing effects on composition of virgin prairie in North-Central Missouri. Ecology 37:389-391.

Küchler, A. W. 1964. Potential natural vegetation of the coterminous United States. American Geographical Society Special Publication , New York.

Larson, F. 1940. The role of bison in maintaining the short grass plains. Ecology 21:113-121.

Lauenroth, W. K. and W. A. Laycock (eds.). 1989. Secondary succession and the evaluation of rangeland condition. Westview Press, Boulder. 163 pp.

Lauenroth W. K. and D. G. Milchunas. 1991. Short-grass steppe. pp 183-226. *In*: R. T. Coupland (ed.) Natural grasslands: Introduction and western hemisphere. Vol. 8A, Ecosystems of the world, Elsevier Press, New York. 469 pp.

Leetham, J. W. and D. G. Milchunas. 1985. The composition and distribution of soil microarthropods in the shortgrass steppe in relation to soil water, root biomass, and grazing by cattle. Pedobiologia 28:311-325.

Mack, R. N. and J. N. Thompson. 1982. Evolution in steppe with few large, hooved mammals. Am. Nat. 119:157-773.

Maher, W. J. 1979. Nesting diets of prairie passerine birds at Matador, Saskatchewan, Canada. Ibis 121:437-452.

McNaughton, S. J. 1976. Serengeti migratory wildebeest: facilitation of energy flow by grazing. Science 191:92-94.

Milchunas, D. G., O. E. Sala, and W. K. Lauenroth. 1988a. A generalized model of the effects of grazing by large herbivores on grassland community structure. Am. Nat. 132:87-106.

Milchunas, D. G., W. J. Parton, D. S. Bigelow, and D. S. Schimel. 1988b. Factors influencing ammonia volatilization from urea in soils of the shortgrass steppe. J. Atmos. Chem. 6:323-340.

Milchunas, D. G. and W. K. Lauenroth. 1989. Three-dimensional distribution of plant biomass in relation to grazing and topography in the shortgrass steppe. Oikos 55:82-86.

Milchunas, D. G. and W. K. Lauenroth. 1993. A quantitative assessment of the effects of grazing on vegetation and soils over a global range of environments. Ecol. Monogr. 63: 327-366.

Milchunas, D. G., W. K. Lauenroth, and P. L. Chapman. 1992. Plant competition, abiotic, and long- and short-term effects of large herbivores on demography of opportunistic species in a semiarid grassland. Oecologia 92:520-531.

Milchunas, D. G., W. K. Lauenroth, P. L. Chapman, and M. K. Kazempour. 1989. Effects of grazing, topography, and precipitation on the structure of a semiarid grassland. Vegetatio 80:11-23.

Milchunas, D. G., W. K. Lauenroth, P. L. Chapman, and M. K. Kazempour. 1990. Community attributes along a perturbation gradient in a shortgrass steppe. J. Veg. Sci. 1:375-384.

Mitchell, J.E. 1988. Impacts of the conservation reserve program in the Great Plains. USDA Forest Service Gen. Tech. Rep. RM-158.

Naeth, M. A., R. L. Rothwell, D. S. Chanasyk, and A. W. Bailey. 1990. Grazing impacts on infiltration in mixed prairie and fescue grassland ecosystems of Alberta. Can. J. Soil Sci. 70:593-605.

Newman, J.B. 1988. Overview of the present land-use situation and the anticipated ecological impacts of program implementation. pp 55-59. *In*: Mitchell, J.E. (ed). Impacts of the conservation reserve program in the Great Plains. USDA Forest Service Gen. Tech. Rep. RM-158.

Olendorff, R. R. 1972. The large birds of prey of the Pawnee National Grassland: nesting habits and productivity, 1969-1971. Tech. Rep. 151, US IBP Grassland Biome, Natural Resource Ecology Lab., Colorado St. Univ., Ft Collins, Colorado. 59 pp.

Olendorff, R. R. 1973. The ecology of the nesting birds of prey of northeastern Colorado. Tech. Rep. 211, US IBP Grassland Biome, Natural Resource Ecology Lab., Colorado St. Univ., Ft Collins, Colorado. 233 pp.

Pase, C. E. and J. F. Thilenius. 1968. Composition, production, and site factors of some grasslands in the Black Hills of South Dakota. USDA Forest Serv. Res. Note RM-103.

Penfound, W. T. 1964. The relation of grazing to plant succession in the tall grass prairie. J. Range Manage. 17:256-261.

Rasmussen, D. I. 1941. Biotic communities of Kaibab Plateau, Arizona. Ecol. Monogr. 11:229-275.

Rauzi, F. 1963. Water intake and plant composition as affected by differential grazing on rangeland. J. Soil Water Cons. 18:114-116.

Rauzi, F. and F. M. Smith. 1973. Infiltration Rates: Three soils with three grazing levels in Northeastern Colorado. J. Range Manage. 26:126-129.

Risser, P. G., E. C. Birney, H. D. Blocker, S. W. May, W. J. Parton, and J. A. Wiens. 1981. The true prairie ecosystem. Hutchinson Ross Publ. Co., Stroudsburg, PA. 557 pp.

Rogers, L., R. Lavigne, and J. L. Miller. 1972. Bioenergetics of the western harvester ant in the shortgrass plains ecosystem. Environ. Entomol. 1:763-768.

Russel, J. C. 1929. Organic matter problems under dry farming conditions. Agronomy Journal 21:960-969.

Ryder, R. A. 1980. Effects of grazing on bird habitats. pp 51-66. *In:* Management of western forests and grasslands for nongame birds. USDA Forest Serv. Gen. Tech. Rep. INT-86.

Sala, O. E., W. J. Parton, L. A. Joyce, and W. K. Lauenroth. 1988. Primary production of the central grassland region of the United States. Ecology 69:40-45.

Schimel, D. S. 1986. Carbon and nitrogen turnover in adjacent grassland and cropland ecosystems. Biogeochemistry 2:345-357.

Schimel, D. S., W. J. Parton, F. J. Ademsen, R. G. Woodmansee, R. L. Senft, and M. A. Stillwell. 1986. The role of cattle in volatile loss of nitrogen from a shortgrass steppe. Biogeochemistry 2:39-52.

Schwartz, C. C., J. G. Nagy, and R. W. Rice. 1977. Pronghorn dietary quality relative to forage availability and other ruminants in Colorado. J. Wildl. Manage. 41:161-168.

Schwartz, C. C. and J. E. Ellis. 1981. Feeding ecology and niche separation in some native domestic ungulates on the shortgrass prairie. J. Appl. Ecol. 18:343-353.

Senft, R. L. 1983. The redistribution of nitrogen by cattle. Ph.D. thesis, Colorado State Univ., Ft Collins, Colorado. 198 pp.

Severson, K. E. (ed.). 1990. Can livestock be used as a tool to enhance wildlife habitat? USDA Forest Serv. Gen. Tech. Rep. RM-194. 123 pp.

Shantz, H. L. 1923. The natural vegetation of the Great Plains. Ann. Assoc. Amer. Geographers 13:81-107.

Sims, P. L. and J. S. Singh. 1978. The structure and function of ten western North American grasslands, III. Net primary production, turnover and efficiencies of energy capture and water use. J. Ecol. 66:573-597.

Sims, P. L., B. E. Dahl, and A. H. Denham. 1976. Vegetation and livestock response at three grazing intensities on sandhills rangeland in eastern Colorado. J. Range Manage. 41:282-286.

Sims, P. L., J. S. Singh, and W. K Lauenroth. 1978. The structure and function of ten western North American grasslands, I. Abiotic and vegetational characteristics. J. Ecol. 66:251-285.

Singh, J. S., W. K. Lauenroth, H. W. Hunt, and D. M. Swift. 1984. Bias and random errors in estimations of net primary production: a simulation approach. Ecology 65:1760-1764.

Smith, C. C. 1940. The effect of overgrazing and erosion upon the biota of the mixed-grass prairie of Oklahoma. Ecology 21:381-397.

Smoliak, S., J. F. Dormaar, and A. Johnston. 1972. Long-term grazing effects on stipa-bouteloua prairie soils. J. Range Manage. 25:246-251.

Stoddard, L. A. and A. D. Smith. 1955. Range management. McGraw-Hill Book Co., Inc. New York. 433 pp.

Tans, P. P., I. Y. Fung, and T. Takahashi. 1990. Observational constraints on the global atmospheric CO_2 budget. Science 247:1431-1438.

Tieszen, H., J. W. B. Stewart, and J. R. Betany. 1982. Cultivation effects on the amounts and concentration of carbon, nitrogen, and phosphorus in grassland soils. Agronomy Journal 74:831-835.

Tomanek, G. W. and F. W. Albertson. 1953. Some effects of different intensities of grazing on mixed prairies near Hays, Kansas. J. Range Manage. 6:299-306.

Tomanek, G. W. and F. W. Albertson. 1957. Variations in cover, composition, production, and roots of vegetation on two prairies in Western Kansas. Ecol. Monogr. 27:265-281.

Torell, L. A., W. W. Riggs, and R. H. Hart. 1989. Economic principles for the efficient use of rangelands with intertemporal considerations. pp. 46-80. *In:* F. Wagstaff and D. Reesman (eds.) Multiple users—Multiple products. Soc. Range Management, Denver, CO.

Uresk, D. W. and P. L. Sims. 1975. Influence of grazing on crude protein content of blue grama. J. Range Manage. 28:370-371.

Vesy-Fitzgerald, D. F. 1960. Grazing succession among East African game animals. J. Mammal. 41:161-172.

Vogel, W. G. and G. M. Van Dyne. 1966. Vegetation responses to grazing management on a foothill sheep range. J. Range Manage. 19:80-85.

Weaver, J. E. 1954. North American prairie. Johnsen Publishing Co., Lincoln, NE 348 p.

Weaver, J. E. and F. W. Albertson. 1956. Grasslands of the Great Plains. Johnsen Publishing Co., Lincoln, NE. 395 pp.

Weaver, J. E. and F. E. Clements. 1938. Plant Ecology. McGraw-Hill Book Co. Inc., New York. 601 pp.

Weaver, J. E. and G. W. Tomanek. 1951. Ecological studies in a mid-western range; the vegetation and effect of cattle on its composition and distribution. Univ. Nebr. Conserv. Surv. Div. Bull. 31, Univ. Nebraska, Lincoln. 82 pp.

Webb, W. P. 1931. The great plains. Ginn and Co. Boston. 525 pp.

Whicker, A. D. and J. K. Detling. 1988. Ecological consequences of prairie dog disturbances in a North American grassland. BioScience 38:778-785.

Whittaker, R. H. 1952. A study of summer foliage insect communities in the Great Smoky Mountains. Ecol. Monogr. 22:1-44.

Williamson, S. C., J. K. Detling, J. L. Dodd, and M. I. Dyer. 1989. Experimental evaluation of the grazing optimization hypothesis. J. Range Manage. 42:149-152.

Wood, M. K. and W. H. Blackburn. 1984. Vegetation and soil responses to cattle grazing systems in the Texas Rolling Plains. J. Range Manage. 37:303-308.

Implications of Livestock Grazing in the Intermountain Sagebrush Region: Plant Composition

Richard F. Miller, Tony J. Svejcar, and Neil E. West

Abstract

The Intermountain Sagebrush semi-desert vegetation represents the largest kind of temperate semi-desert in North America. Vegetation throughout this region has been in a continued state of flux during the past 10,000 years. Shifts in climate and fire return intervals have caused altitudinal changes in plant communities, increases and decreases in the abundance of salt desert species and the relative abundance of upland shrubs and grasses. However, since settlement, approximately 150 years ago, changes in plant/animal composition have occurred at unprecedented rates across the region. The introduction of cattle, sheep, horses and aggressive alien plant species, cultivation, elevated CO_2 levels, altered fire frequencies, recreation, mining and demands for water, interacting with a gradual change in climate, have had a cumulative effect on the landscape. The focus of this chapter will be to describe changes that have occurred in the mountain sagebrush region prior to and following settlement with specific emphasis on livestock grazing.

Introduction

The sagebrush steppe and Great Basin-Colorado Plateau sagebrush semi-desert vegetation types are the largest kind of temperate semi-desert in North America, covering 44.8×10^6 ha and 17.6×10^6 ha, respectively (Fig. 1) (Kuchler 1970). Since settlement, approximately 150 years ago, changes in plant and animal composition have occurred at unprecedented rates across the region. The introduction of cattle, sheep, horses and aggressive alien plant species, cultivation, elevated CO_2 levels, altered fire frequencies, recreation, mining and demands for water, interacting with a gradual change in climate, have had a cumulative effect on the landscape. It is well documented that overgrazing by domestic livestock was a major factor in altering this large semi-desert region, causing dramatic changes in vegetation composition. The impacts of light to moderate grazing by domestic livestock, however, are less clear. It is difficult, and at times impossible to separate the effects of livestock on plant succession from other environmental influences. Both abiotic and biotic factors must be considered along with livestock grazing for a more complete picture of cause and effect relationships which have influenced plant succession in the Intermountain Sagebrush Region.

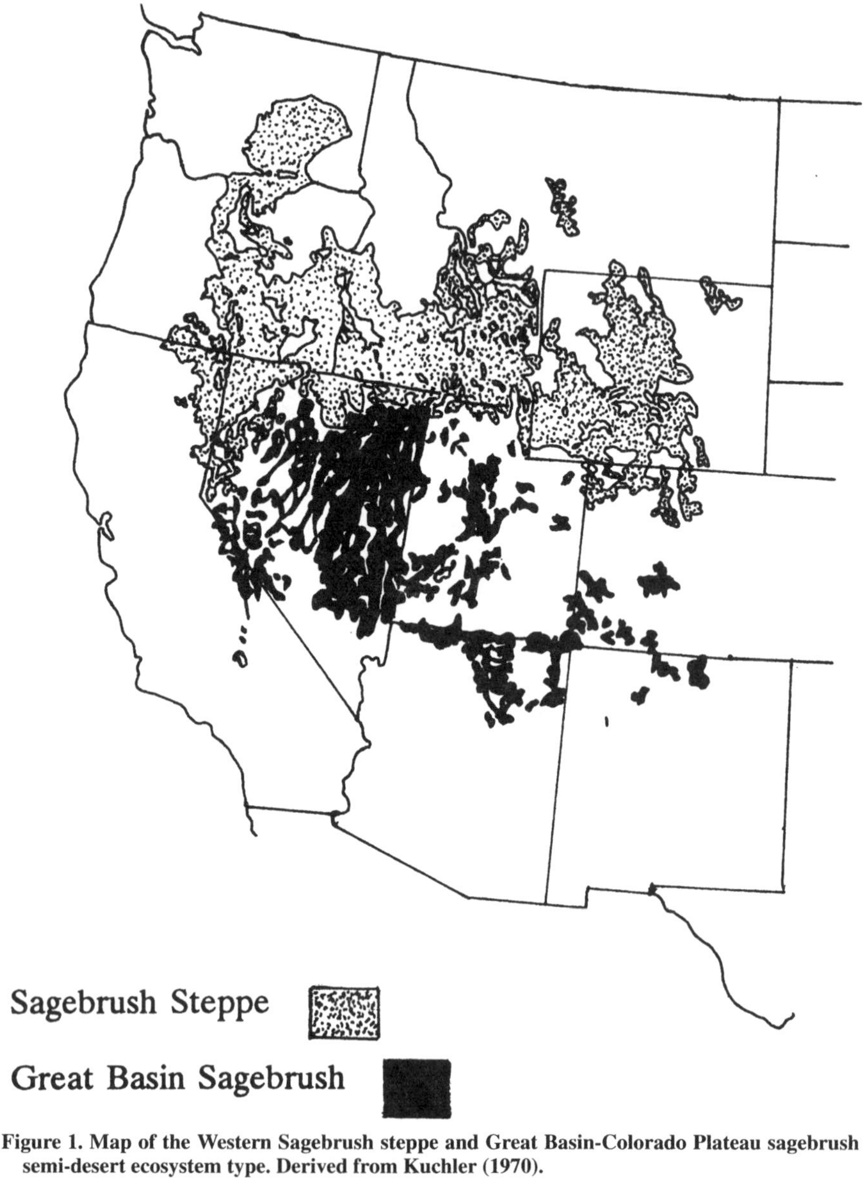

Figure 1. Map of the Western Sagebrush steppe and Great Basin-Colorado Plateau sagebrush semi-desert ecosystem type. Derived from Kuchler (1970).

The focus of this chapter will be to describe changes that have occurred in the Intermountain Sagebrush Region since the time of settlement with specific emphasis on livestock grazing. However, to fully understand landscape changes during the past 150 years we will first focus our discussion on the dynamics of these landscapes and the environmental factors influencing vegetation composition during the past 10,000 years. We will attempt to describe plant community composition just prior to settlement based on early observations by explorers, trappers and settlers in the Intermountain Sagebrush Region. The chapter will then focus on vegetation changes which occurred during settlement, the introduction of Eurasian plant species, the possible movement of plant succession to new steady states, and the implications of livestock grazing on plant succession in the sagebrush region.

The Setting: Sagebrush Region

The Intermountain Sagebrush Region lies between the Sierra Nevada and Cascade Mountains and the Rocky Mountains. The ecosystem type is usually separated into the sagebrush steppe and the more arid sagebrush semi-desert (West 1983). The sagebrush steppe, which is more similar to a semi-arid grassland, is dominated by sagebrush (*Artemisia* spp.) and perennial bunchgrasses. This ecosystem is bordered on both its eastern and western flanks by various conifer forests. The sagebrush steppe gives way to salt desert shrublands on the poorly drained valley bottoms, where salty soils have developed. The sagebrush semi-desert, more closely related to a true desert, is also dominated by sagebrush but has a lower abundance of perennial grasses. It is located south of the sagebrush steppe and generally north of the creosote and blackbrush types (West 1983). It lies in elevation above the salt-desert shrublands and below the semi-arid conifer woodlands.

The salt-desert shrublands and semi-arid woodlands are two major vegetation types of considerable size, encompassed by both sagebrush ecosystem types. The salt-desert shrublands typically occupy saline soils but many of these salt-tolerant species such as shadscale (*Atriplex confertifolia*) are also found in areas apparently too arid for sagebrush (Billings 1949). The semi-arid woodlands are typically found at the higher elevations of the sagebrush type and consist primarily of juniper species (*Juniperus* spp.) and single-needle pinyon pine (*Pinus monophyla*).

Climate in the Intermountain Sagebrush Region is characterized by cold wet winters and hot dry summers. The primary source of moisture for this region, of which about 60% falls as snow in the north, is the Pacific Ocean, with low pressure centers moving inland during the fall, winter and spring. The frequency of summer thunderstorms increases as one travels from the west to the east within the sagebrush zone (Stoddart 1941, Houghton et al. 1975). Mean annual precipitation for the region is 280 mm, commonly varying between 150 and 400 mm (Flaschka et al. 1987). Elevation varies from 150 m in the Columbia River drainage to 2200 m in the Great Basin and Wyoming Basin (West 1983).

Pre-Settlement Vegetation: The Last 10,000 Years

Since the end of the last ice age, 10,000 to 12,000 B.P. (before present), vegetation composition across the sagebrush region has been dynamic (Mehringer and Wigand 1990, Thompson 1990). Prior to European settlement, climate and frequency of fire were probably the primary forces influencing change in plant composition throughout this region. The relative abundance of sagebrush and perennial grasses has shifted, coincident with changes in climate and fire frequency (Morrison 1964, Aikens 1986, Mehringer 1986, Mehringer and Wigand 1990, Thompson 1990). Adjacent salt-desert shrub communities and juniper woodlands also expanded and contracted during this period (Wigand and Mehringer 1985, Mehringer 1987, Thompson 1990).

Climate and Vegetation

During the past 10,000 years climate has fluctuated with periods of cooler and wetter, warmer and drier, and warmer and more humid weather patterns than present (Fig. 2) (Russell 1885, Antevs 1938, 1948, Morrison 1964, Davis 1982, Aikens 1986, Mehringer 1986). There is also evidence that the season of maximum precipitation has varied across the Intermountain Sagebrush Region (Davis 1982, Thompson and Hattori 1983). The duration of these periods extended from several hundred to several thousand years. Fluctuations in lake levels, glaciers, and snow lines; soil deposition and lake salinity levels; tree ring widths, pollen cores from lake bottoms, rat and Indian middens, and the relative abundance of various plant and animal fossils provide evidence for these climatic changes. In historic time weather measurements, runoff and lake levels provide additional data.

During the past 10,000 years, prior to European settlement, flora in the intermountain sagebrush steppe has essentially remained unchanged (Butler 1976). However, changes that did occur in vegetation during this period were altitudinal shifts (Butler 1976, Madsen 1976), and the relative abundance between sagebrush

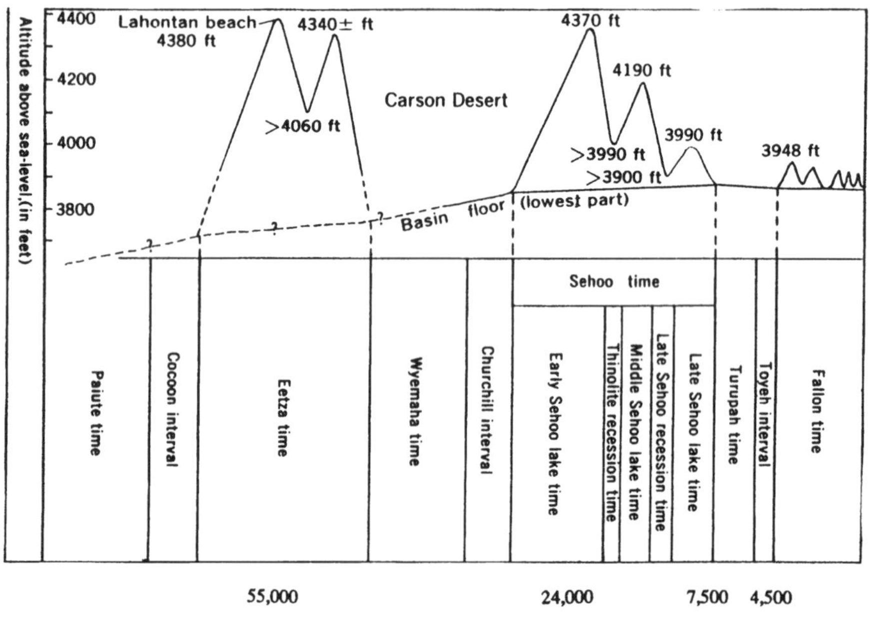

Figure 2. Elevation of Lake Lahontan during the past 100,000 years. The Wyemaha and Turupah time periods were warmer and drier than current conditions and represented periods of accelerated soil erosion. The Cocoon, Churchill and Toyeh intervals were warmer and wetter than current conditions and were important periods of soil formation in the Great Basin. Elevated lake levels suggest cooler and wetter weather than at present (Morrison 1964).

species, perennial grasses and salt desert species (Morrison 1964, Mehringer 1985, Mehringer and Wigand 1987, 1990, Beiswenger 1991).

Though rising temperatures brought the last glacial period to an end about 10,000 years B.P., the climate was generally cooler than present until about 8,000 years B.P. During this period, elevation of vegetation zones were 400-500 m lower than present (Wigand and Mehringer 1985, Thompson 1990). The Snake River Plain, now dominated by sagebrush species and introduced annual grasses, was dominated by pine with only a scattered presence of sagebrush (Butler 1976). The sagebrush steppe and semi-desert were well below their current elevational limits (Mehringer 1967, Thompson 1990) and lake levels in the Great Basin were still higher than present (Antevs 1948, 1952). For example, at Connely Caves in eastern Oregon, where western juniper (*Juniperus occidentalis*) and big sagebrush (*Artemisia tridentata*) characterize the present day landscape, pine ashes in the fire rings and remains of waterfowl and mammals which occupy cooler and moister habitats were common in deposits dated 8,000 years B.P. (Bedwell 1973, Aikens 1986). Evidencc of pika in this area also suggests cooler temperature regimes (Grayson 1987).

Following this period of cooler and moister conditions, climate became warmer and drier than present in the Intermountain Sagebrush Region and lakes occupying the basins fell below current levels, and in most instances were totally desiccated (Fig. 3) (Antevs 1938 1948, Hansen 1947a, Morrison 1964, Reveal 1979, Allison 1982, Wigand 1985, Thompson 1990). This Hypsithermal or Altithermal (meaning warmer and/or drier) interval occurred between 7,000 and 4,500 years B.P (Fig. 2). Summers became wetter, but winters were dry (Davis 1982, Thomp-

Figure 3. Looking east from Wheeler Peak, Nevada with desiccated Pleistocene Lake Bonneville in the background. Once covered by hundreds of feet of water, the area is currently occupied by an extensive salt desert.

son and Hattori 1983). Treelines, sagebrush steppe and semi-desert communities rose in elevation. At lower elevations the sagebrush steppe and semi-desert gave way to salt desert species (Hansen 1947a, b, Byrne et al. 1979, Wigand and Mehringer 1985, Thompson 1990).

By 7,000 years B.P., sagebrush replaced pines as the dominant plant genus on the Snake River Plain (Butler 1976). During this warm dry period total plant cover probably declined, and shrubs dominated the landscape (Fig. 3) (Morrison 1964, Mehringer 1985, 1987). Little soil was formed and massive erosion, as evidenced by deposition of vast quantities of eolian sand, took place during this period (Morrison 1964). At Connely Caves, fossil evidence in the middens, of plants, mammals and birds requiring moist habitats steadily declined while evidence of animals living in dry habitats increased (Bedwell 1973, Aikens 1986). Pine ashes in the fire rings were replaced by juniper and sagebrush. Eventually the Indians abandoned many of these caves throughout the Great Basin during the Hypsithermal period (Morrison 1964, Bedwell 1973).

Shifts from cool moist to warm dry periods between 10,000 to 2,000 years B.P. do not coincide with changes in climate for the eastern portion of the Intermountain Sagebrush Region (Beiswenger 1991). In the Grays Lake area of eastern Idaho and western Wyoming pollen from Chenopodiaceae, Amaranthaceae and Compositae remained stable and *Pinus* pollen increased between 7,100 to 2,000 years

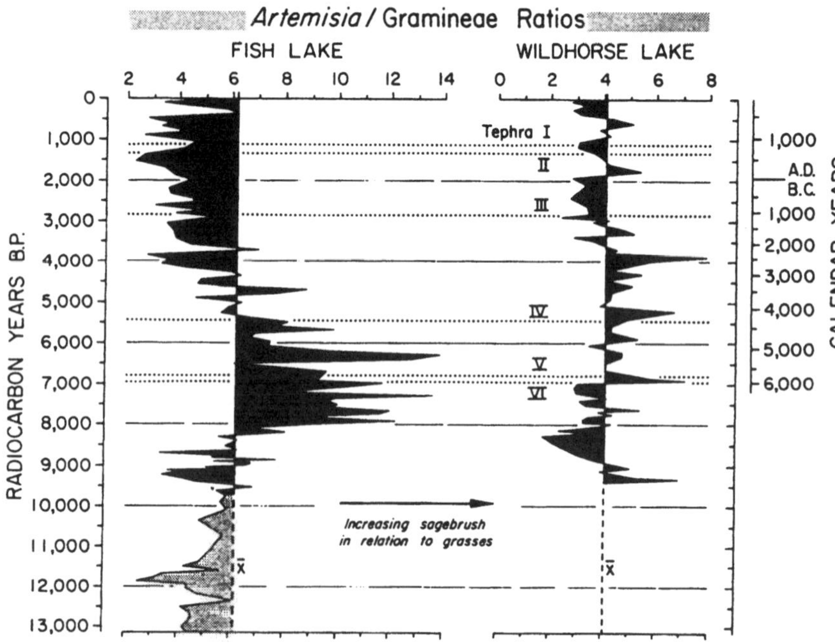

Figure 4. The ratios of sagebrush/grass pollen plotted about their means for the last 9,700 and 9,300 radiocarbon years for the Steens Mountain, Oregon. Increasing sagebrush in relation to grass at Fish Lake (2,250m) indicates less effective moisture. A similar but somewhat delayed trend at Wildhorse Lake, which lies 315 m higher. Wildhorse Lake which is at the current elevational limit of sagebrush, suggests the upward advance of sagebrush owing to warmer conditions (Mehringer 1985).

B.P. These contrasting climate and vegetational histories may be attributed to changes in timing of summer high temperatures from early to late summer during the Holocene (Davis et al. 1986).

Approximately 4,000 years B.P. the climate in much of the sagebrush region became moister, and then cooler about 3,500 years B.P. than present (Antevs 1938, 1948, Morrison 1964, Mehringer and Wigand 1987, Thompson 1990). Both winter and summer precipitation were abundant (Davis 1982). This period was characterized by at least five shallow lake cycles (Morrison 1964) and three episodes of montane glacial buildup (Miller 1973). The most recent glacial buildup, the Little Ice Age, was associated with high volcanic activity around the globe (Bryson 1989). It began in the late 1500s and culminated during the mid 1800s. During the Little Ice Age, winters were colder, summers cooler and precipitation greater (Pielou 1991).

Following the end of the Hypsithermal period tree lines descended down the mountains (LaMarche and Mooney 1972, LaMarche 1973) but did not regain earlier post glacial abundance. Sagebrush communities also advanced down slope into the valley bottoms replacing chenopods and the proportion of grass pollen increased relative to that of sagebrush pollen (Fig. 4 and 5) (Hansen 1947b, Mehringer 1985, Mehringer and Wigand 1990, Thompson 1990). The prehistoric expansion of juniper began during this period at the lower elevations (Wigand 1987). Connely cave was reoccupied by humans, but at a lower level of activity than 8,000 years B.P. (Cressman 1981).

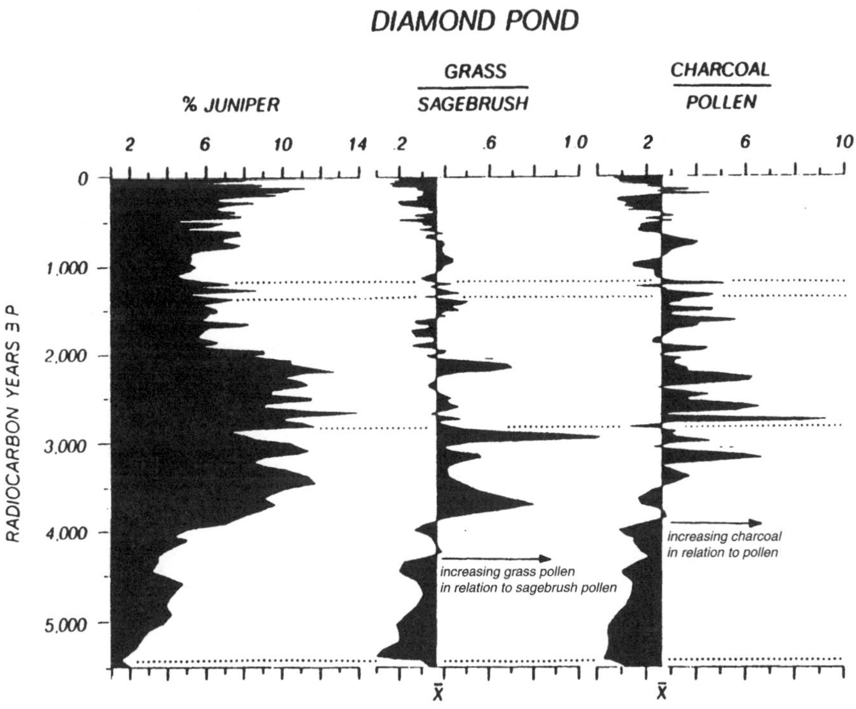

Figure 5. The ratios of grass/sagebrush pollen and charcoal/pollen plotted about their means during the last 5,500 radiocarbon years for Diamond Pond in Diamond Crates, Oregon (Mehringer 1987).

During the last 2,000 years the general trend in climate has been erratic but moving towards increasing aridity and an increased proportion of precipitation occurring during winter (Antevs 1948, Morrison 1964, Davis 1982). Since the culmination of the Little Ice Age in the mid 1800s temperature has been rising (Antevs 1948, Easterling 1990, Ghil and Vautgard 1991).

Short term climatic cycles have also affected plant community composition. Short term climatic regimes may last from 2-21 years in the Intermountain Sagebrush Region (Neilson 1986, Ghil and Vautgard 1991). Tree ring chronology in the northern part of the Great Basin reflects the cyclic nature of these wet and dry cycles (Fig. 6) (Antevs 1938). Periods of increased precipitation in the Great Basin, lasting between 5 and 6 years, are probably associated with El Nino (periods of elevated sea surface temperatures and reduced barometric pressure in the eastern Tropical Pacific)(Ropelewski and Halpert 1986, Nicholls 1988, Ghil and Vautgard 1991). Interdecadal periods of increased or decreased precipitation may be associated with changes in the ocean circulation outside the tropics (Ghil and Vautgard 1991). There have been exceptions to the duration of these cycles in recent time: the great drought of A.D. 1276-1299 and above average precipitation from 1850 to 1916 in the northwestern portion of the Great Basin (Antevs 1948, Graumlich 1985).

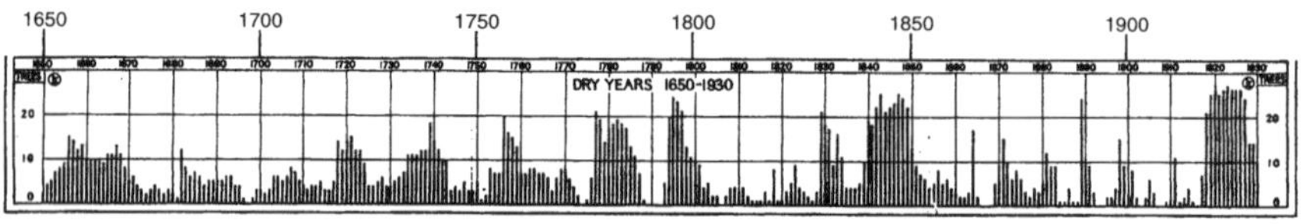

Figure 6. Estimated relative available moisture based on tree growth at various locations throughout the northern Great Basin (Antevs 1938). The Y axis is an index developed using tree ring growth representing wet (0) to dry (>20) conditions; the x axis represents the years 1650–1930.

Neilson (1986, 1987) proposed that these wet and dry cycles influence plant composition by controlling seedling establishment and physiological processes. He said that the effects of climate on community patterns are the result of: (1) weather cycles selecting for specific species, (2) spatial interplay of environmental gradients, (3) temporal pattern of weather oscillation producing episodic establishment and extinction, and (4) species interaction. For example, big sagebrush (Pechanec et al. 1937) and threetip sagebrush (*Artemisia tripartita*) (West et al. 1979) recover more rapidly than perennial bunchgrasses following drought. Drought may be an important environmental factor that allows sagebrush species to increase in dominance. Once established, vegetation may survive under a different climatic regime (Neilson 1986, Archer and Smeins 1991). The ability of plants to persist under different environmental conditions than which they initially established is known as biological inertia.

Sagebrush, a well adapted species in this environment, is an aggressive competitor for resources allowing it to attain dominance throughout the intermountain shrub region. The concentration of precipitation in the winter, throughout the region, probably favors the growth of sagebrush over the perennial grasses. Woody plants typically predominate in areas where most of the precipitation comes in winter as opposed to spring (Stoddart 1941, West and Tueller 1972, West 1983). The accumulation of winter snow allows percolation of moisture deep into the soil profile (EOARC file data), favoring the deeper rooted woody species (Walker et al. 1981). As the surface soils dry, big sagebrush continues to use soil water below one meter in the soil profile during the summer (Campbell and Harris 1977, Sturges 1977). Spring and summer precipitation primarily recharge the upper 300 mm of the soil profile (West 1983) where grasses are more efficient than shrubs in extracting water (Walker et al. 1981). Moisture distribution in the soil profile influences the effort plants allocate towards root development. Root:shoot ratios, which average around four:one in the Intermountain Sagebrush Region are typically one or less for plant species in the hot deserts to the southwest (Caldwell 1985, Noy-Meir 1985).

The presence of a relatively large leaf area early in the spring allows plant species in these winter precipitation areas to take full advantage of favorable growing conditions early in the growing season (Slayter 1970, Caldwell 1978). It is important for a plant to maximize its leaf area as early as possible during the growing season to make the most beneficial use of water (Ritchie 1974). The persistence of winter leaves and the early development of the larger ephemeral leaves in big sagebrush (Miller and Shultz 1987), is a mechanism allowing it to capture soil resources more readily, early in the spring, than associated deciduous species (Miller 1988).

Big sagebrush also appears to be more opportunistic than perennial grasses in nutrient acquisition. Aboveground growth of big sagebrush has a greater capacity to respond to increased levels of nitrogen than the associated perennial grass Thurbers needlegrass (*Stipa thurberiana*) (Miller et al. 1991). Big sagebrush can also absorb a larger proportion of phosphorus than bluebunch wheatgrass (*Agropyron spicatum*) (Caldwell et al. 1985). The ability of sagebrush to compete more effectively for soil resources may also be partially attributed to its potential growth rate being greater than perennial tussock grasses (Eissenstat and Caldwell 1987).

Fire

Prior to European settlement, fire played an important role in the sagebrush steppe, enabling herbaceous species to retain a codominant position with the shrubs in the landscape (Burkhardt and Tisdale 1976, Gruel 1985). Besides natural caused fires, North American Indians frequently burned portions of the landscape (Stewart 1951, 1963, Pyne 1988). Indian-set fires greatly added to lightening fires (Barrett and Arno 1982). Lewis and Clark (Thwaites 1959) and John Fremont (Jackson and Spence 1970) observed frequent fires set by the natives. Over the past 5,500 years, the abundance of charcoal has been inversely related to the abundance of sagebrush pollen and directly related to Graminae pollen (Fig. 5) (Mehringer and Wigand 1990). During peak prehistoric juniper expansion, based on relative pollen abundance (Wigand 1987), frequent fires probably maintain open savannah like stands. In recent times, sagebrush species have been observed to increase or maintain dominance in the absence of fire regardless of the presence or absence of livestock grazing (Robertson 1971, Anderson and Holte 1981, Sanders and Voth 1983, Sneva et al. 1984, West et al. 1984, Daddy et al. 1988). The dramatic increase in juniper species density in sagebrush steppe communities throughout the west has been partially attributed to reduced fire frequencies (Johnsen 1962, Blackburn and

Tueller 1970, Burkhardt and Tisdale 1976, Young and Evans 1981). However, some plant communities throughout the sagebrush region have maintained a dominance of perennial grasses in the absence of fire. Site characteristics interacting with a variable climate are likely to determine the ability of woody species to dominate a plant community in the absence of fire. Also, some plant communities may be maintained through biological inertia in the absence of disturbance.

Fires were thought to occur every 50 to 100 years in the intermountain sagebrush steppe, prior to settlement (Wright and Bailey 1982). However, on the more productive sagebrush sites, fires have been found to be more frequent (Burkhardt and Tisdale 1976). On mountain big sagebrush sites (*Artemisia tridentata* ssp. *vaseyana*) fire frequencies have been reported to be 20-25 years in Yellowstone (Houston 1973), 20 years in Great Basin National Park (Eddleman personal communication), and 17 years in southwestern Idaho (Burkhardt and Tisdale 1976). Burkhardt and Tisdale (1976) concluded, fire frequency intervals of 30 to 40 years would be adequate to keep juniper from invading a sagebrush steppe community. Many of the herbaceous species in the intermountain sagebrush steppe are well adapted to fire whereas big sagebrush and young juniper species, less than 50 years old, particularly the juvenile and seedling trees, are easily killed by fire (Blaisdell 1953, Wright and Klemmedson 1965, Conrad and Poulton 1966, Burkhardt and Tisdale 1976, Wright and Bailey 1982, Young and Miller 1985). In grasslands, fire maintains vigor of the herbaceous layer, while fire exclusion leads to its deterioration (Medina and Silva 1990). Net primary productivity of the herbaceous component usually increases greatly following a fire (Daubenmire 1968).

Volcanic and Tectonic Activity

Probably the most abrupt change in plant and animal distribution to occur across a large area in the sagebrush region was caused by volcanic activity (Mehringer 1986). During the past 10,000 years, deposits of ash and lava flows in the western portion of this region originated from the Cascades and Sierra Nevadas. Evidence of at least six volcanic events during the past 7,000 years are found on Steens Mountain in southeastern Oregon. One of the largest eruptions was Mount Mazama, approximately 6,600 years ago (Allison 1966), which produced 7 cubic miles of material distributed across 900,000 km^2 including a large portion of the sagebrush steppe (Fryxell 1965). Newberry Craters erupted only 2,000 years B.P. in central Oregon sending its ash plume east into the sagebrush steppe (Cressman 1981). Following eruptions of Mount Mazama and Newberry Craters, parts of eastern Oregon lay barren (Cressman 1981, Mehringer 1986). With passing time, vegetation and animal life returned to the landscape.

Herbivores

Abiotic factors alone do not always fully predict changes in plant composition or favorable growing conditions. Mychorrhiza, herbivore populations or human activities may also influence plant community succession. For example, although plant survival and production is usually associated with wet growing conditions, Allen et al. (1987) reported a decline in big sagebrush growth during a wet cycle. They associated the decline in sagebrush growth to a decrease in mycorrhizal reproduction and an increase in snow mold. Fluctuations in herbivore populations also can influence interspecific plant competition through selective grazing.

Although the biomass of large herbivores prior to European settlement was less in the intermountain sagebrush steppe than in many grasslands, herbivory by the native grazers was still an important process in this ecosystem. Besides the large hoofed mammals, native herbivores important in this ecosystem are small rodents, lagomorphs, birds, insects, mites, and plant-parasitic nematodes. Below-ground grazers, such as *Cicadidae*, are also thought to have a large effect on natural community structure and function (Hugie and Passey 1963, Anderson 1987). The smaller herbivores, due to their abundance, may have had a larger impact on this ecosystem than the large animals.

There was a mass extinction of the megafauna approximately 10,000 years ago, attributed to a combination of a changing climate and the arrival of humans to the sagebrush region (Martin 1967, Aikens 1986). Since then, large herbivores present in this ecosystem have been deer (*Odocoileus* spp.), elk (*Cervus elaphus*), antelope (*Antilocarpa americana*), bighorn sheep (*Ovis canadensis*) and bison (*Bison bison*) (Bedwell 1973, Aikens 1986, Van Vuren 1987, Urness 1989). Bison were widely distributed throughout the sagebrush steppe, but most abundant in southwestern Wyoming and southeastern Idaho (Van Vuren 1987, Urness 1989). In 1825, Ogden wrote in his journals that buffalo were plentiful, as he headed northwest from Henry's Fork, Idaho (Rich et al. 1950). Throughout most of the sagebrush steppe, however, bison herds were probably small and isolated (Van Vuren 1987). Bison skulls found in Malheur Lake in eastern Oregon were from an isolated inbred herd, based on a high incidence of dental anomalies. By the 1830s bison had disappeared west of the Rocky Mountains. Big horn sheep populations were also probably higher than present based on the frequency of remains found in Indian caves (Aikens 1986).

The relative abundance of deer, elk and antelope of the past compared to present is difficult to determine. Some of the early records suggest pre-settlement populations may have been low (Rich et al. 1950, Williams et al. 1971, Urness 1989). Odgen frequently mentions the scarcity of game as they traveled through the sagebrush steppe in the 1820s. The killing of horses to supply his group with meat during the winter of 1825, is frequently cited as evidence suggesting the scarcity of big game. However, a the time Ogden describes winter conditions as very severe. Ogden writes, "And as the weather continues, starve we must." In March, Ogden reflects over the past winter and states, ". . . not one horse would have fallen for the kettle but who can contend against the elements." In the winter of 1827-28, traditionally thought as the worst winter during the fur trapping era, Ogden noted hundreds of buffalo and antelope dying near Pocatello. Was game scare in the early 1800s due to severe winters or was habitat limiting? Some attribute the scarcity of game to year round hunting by Indians (Urness 1989). In Ogden's journals he frequently mentioned that game were scarce due to the abundance of Indians in the area (Rich et al. 1950, Williams et al. 1971). In the Great Basin, antelope trapping by the Indians appeared to be so successful it took several years to regenerate the population (Frison 1978). In Jedediah Smith's travels through the Great Basin during the 1820s, he mentions where he found water he found Indian sign (Strong

1969). Limited watering sites throughout much of the Great Basin would have made animal populations in this semi-arid region vulnerable to predation.

The early historic journals, however, do not always paint a picture of low numbers of large herbivores in the region. In other parts of Ogden's journals he mentions times when hunting was good. In July of 1826 he notes that deer and elk appear abundant on the main branch of the Crooked River in central Oregon (Rich et al. 1950). In the early to mid 1800s, other early explorers and trappers mention the abundance of deer, elk, and antelope in eastern Oregon (Stuart 1812, Clark 1932). Although mule deer populations were scarce in the arid portions of the sagebrush region, they were apparently abundant in the high mountain ranges (Young and Sparks 1985). One hunting party, consisting of two individuals killed 45 mule deer on a trip to southwestern Idaho in 1883.

The increase in mule deer and elk populations in the 20th century has been attributed to regulated hunting, successional changes favoring shrubs (Urness 1989), water developments and the elimination of large predators such as grizzly bears and wolves. Evidence, however, supporting arguments that 20th century populations of big game are greater or smaller than 19th century populations is limited.

Jackrabbits were abundant in the sagebrush steppe prior to settlement. In Escalante's (1776) journal he states; "The indian hunts hares, rabbits and sage-hens of which there is a great abundance." Early settlers and gold seekers in Utah also frequently mention the abundance of hares and sagehens on the landscape (Christensen and Johnson 1964). The black-tailed jackrabbit and white-tailed jackrabbit were the two most common lagomorphs in the Intermountain Sagebrush Region. Black-tailed jackrabbits graze grasses, forbs and shrubs in the early spring, switching to a shrub diet in the summer, fall and winter (Currie and Goodwin 1966, Uresk 1978). Cheatgrass (*Bromus tectorum*) is seldom found in their diets (Uresk 1978, Rogers et al. 1988). White-tailed jackrabbits feed primarily on grasses and forbs, switching over to shrubs during the winter (McAdoo and Young 1980). The abundance of the white-tailed jackrabbit has declined since settlement while the black-tailed jackrabbit is still abundant (McAdoo and Young 1980). White-tailed jackrabbits are usually found in higher densities than the black-tailed jackrabbit in grass dominated communities. The shift in relative abundance between the two species may reflect a change in plant composition in the sagebrush steppe. Changes in shrub architecture and food resources, however, did not appear to influence rodent populations in the sagebrush steppe (Parmenter and MacMahon 1983).

Another group of important herbivores in the Intermountain Sagebrush Region, of which we know little about, are the insects such as grasshoppers and Mormon crickets (Orthoptera), caterpillars (Lepidoptera), and ants (Hymenoptera) (Hewitt et al. 1974). Outbreaks of grasshoppers (Acridae), Mormon crickets (Tettigoniidae) and cutworms have been reported to intensively defoliate numerous plant species in the sagebrush region. Occasional outbreaks of the Aroga moth larvae have defoliated large areas of big sagebrush. Big sagebrush was also the most frequently selected plant species in the diet of the migratory grasshopper (Sheldon and Rogers 1978). However, insects also influence native perennial grasses. A cutworm infestation caused extensive mortality of Idaho fescue (*Festuca idahoensis*) following 4 years of livestock exclusion (Ellison 1960). Grazed plants outside of the exclosure, unaffected by cutworms, were alive and healthy. Another group of insects common to the sagebrush steppe are the Cicadidae (Hugie and Passey 1963). The newly hatched nymphs attach themselves to plant roots sucking the plant juices. Besides directly influencing the plants, the extensive burrowing by the nymphs in some soils influences soil structure.

The evolutionary history of grazing in an ecosystem has a large impact on its physiognomy (Milchunas et al. 1988). Grasses that have evolved with large numbers of herbivores are often characterized as small in stature with a large proportion of basal meristems, have minimum supportive tissue, high shoot density and rapid leaf turnover, and can reproduce vegetatively. In the absence of grazing, plant characteristics which allow a species to be more competitive in its environment may preclude morphological, physiological and biochemical mechanisms that make it grazing tolerant (Archer and Tiezen 1986). In the sagebrush steppe the tussock growth form of grasses, which makes them less resistant to grazing, is thought to be more resistant to heat transfer, to retard water vapor transfer during dry hot summers and to increase soil heat flux in the early spring accelerating the rise of soil temperature (Mack and Thompson 1982).

In the Intermountain Sagebrush Region, environmental conditions and hunting pressure by Indians appeared to keep large herbivore populations low. Grazing impacts by large herbivores in the sagebrush steppe was probably light, with heavy grazing limited to localized areas. Grazing was also probably seasonal, with animals moving up in elevation as the warm season ensued. Grazing by large and small mammals, birds and insects was also likely characterized by cyclic extremes of heavy and light grazing. Jackrabbit populations peak at approximately 10-year intervals (Gross et al. 1974, Johnson and Peek 1984). Total plant cover was probably reduced during peak population densities of jackrabbits in the sagebrush steppe (Anderson and Shumar 1986). No single plant species, however, appeared to be irreparably impacted since jackrabbits grazed nearly all the plant species within these communities. Due to the lack of differential grazing by jackrabbits, they had little impact on vegetation composition during periods of high population densities. However, Ellison (1960) concluded that rabbits and small mammals influence plant succession by destroying or disseminating plant seeds of various plant species. A population outbreak of voles in southwestern Montana during the winter of 1963-1964 caused a high mortality of big sagebrush (Mueggler 1967a). Areas affected by vole damaged ranged from 20 to 1,200 acres, killing from 10 to 84% of the sagebrush on these sites. Severe outbreaks of vole populations across large areas in the sagebrush steppe also occurred in 1907-1908 (Piper 1909) and 1956-1958 (Jellison et al. 1958).

Selective grazing pressure by the native herbivores as a group in the sagebrush steppe may have been minimal, reducing their affect on interspecific competition. As a group, the native herbivores in this region readily grazed forbs, grasses and shrubs. During the winter pronghorn, deer and sage grouse browse on sagebrush. In the spring the aroga moth larvae browse the new growth on the twigs. Spring forbs are grazed by many classes of animals including large ungulates, lagomorphs, small rodents, birds and insects. Grasses were consumed by elk, bison, and to a lesser degree pronghorn and deer. Lagomorphs and insects such as the grasshopper also readily consumed many grass species. If preferential grazing was minimal, the physiognomy of the sagebrush steppe may have evolved primarily to be adapted to cold wet springs and hot dry summers rather than herbivory.

Anthropogenic Disturbance

The earliest uncontested evidence of humans in the Intermountain Sagebrush Region is about 13,000 years B.P. (Aikens 1986). Besides using fire, early humans influenced their landscape through seed gathering and transport, root digging, farming and hunting (Strong 1969, Cressman 1977, Aikens 1986). However, by 1800 their influence on the landscape was already declining. By the close of the eighteenth century, Indian populations throughout this region were reduced eighty percent by such European diseases as smallpox, measles, venereal disease and possibly typhus (Thompson 1916, Cressman 1981). Their influence was further reduced when they were moved onto Indian reservations in the 1870s.

Man's earliest impacts on herbivory may have been the reduction of browsers (West 1987). Some ecologists feel the low numbers of large herbivores present during the 1800s was due to year-round hunting by the American Indian (Urness 1989). One of the greatest impacts of prehistoric humans on landscape processes may have been their contribution to the extinction of the megafauna (Martin 1967). The sudden extinction of large herbivores and their ecologically dependent carnivores and scavengers is not clearly related to climate. Although the extinction of most animals occurred during the warming and drying period at the end of glaciation, it also closely correlates with the spread of prehistoric man. The megafauna had survived several glacial and interglacial phases in the past. However, the last glacial and interglacial transition occurred for the first time when humans were present on the North American continent.

Pre-Settlement: Early-Mid 1800s

Shrub-Grass Communities

It is difficult to reconstruct an accurate picture of pristine vegetation in the Intermountain Sagebrush Region just prior to settlement. Early observations describing the vegetation in the sagebrush steppe are often contradictory. Father Escalante (1776), John Fremont (Jackson and Spence 1970), Russell (1885) and early pioneers traveling through Utah (Christensen and Johnson 1964) describe a picture of abundant grasslands. A buckaroo working for Pete French in the late 1800s describes the abundant grasses on Steens Mountain in southeastern Oregon brushing the stirrups as he road across the country side (John Scharff, personal communication). However, Peter Skene Ogden mentions the scarcity of grasses for his horses in central Oregon (Rich et al. 1950). Based on a summarization of early observations for the Snake River Plain, Vale (1975) concluded the dominant vegetation was sagebrush species with little grass present in the understory. In part, this may be attributed to locations where and when visits occurred. For example, vegetation was probably altered along the Oregon Trail by the late 1800s due to the large number of livestock moving across this route since the 1840s. Settlers also most frequently traveled through the grasslands east of the Rockies in the early spring when the country looked its best and entered the sagebrush steppe during its driest period in the late summer and early fall (Young 1989). Also the time since the last major fire would influence how the landscape looked to the European settlers.

Careful interpretation of past observations, relic areas, old exclosures and current successional trends provides us with a general picture of vegetation composition prior to settlement. In the early to mid 1800s, much of the sagebrush steppe was probably composed of open stands of shrubs with a strong component of long-lived perennial grasses and forbs in the understory (Fig. 7) (Hanson and Stoddart 1940, Blaisdell 1953, Christensen and Johnson 1964, Daubenmire 1970, Laycock 1978, Rogers 1982). Shrub canopy cover probably ranged between 5-10% in the drier Wyoming big sagebrush (*Artemisia tridentata* ssp. *wyomingensis*) communities (Cooper 1953, Young et al. 1976, Winward 1991), to 10-20% on the more mesic sites, occupied by mountain big sagebrush (Tisdale et al. 1965, Winward 1991). In eastern Oregon, near the confluences of the Deschutes and Crooked Rivers, a relic vegetation stand consists of abundant Idaho fescue and bluebunch wheatgrass with scattered big sagebrush and western juniper (Driscoll 1964). Total herbaceous cover was 14.2% and sagebrush cover 9.6%. In an ungrazed kipuka (an area of older land surround by a more recent lava flow) in southern Idaho, cover of big sagebrush and threetip sagebrush ranged from 10 to 13% and perennial grass cover 7 to 9% (Tisdale et al. 1965). The abundance of woody versus herbaceous species across the sagebrush steppe landscape varies, depending on climate, soils, aspect (Passey et al. 1982) and fire frequency. Soil fertility, temperature, water holding capacity, depth to stored water, presence of restricting soil layers, aeration and other soil related factors affect both plant growth and plant composition (Passey et al. 1982).

Figure 7. A mountain big sagebrush/Idaho fescue stand, occupied by a mixture of deep rooted perennial grasses and forbs, with about 5% cover of sagebrush.

Early explorers in Utah describe hillsides covered with abundant grasses. The primary species present were bluebunch wheatgrass and sandbergs bluegrass (*Poa sandbergii*) with scattered big sagebrush and Utah juniper (*Juniperus osteosperma*) (Christensen and Johnson 1964). On the level areas, western wheatgrass (*Agropyron smithii*) was the dominant species, forming a heavy sod. The grass dominated slopes lead to the valley bottoms and graded into shrub dominated communities composed of sagebrush species, *Atriplex* species and winterfat (*Ceratoides lanata*), and at the lowest points, greasewood (*Sarcobatus vermiculatus*). In the foothills, juniper species were scattered, giving an appearance of an open savanna. Abundance of juniper pollen prior to settlement was considerably less than amounts produced in present day woodlands (Wigand 1987). In eastern Oregon, western juniper was also found in open scattered stands across the landscape (Griffiths 1902, Rich et al. 1950). In the mid 1800s, topographic engineer Lieutenant Joseph Dixon, described vegetation occupying a flat valley fed by the Silvies River in the northwestern corner of the Great Basin, as "carpeted with bunch grass, wild pea vines and red clover, interspersed with fields of camas. . . ." (Clark 1932). The most frequent mention of a grassland appearance of the landscape by early pioneers and settlers, occurs in the eastern portion of the sagebrush steppe. Stoddart (1941), concluded that large areas in northern and central Utah were similar to the Palouse grassland association. The predominance of grass species in the eastern portion of the sagebrush steppe may be due to the increased proportion of April-September precipitation as compared to the western portions of this region (Stoddart 1941).

Microphytic Crusts

An important but often ignored component of vegetation of the Intermountain Sagebrush Region type is microphytic crust (West 1990). The microphytic component of sagebrush vegetation may play a large role in nutrient cycling, erosion and successional dynamics. Microphytes are represented by a complex surface mat of mosses, lichens, liverworts, algae, fungi and bacteria. Prior to settlement of the semi-arid sagebrush region, soil surfaces free of vascular plant litter that were not frequently flooded, burned, excessively stony, nor characterized by mobile sand or intensively traversed by hoofed animals were probably covered by these communities of small, non-vascular plants. Cryptogamic cover in Canyonlands National Park was 38% in the ungrazed and 5% in the grazed areas (Kleiner and Harper 1972). Microphyte distribution apparently is heavily influenced by micro-climate, forming patterns around individual shrubs and rocks. Mosses and liverworts seem to dominate the more mesic sites. Cyanobacteria favor the harshest sites and lichens dominate in intermediate circumstances (Eckert et al. 1986b). Microphytic crusts may interact with physical and chemical conditions of the soil (e.g. laminar crusts, rain crusts, vesicular crusts) and influence hydrologic and nutrient cycles. For instance, cyanobacteria are noted for their production of mucilaginous materials as well as N-fixation (West 1990).

Nitrogen fixation rates for microphytes have been estimated to range from 4 to 41 kg N ha^{-1} $year^{-1}$ (Rychert and Skujins 1974, Rychert et al. 1978, Skujins and Klubek 1978a b). Carbon inputs of microphytes may also enhance heterotrophic microbial N fixation (Klubek and Skujins 1980). Linkages between microphyte-fixed N and greater vascular plant growth, however, has yet to be demonstrated in the field. High rates of denitrification (Skujins and Klubek 1978b) and volatilization of ammonia (Klubek et al. 1978) warn us that a high N fixation rate does not automatically mean that productivity of nearby vascular plants will be enhanced.

Microphytic crusts may also influence the hydrologic cycle through increased infiltration rates and reduced overland flow connected to increased surface rugosity where mosses and lichens prevail (Blackburn 1975). Johnson and Gordon (1986) and Johnson et al. (1984, 1985) have reported wide variations in infiltration rates, overland flow and sediment production obtained with infiltrometer measurements on sagebrush steppe in southwestern Idaho. The variations could not be adequately explained by physical and chemical characteristics of the soil, vascular plant cover or litter. Because microphytes were abundant, but also variable in composition, they may have contributed to the variable hydrologic responses (Johnson and Blackburn 1987).

Once the water is under the crust, evaporation could be reduced, although microphytes with dark pigments could produce increased temperatures. Harper and Marble (1988) found that dark microphytes led to significantly more water loss from the upper 7.5 cm of soil than intermixed, scalped soil. Johansen (1984) demonstrated that extreme surface flow from convectional storms temporarily destroyed algal crusts in sagebrush communities, with Cyanophyceae least damaged.

Microphytes may influence the establishment of vascular plants by increasing the availability of "safe sites" for seed germination and establishment which is usually more limiting to plant establishment than amount of seed or soil resources (Crawley 1983). Although litter of the moss tortula (*Tortula ruralis*) in southern Idaho can retard germination and early growth of perennial bunchgrasses Thurbers needlegrass, squirreltail (*Sitanion hystrix*) and bluebunch wheatgrass, growth of these same species was stimulated by the moss litter four weeks after germination (Schlatterer and Tisdale 1969). Similar studies in Nevada by Eckert et al. (1986a) showed better establishment of big sagebrush, Thurbers needlegrass, squirreltail and cheatgrass on micro-sites where microphytes were abundant than in the bare interspace. In greenhouse experiments, establishment of the perennial bunchgrasses tall wheatgrass (*Agropyron elongatum*) and Russian wildrye (*Elymus junceus*) were enhanced by microphytic crusts (St. Clair et al. 1984).

Post-Settlement Vegetation

A large portion of the sagebrush region was the last area to be explored and settled in North America (Cline 1963). By 1750 only the Great Basin remained unexplored. It wasn't until 1844 that Fremont concluded the Great Basin did not drain into an ocean or sea (Jackson and Spence 1970). Although early settlement began during the 1840s in Utah, active settlement in much of this region did not begin until the 1860s.

Settlement

The first major impacts by Eurasians on the landscape in the sagebrush ecosystem came prior to settlement. Not only did they effect Indian populations in the late 1700s, but in the 1820s, thousands of beaver (*Castor canadensis*) were trapped from drainages throughout this region as the Hudson Bay Company attempted to ruin the country for the Americans (Williams et al. 1971). Governor George Simpson, head of the Hudson Bay Company, felt the best protection against the Americans taking over the Oregon Country was to keep it closely hunted. Ogden states, "It is scarcely credible what a destruction of beaver by trapping at this season, within the last few days upwards of fifty females have been taken and on an average each with four young ready to litter. Did not we hold this country by so slight a tenure it would be most to our interest to trap only in the fall, and by this mode it would take many years to ruin it." It did not take long for the various fur companies to trap out the beaver. The sudden reduction in beaver populations must have had a major impact on riparian and aquatic systems. Numerous beaver dams would have washed out and few new ones constructed.

Settlement of the Intermountain Sagebrush Region occurred shortly after the end of the Little Ice Age, in the mid 1880s. During the next 60 years, 1850-1913, the climate in much of the sagebrush steppe was characterized by increased warming, accelerated glacial melt and above normal precipitation (Antevs 1938, 1948, 1952, Graumlich 1985). Had settlement occurred during another time period, expectations of farming and livestock-carrying capacities would have been lower. The settlers were fortunate to have missed the worst drought in the 19th century, the decade of the 1840s. The period of above normal precipitation ended with the drought of the 1930s, the severest drought since the 13th century.

Settlement introduced several new factors that caused dramatic changes to vegetation in the sagebrush ecosystem; livestock, aggressive alien plant species, cultivation and water diversion. In addition, unregulated hunting during this period caused deer, elk, antelope and sage grouse populations to decline to low levels (Klebenow 1985, Urness 1989). Predators such as wolves and grizzly bears were aggressively hunted (Oliphant 1968), fire frequencies were dramatically altered (Houston 1973, Burkhardt and Tisdale 1976, Johnson and Larsen 1991) and atmospheric CO_2 levels would soon begin to increase from 275-285 ppm in the late 1800s (Gammon et al. 1985) to the present level of 350 ppm (Keeling et al. 1989). All of these factors, interacting with a gradual change in climate, have had a cumulative effect on the landscape since settlement.

As early as 1861, livestock were reported to be moving into eastern Oregon and northern Nevada (Oliphant 1968). From 1870 through the early 1900s, the numbers of cattle, sheep and horses rapidly increased, peaking in the early 1900s (Oliphant 1968, Young et al. 1976). By 1873, most all of the range in southeastern Oregon was occupied (Oliphant 1968). In 1900 on Steens Mountain in eastern Oregon, over 182,000 sheep grazed for a four to five month period during the growing season (Griffiths 1902). Griffiths estimated densities of about 450 animals per square mile. In addition to sheep, horses and cattle also grazed the mountain. In August 1901, some of the flocks were moved from summer range to lower elevation winter pasture due to a shortage of feed caused by overstocking. Griffiths states, ". . . overstocking was very conspicuous. The Whitehorse Mountains were being pastured by sheep the second time this season. . . . The most closely pastured region visited was Steins Mountains (currently spelled Steens). On the whole trip of three days we found no good feed except in the very steep ravines. . . ."

Many exotic plant species were also introduced during this period, some imported with wheat seed. The general trend in vegetation change on overgrazed landscapes in the sagebrush steppe has been either towards the increased dominance of woody species at the expense of herbaceous species, or the increased dominance of introduced annuals. By the 1930s, grazing capacity was estimated to have declined by 60 and 90% (McArdle et al. 1936). However, areas moderately grazed by domestic sheep in central Washington, from the mid-1800s to 1940, showed little or no evidence of grazing today (Beedlow et al. 1988).

Shrub-Grass Communities

Woody plants frequently increase or invade overgrazed shrub-grasslands and semi-arid savannas (Walker et al. 1981). In the sagebrush steppe, where fire frequencies have decreased, sagebrush species and juniper species have increased in dominance. Wyoming big sagebrush cover has increased from less than 10% to 20%, and mountain big sagebrush cover from less than 20% to 30 and 40% (Winward 1991). In Utah, the rapid decrease of perennial grasses and increase of shrubs occurred between 1870 and 1900 (Cottam and Stewart 1940, Christensen and Johnson 1964). Reports of abundant grass covering the hillsides between 1850 and 1870, changed to reports of declining grass, bare hillsides and increasing sagebrush by the 1880s. The rapid decline in giant wild ryegrass (*Elymus cinereus*) (Christensen and Johnson 1964) and palatable shrubs such as winterfat (Griffiths 1902) had occurred by the late 1800s. In the Lahontan Basin, dense stands of bunchgrass had been destroyed by overgrazing by 1880 (Russell 1885). The decline of palatable forage species and an increase in plant species of low palatability and bare ground took only 10 to 15 years at any given site under heavy uncontrolled grazing (Hull 1976).

The increase in woody vegetation during the late 1800s and early 1900s was due to reduced fire frequency, intensive season long grazing and climatic change. Fire, which maintained the dominance of herbaceous vegetation on many areas, declined with the reduction of fine fuels through grazing, building of roads and fire suppression. The influence of North American Indians on fire frequency was also reduced in the 19th century due to a decline of American Indian populations caused by European diseases (Thompson 1916, Cressman 1981) and movement of Indians to reservations in the early 1870s. The overall trend on a large portion of the sagebrush steppe has been a reduction in palatable grasses and forbs and an increase in sagebrush species (Ellison 1960, Young et al. 1976). The frequency of disturbance also influences populations of green rabbitbrush (*Chrysothamnus viscidiflorus*). As frequency of disturbance declined to more than 40 or 50 years, density of green rabbitbrush decreased (Young and Evans 1974). In eastern Oregon, grazing management that allowed the reestablishment and maintenance of perennial herbaceous species also maintained a higher density of green rabbitbrush as compared to adjacent ungrazed exclosures (EOARC 1991).

The increase in juniper species throughout much of the sagebrush steppe and semi-desert over the past 100 years has been dramatic (Fig. 8). Overgrazing alone, however, is not sufficient to explain the increase in juniper throughout the sage-

Figure 8. A juniper (*Juniperus occidentalis*) woodland in the early stages of development on Steens Mountain. Almost all the trees in the picture are less than 100 years old, and half are not yet above the sagebrush canopy.

brush region (Neilson 1986). Juniper species can readily invade sagebrush communities (Leopold 1924, Cottam and Stewart 1940, Jameson 1967, Johnsen 1962, Arnold et al. 1964, Blackburn and Tueller 1970, Eddleman 1987, Sparks et al. 1990). In Nevada, juniper densities slowly increased during the late 1800s (Blackburn and Tueller 1970). By the 1920s, the rate of invasion accelerated probably due to an increase in seed source. Western juniper begins to produce cones by 25-years of age (Eddleman personal communication). On Steens Mountain in eastern Oregon, western juniper densities were low prior to 1870, with the old trees restricted to rocky outcrops which produce little fine fuel (Miller unpublished data). Invasion of western juniper began in the late 1800s and rapidly accelerated through the 20th century (Fig. 9). Approximately half of the trees on Steens Mountain established between 1975 and 1990. The abundance of western juniper pollen has increased five fold since 1900 in this area (Mehringer and Wigand 1990). This species reaches its peak potential pollen production at around 70 to 80 years of age (Eddleman personnel communication, Miller unpublished data).

During the past several thousand years, the abundance of juniper species has been closely tied to fire frequency and climate (Mehringer and Wigand 1990). The primary influence of grazing, which has allowed the encroachment of juniper species into sagebrush communities, has been the reduction of fire frequency through the removal of fine fuels. The reduction of competition from perennial grasses through overgrazing in the late 1800s and 1900s may have enhanced seedling establishment of some juniper species. Pearson (1942) reported that cool season grasses are effective in competing for soil resources during the critical time of conifer seedling establishment. However, the effects of livestock grazing on interspecific plant competition probably has had minimal impact on the increase

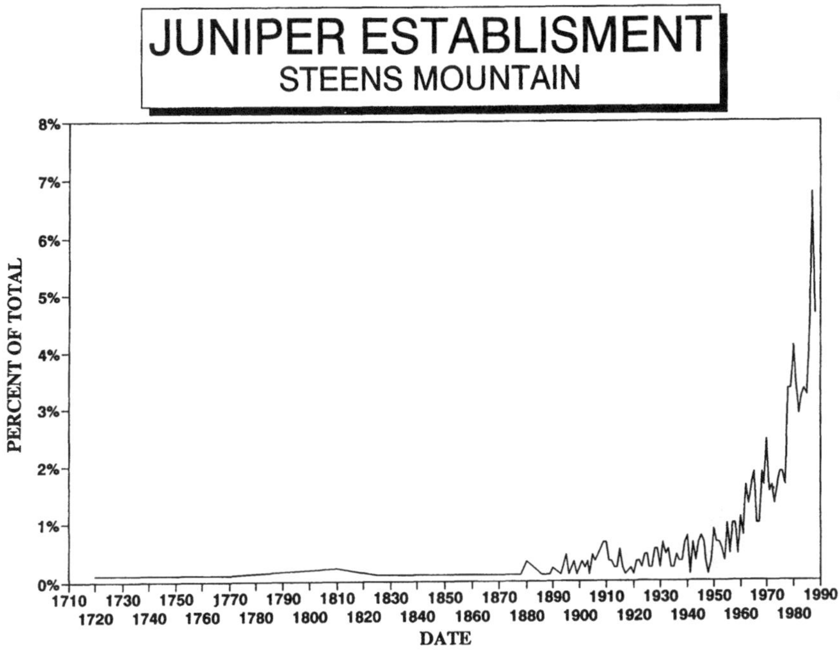

Figure 9. Years of establishment for the current population of *Juniperus occidentalis* on Steens Mountain, Oregon (n=1400).

in western juniper in the sagebrush steppe. Ecological condition of the site appears to have little influence on the ability of this species to invade sagebrush steppe communities (Burkhardt and Tisdale 1976, Miller unpublished data). Although seed dispersal of woody species by livestock has also been attributed to the increase of shrubs in other ecosystems (Brown and Archer 1989), dispersal of juniper species seed by livestock has probably been minimal (Johnsen 1962). Native herbivores such as the Townsend solitare (*Myadestes townsendii*) (Lederer 1977, Poddar and Lederer 1982), the American robin (*Turdus migratorius*), Steller's jay (*Cyanocitta stelleri*) and scrub jay (*Aphelocoma coerulescens*) are primary vectors of juniper seed dispersal (Gabrielson and Jewett 1970).

In addition to overgrazing and reduced fire frequencies, the increase in juniper species has also been attributed to climate (Johnsen 1962). The relatively more mild and mesic climate between 1850 and 1923 may have caused the initial increase in juniper species (Arnold et al. 1964, West et al. 1975). Wide annual growth rings in western juniper generally occur with mild wet winters and cool wet springs (Earle and Fritts 1986, Fritts and Xiangding 1986, Mehringer and Wigand 1990). In the late 1800s and early 1900s, mild wet conditions, reduced fire frequencies and reduced competition from perennial grasses created an ideal environment to initiate succession of shrub-steppe communities towards juniper woodlands. Formed under different climate and fire regimes, these woodlands are denser and contain a less abundant understory than prehistoric stands. In the southwest, Savage and Swetman (1990) associated these same environmental conditions with the increase in ponderosa pine (*Pinus ponderosa*) densities.

Introduced Eurasian Plants

Everyone is probably familiar with specific examples of biological invasions. However, the extent to which species have been introduced into North America during the past several hundred years may be surprising to many. For example, nearly 16% of the vascular plant species in California are non-native (Johnson 1985). The addition or deletion of individual exotic species may have relatively little impact on a plant community or ecosystem (Johnson 1985, Westman 1990). However, there are also examples of exotic species that can be very disruptive and actually displace native species (Westman 1990). In the Great Basin there is a relatively long list of exotic species that have been introduced within the past 150 years. Some of the more notable exotics include cheatgrass, red brome (*Bromus rubens*), medusahead (*Teniatherum asperum*), Russian thistle (*Salsola kali*), halogeton (*Halogeton glomeratus*) and a number of knapweeds (*Centaruea* spp.) We will center our discussion around cheatgrass which potentially will have an extensive and long-lasting influence on vegetation of the Great Basin (Billings 1990).

Cheatgrass was apparently introduced into North America from Europe some time during the mid-1800s, and was first documented in the Great Basin or Intermountain Region during the late-1800s (Klemmedson and Smith 1964). The spread of cheatgrass during the early part of this century was dramatic (Billings 1990), and this species is now thought to be dominant on more than 40 million hectares in the Intermountain West (Mack 1981). Although cheatgrass is probably most notable as an invader of rangelands, it also invades crop fields such as winter wheat (Rydrych 1974) and alfalfa (Kapusta and Strieker 1975). It is interesting to note that cheatgrass was described as a serious weed problem in Michigan in the 1930s (Megee 1938).

Many exotic species have developed broad distributions since their introduction into North America. However, few have the potential to impact entire ecosystems as does cheatgrass. Upadhyaya et al. (1986) summed up the situation as follows, "Seldom in recent history has the vegetation of such a large area been transformed so rapidly, and probably so permanently, as has been demonstrated by the invasion and spread of cheatgrass in the Great Basin and Columbia Basin areas." Billings (1990) suggests that cheatgrass may reduce the biotic diversity of the sagebrush biome by eliminating some plant and animal species, and potentially altering entire ecosystems. And it has recently been suggested that cheatgrass may be expanding its range to include salt desert and higher elevation woodland ecosystems (Young et al. 1987). What then are the mechanisms that allow one chance introduction to have such a dramatic impact?

The success of cheatgrass relates to both individual and population level characteristics. This species is very competitive with seedlings of perennial grasses (Evans 1961, Hull 1963, Harris 1967), and thus may limit recruitment of native perennial grass seedlings (Young and Evans 1973, 1978). The ability to germinate and extend roots at relatively low soil temperatures gives cheatgrass a major advantage over the native perennials in the Great Basin; it germinates and sets seed months earlier than native bunchgrasses. Climatic patterns in the Great Basin tend to favor species that can actively grow during the cool portion of the year when moisture is most available (Evans et al. 1970).

Harris (1967) has clearly demonstrated that cheatgrass is able to establish an extensive root system more quickly than the native bluebunch wheatgrass. The explanation for rapid growth may be: 1) the ability to extend roots at low soil temperatures, and 2) the inherent growth rate and efficiency of cheatgrass. Under favorable growing conditions, cheatgrass produced about twice the root length and leaf area compared to the very competitive seedlings of introduced crested wheatgrass (*Agropyron desertorum*) (Svejcar 1990). In this study, cheatgrass produced nearly twice as much root length per unit mass as crested wheatgrass. Crested wheatgrass seedlings are more aggressive than seedlings of most native species. Thus, the ability of cheatgrass to quickly establish and exploit soil moisture and nutrient reserves may severely limit the potential of perennial grasses to become established before the typical summer drought period. Cheatgrass may effectively extend the drought period experienced by perennial seedlings by using soil moisture reserves in the late spring period.

In addition to being very competitive as an individual, cheatgrass exhibits reproductive characteristics that ensure success at the population level. After studying cheatgrass, Mack and Pyke (1983) concluded that "Characterizations of any species as a colonizer, etc., on the basis of life history traits alone may be erroneous; knowledge of the variation in such population attributes is also necessary." In the Columbia Basin, Mack and Pyke (1983) found that cheatgrass recruitment generally occurred in late summer or autumn, but could occur at any time until mid-May of the following year; individuals less than 45 days old often produced at least one viable seed by June; and most plants survived to produce seed. Young and Evans (1985) also concluded that cheatgrass would behave as a winter annual if there is sufficient late-summer and fall precipitation, but was very successful as a spring annual when fall germination doesn't occur. They further state that, "The phenotypic plasticity of this species allows a dynamic rebound in seed production even at population densities as low as 0.1% of normal." On cheatgrass-dominated sites there are generally sufficient seed numbers to restore the stand and fully exploit the environmental potential even after 2 years of extreme drought (Young and Evans 1985). The population dynamics of this species after wildfire provides an example of its plasticity. In a case where cheatgrass density was 10,000 plants per m^2 before burning and 10 plants per m^2 after burning, herbage and seed production per unit area were not changed (Young 1991).

In addition to its competitive nature, cheatgrass can influence ecosystem processes in another important way — by increasing the fire frequency on a site. The initial establishment of cheatgrass on a site, and subsequent reestablishment after wildfire, predisposes vegetation to recurring wildfires (Young and Evans 1978). Whisenant (1990) provided an example from what might be the extreme case, the Snake River Plain in southern Idaho. In this area the fire return interval may have been as low as 50 to 100 years, and wildfires now occur at 3 to 5-year intervals. In addition, the fires are more extensive and there are fewer unburned patches remaining within the burned area. As Whisenant pointed out, the continuity of fine fuel is much greater in a cheatgrass-dominated community than it is a native bunchgrass community. As fire frequency increases the non-sprouting woody species are lost and, at the very high frequencies experienced in the Snake River Plain, virtually all perennials are lost. The result is a decline in species diversity at both the patch and landscape-level (Whisenant 1990). Another potentially important but often overlooked factor is the change in timing of burns. Because cheatgrass matures several months earlier than the native bunchgrasses, wildfires may occur at a time when the native plants are still actively growing. As West and Hassan (1985) pointed out, cheatgrass dominance leads to potentially harmful earlier reburns.

The combination of frequent fires and severe competition has led Billings (1990) to conclude that cheatgrass may place some native species and even entire ecosystems at risk in the Great Basin. Excessive grazing during the late 1800s and the early part of this century probably hastened the rapid spread of this species across the intermountain west. However, grazing may not have been a prerequisite for the successful introduction of cheatgrass into the Great Basin. Daubenmire (1940) concluded that the cheatgrass stands in southeastern Washington were not a direct result of overgrazing. Daubenmire (1940), Tisdale et al. (1965), Harris (1967) and Passey et al. (1982) cited examples of cheatgrass invasion into climax communities that had not been burned or grazed for at least 50 years. Cheatgrass, and its close relative red brome have come to dominate areas that previously had little or no grazing and no record of recent wildfires (Hunter 1990, Svejcar and Tausch 1991). In such cases the influence of annual *Bromus* species on the native vegetation may be more subtle. Svejcar and Tausch (1991) suggest that the annual *Bromus* species may inhabit safe sites and thus limit recruitment of native bunchgrasses into the population. Given the long life span of some native bunchgrasses (West et al. 1979), the loss of these species from the stands may occur over a relatively long time-frame.

The debate of grazing effects on annual *Bromus* invasion will no doubt continue. There are probably examples when a disturbance such as grazing is necessary for *Bromus* establishment and examples where these species are "preadapted" and can dominate the native vegetation without a major disturbance. Cheatgrass in particular seems well adapted to areas with rather mild winters and precipitation concentrated in the cool winter-early spring period. At this point the removal of grazing pressure may serve to exacerbate the problem of increased fire frequencies. Young and Evans (1978) felt that the evidence was fairly good that removal of livestock would accelerate conversion to annual dominance by allowing fuel accumulations. Given the current situation, it appears that more research is needed to evaluate whether grazing might be used to favor native perennials over the annual invaders. Might grazing pressure during winter and early spring serve to shift the competitive advantage toward the native perennial vegetation? Can controlled burning (during spring or fall) be used to favor the perennial bunchgrasses that suffer during summer burns? We need answers to these questions to better manage our rangelands in the Great Basin.

Microphytic Crusts and Disturbance

Mack (1981) and Rickard and Vaughan (1988) think that microphytic crusts have been largely lost from the entire northern Intermountain Region of Eastern Washington due to a combination of livestock grazing, fire, cultivation and recent volcanic deposits. This region is where the introduced annual grass cheatgrass has become especially abundant, allowing a pronounced increase in fire frequencies (West 1988).

Many researchers in the Intermountain Sagebrush Region (Mack and Thompson 1982, Johansen 1986, Harper and Marble 1988, Richard and Vaughan 1988) have observed apparent reductions in microphytic crusts following livestock trampling. Marble and Harper (1989) feel that microphytic crusts are especially susceptible to this mechanical damage when they are dry. Theirs and other studies, however, used natural, fence line contrast "experiments." Mosses and foliose lichens appear to be more susceptible to livestock trampling than crustose lichens and microscopic forms. Moderate mechanical disturbance may stimulate these crustose lichens and microscopic forms which are more important in N-fixation (West 1990).

Off-road vehicles are another modern source of disturbance to microphytic crust that has been scarcely investigated. The Webb et al. (1988) study of recovery of plants, including microphytes, on roads in a desert ghost town indicate a long-time requirement for full recovery.

Several recent studies in sagebrush communities have demonstrated that microphytic soil crusts can be at least temporarily damaged by wildfire (West and Hassan 1985, Johansen and Rayburn 1989). Wildfires in the sagebrush steppe and semi-desert dominated by Eurasian annuals, have recently become larger, more frequent and earlier in the year (Whisenant 1990). This change in fire pattern is largely related to the invasion of Eurasian annuals creating more continuous, earlier maturing fine fuels (West 1988). Rosentreter (1986) and Whisenant (1990) described how fire reduced lichen cover on the sagebrush steppe of southern Idaho where deeper soils allow the production of more fuel. Johansen et al. (1984) showed all microphytes were damaged from a fire on an sagebrush steppe in central Utah. The reduction lasted up to 3 years and recovery was not complete after 5 years, even though the years involved were unusually wet.

Photosynthesis-inhibiting herbicides (e.g., tebuthiuron) have been widely used in the Intermountain Sagebrush Region, yet the effects of these herbicides on microphytic organisms has not yet been considered. We could speculate that because cyanobacteria are sensitive to insecticides (Pipe and Shubert 1983), the application of materials such as malathion to control grasshoppers may be causing reductions of microphytic crusts. Although soil conditioners and fertilizers are not widely used in sagebrush communities, slurries and explosives are widely used in fire control. We do not know how microphytes respond to such influences.

Leachates of sagebrush leaves reduce the N-fixing abilities of microphytes by as much as 80% (Rychert and Skujins 1974). Thus the two to four fold increase in shrub cover in many sagebrush communities may have reduced the amount of N fixed in these sites. Skujins and Trujillo Y Fulgham (1978) discovered that sagebrush litter can also inhibit nitrification. This results in a considerable proportion of organic N being denitrified and put into the atmosphere. However, fresh chopped leaf material was used in their study which may have magnified their results.

New Steady States

Changes in plant composition on rangelands aren't always continuous, reversible or consistent (Westoby et al. 1989, Laycock 1991, Friedel 1991) and is thus highly indeterminate (Van Hulst 1979). Past events can cause a plant community to succeed towards a new steady state (Sutherland 1974). This scenario appears to describe vegetation in the Intermountain Sagebrush Region. Heavy grazing, altered fire frequencies, variable climate, introduction of exotic plant species and other perturbations have changed the composition and structure of vegetation on many parts of the landscape to new steady states. Removal of livestock from altered sagebrush communities may not return these communities to the pregrazed state (Robertson 1971, Anderson and Holte 1981, Holechek and Stephenson 1983, West et al. 1984). Examples of vegetation remaining in a new steady state following livestock removal have been noted in the southwestern U.S. (Archer et al. 1988), Australia, Africa and India (Walker et al. 1981). In some cases, woody plants continue to increase following the removal of livestock (Robertson 1971, Anderson and Holte 1981, Sneva et al. 1984). As woody plants increase in savannas and other grass-shrub plant ecosystems, resilience (the ability of a natural ecosystem to restore its structure following acute or chronic disturbance) declines and their stability increases (Walker et al. 1981). Removal of livestock from plant communities dominated by Eurasian annuals usually does not lead towards the reintroduction of native perennial species and the decline of the introduced species (West et al. 1984, Billings 1990, Young and Tipton 1990, Sparks et al. 1990). These recently established annual plant communities appear to have created their own, new equilibrium (Hanley 1979). Plant communities dominated by cheatgrass resist the reestablishment of native species (Rickard and Sauer 1982). The 3 to 5 year fire return intervals in cheatgrass dominated communities in the Snake River Plain (Whisenant 1990), probably reduces the establishment of early seral perennial grasses such as squirreltail. Elevated CO_2 levels may also enhance the competitive ability of cheatgrass over other native grasses (Smith et al. 1987).

In the sagebrush steppe, succession usually moves in one of three general directions: (1) dominance of woody species and a decline in fire frequencies, (2) increase in Eurasian weeds and an increase in fire frequencies, or (3) return to a codominance of perennial native herbaceous species with shrubs (Fig. 10). Once sagebrush species dominate a site there is little evidence the site will return to a codominance of perennial native herbaceous species (West et al. 1984, Winward 1991) unless some perturbation selectively reduces sagebrush. In savannas and shrub steppe ecosystems, the woody plant dominated state is very stable (Niering and Goodwin 1974, Walker et al. 1981, Anderson and Holte 1981). However, while vegetation may appear stable, soil erosion often accelerates degrading the future productivity of the site.

Increases of native herbaceous species have been observed on portions of the sagebrush steppe where livestock have been removed or intensive management implemented (Ellison 1960, Sneva et al. 1984). The potential for increased density and cover of perennial native herbaceous species generally occurs when sagebrush species cover is below the maximum potential level, a residual of perennial herbaceous species is present and competition from Eurasian weeds is absent or minimal (Sneva et al. 1984). Although the perennial native herbaceous species increase in this situation, shrubs frequently increase by a greater magnitude in the absence of fire (Robertson 1971, Anderson and Holte 1981, Sneva et al. 1984).

Two readily identifiable transition thresholds in arid and semiarid rangelands exist, separating grasslands from shrublands and stable from degraded soil (Friedel

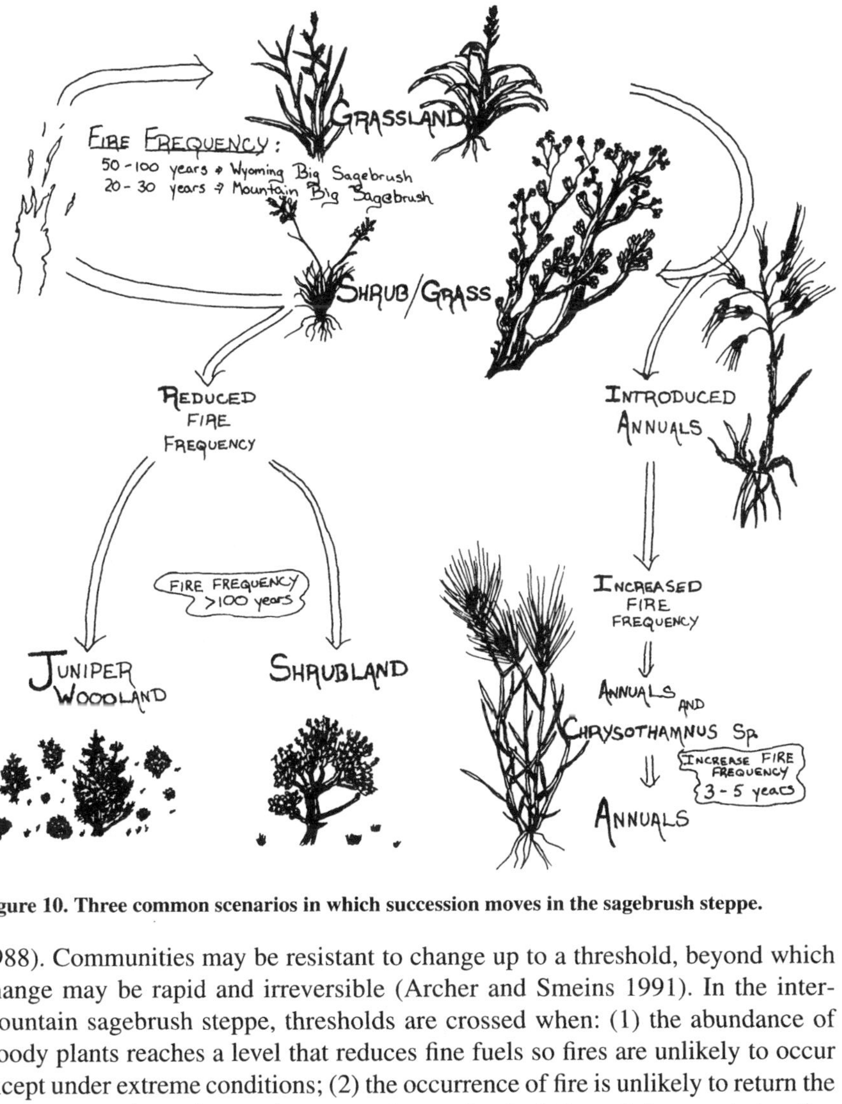

Figure 10. Three common scenarios in which succession moves in the sagebrush steppe.

1988). Communities may be resistant to change up to a threshold, beyond which change may be rapid and irreversible (Archer and Smeins 1991). In the intermountain sagebrush steppe, thresholds are crossed when: (1) the abundance of woody plants reaches a level that reduces fine fuels so fires are unlikely to occur except under extreme conditions; (2) the occurrence of fire is unlikely to return the plant community to its previous state due to the depletion of deep rooted native perennial grasses and forbs, or (3) the herbaceous understory is replaced by introduced annuals and short lived perennials, thereby increasing fire occurrence to greater frequencies than during presettlement times.

Impacts Of Domestic Herbivores On Plant Composition

Herbivory was not a new process in the Intermountain Sagebrush Region at the time of introduction of livestock during the 1860s. The new species of herbivores introduced, however, altered the process of herbivory at the species, community and landscape levels. Numbers of domestic herbivores increased rapidly in the late

1800s and early 1900s (Oliphant 1968, Young and Sparks 1985). Their impact on plant composition in sagebrush communities was not only due to the increased intensity of herbivory caused by the large numbers of livestock usually grazed throughout the growing season, but also selective feeding pressure on specific plant species and their preference for certain portions of the landscape. Livestock also differed from native herbivores in that their populations were not cyclic, so landscapes were often grazed year after year throughout the growing season until no forage remained (Griffiths 1902). Heavy grazing and light grazing across years by native herbivores was replaced by year after year season long heavy grazing by livestock in the late 1800s.

Grazing Methods

Early in this century Sampson (1914) pointed out the importance of knowing the growth requirements and life histories of principal forage plant species to properly manage rangelands. Forty years later, Stoddart and Smith (1955) stated that little is known concerning how much grazing use a plant can withstand without injury. More recently, Eckert and Spencer (1986) said, "The effects of grazing management systems on plant communities in the Great Basin are largely unknown." Some studies have evaluated the effects of grazing versus no grazing on plant community composition. However, grazing treatments, before and after exclosure establishment, and ecological conditions at the time of exclosure establishment tend to be variable and poorly quantified.

Defoliation during the mid growing season, particularly during early flower development, usually has the most negative impact on cool season herbaceous plants growing in the intermountain sagebrush steppe (Daubenmire 1940, Stoddart 1946, Blaisdell and Pechanec 1949, Heady 1950, Wilson et al. 1966, Mueggler 1967b, Trlica and Cook 1971, Harris and Goebel 1976). Recent work has also evaluated the physiological and morphological response of cool season native grasses to defoliation (Caldwell et al. 1981, Richards 1984, Richards and Caldwell 1985, Nowak and Caldwell 1985, Mueller and Richards 1986, Richards et al. 1988). It is apparent, however, that there is little long-term research evaluating the effects of various grazing systems and intensities on native plant communities in the intermountain sagebrush steppe.

The lack of grazing system comparisons is probably due to the land area required, the variability of terrain and climate, the relatively low number of experimental rangeland areas in the intermountain sagebrush ecosystem and the long term commitment required to evaluate vegetation trend. We know of only a few such studies in the sagebrush steppe. Hyder and Sawyer (1951) compared season-long with rotation-deferred grazing in eastern Oregon. They found a slightly better plant response in the season long pasture. Season long use, however, generally does not work well when animal distribution is a problem (Holechek et al. 1989). This is frequently the case in the intermountain sagebrush steppe, where rugged topography, riparian zones, limited watering areas and large pastures (often exceeding 4,000 ha) influence preferred areas for grazing. In one other study which evaluated 10 years of rest-rotation grazing, desirable deep rooted perennial grasses did not increase while Sandbergs bluegrass and big sagebrush did increase (Eckert and Spencer 1986,

1987). However, plant utilization was relatively high, ranging from 65% to 80%. Poor results from rest-rotation are usually caused by heavy grazing (Holechek et al. 1989). One or several years of rest are usually not sufficient to compensate for intermittent years of severe use (Cook and Child 1971, Trlica et al. 1977, Eckert and Spencer 1987). Under lower levels of use, Holechek et al. (1989) concluded rest-rotation was a good system to consider in rugged terrain. Long term research evaluating plant composition changes to spring and fall sheep grazing and total protection on threetip sagebrush range has also been conducted in eastern Idaho (Mueggler 1950, Laycock 1967). Both fall grazing and total protection maintained plant communities in good ecological condition while heavy spring use allowed sagebrush to increase and perennial grasses and forbs to decline. Palatable forbs such as arrowleaf balsamroot (*Balsamorhiza sagittata*) were reduced as much as 80% by spring sheep grazing (Laycock 1967). Threetip sagebrush, however, is more readily browsed by sheep in the fall than is big sagebrush.

Results comparing no grazing with grazing are usually mixed and often hard to interpret. Although a number of exclosures have been established throughout the Intermountain Sagebrush Region, measurements quantifying grazing use outside are often not available. Condition of the site when the exclosure was established will also affect plant succession. Most of these exclosures were established following heavy livestock use during the late 1800s and early 1900s. In general, where proper grazing management was applied, desirable perennial herbaceous vegetation increased both with and without grazing (Gibbens and Fisser 1975, Sneva et al. 1984, Kindschy 1987). However, on sites where competition from woody plants or exotic annuals limited the reestablishment of desirable plant species, little change or a decline in desirable species occurred both inside and outside the exclosure (Hughes 1980, Anderson and Holte 1981, Sanders and Voth 1983, West et al. 1984).

Due to the cyclic nature of climate, studies evaluating grazing effects on plant composition should be done over a period of a minimum of 10 years. The importance of long term studies is emphasized by Branson's (1985) summary of changes in plant cover from 1932 to 1943. Plant cover declined and increased with climate, regardless of grazing intensity or the absence of grazing. Due to the limited amount of work evaluating long term effects of various grazing strategies on plant composition of native ranges, managers must base their decisions on principles of plant growth requirements, interspecific competition and animal behavior.

Defoliation and Plant Composition

The degree of influence domestic herbivores have on plant community composition depends upon intensity, season, duration and frequency of use, level of selectivity and site characteristics. How these factors, in combination with climate, influence plant longevity and recruitment will affect plant composition over time. Also, the evolutionary history of the Intermountain Sagebrush Region, in relation to herbivory, influences ecosystem response to livestock grazing (Stebbins 1981, Milchunas et al. 1988).

Domestic herbivores influence plant composition in sagebrush communities both directly and indirectly. Direct effects are removal or breakage of aboveground plant tissue influencing both physiological and morphological processes (discussed in Chapter 5). Indirect effects of livestock grazing on plant community composition are alteration of the microenvironment and fire return intervals, and influence on ecosystem processes such as nutrient cycling, energy flow and the water cycle. Heavy grazing has also caused substantial increases in populations of small herbivores (Smith 1940, Hayes 1927, Kamm et al. 1978) such as grasshoppers (Ellison 1960) and jackrabbits (Christensen and Johnson 1964, Jaksic and Fuentes 1980). Heady (1975) and more recently Archer and Smeins (1991) discussed the effects of livestock grazing on ecosystem processes.

One of the primary effects of grazing on plant community composition is its influence on interspecific competition. This is particularly true when certain species are being selectively grazed while associated species are either being lightly or not grazed (Mueggler 1970, 1972). It is frequently state, that the rapid decline of perennial grasses in the sagebrush steppe was due to the sensitivity of these species to defoliation since they evolved in an ecosystem with few large herbivores. However, the sensitivity of the native grasses to grazing may be as much or more due to the competitive interaction with ungrazed species such as sagebrush and juniper species. The effect of selective grazing on interspecific competition may override a plant species tolerance to grazing (Archer and Tiezen 1986). The partial reduction of competition from surrounding neighbors can offset the adverse effects of heavy defoliation on native perennial grasses (Mueggler 1970, 1972).

Although a great deal of research has evaluated individual plant response to season, intensity and frequency of grazing on important range forage species, little work has evaluated the response of the ungrazed neighbor. Research in the sagebrush steppe has shown that defoliation of grasses during rapid growth decreases the depletion rate of soil water (Wraith et al. 1987, Daddy et al. 1988, Miller et al. 1990). Reduced uptake of soil resources by forage species due to decreased leaf area and root growth may enhance growth and or establishment of ungrazed neighboring plants. Grazing avoidance-type plants often gain the competitive advantage over grazed plant species (Archer and Smeins 1991). In the Great Basin, southwestern hot deserts and the southern plains, shrub-steppe and grassland communities have become dominated by unpalatable grazing avoidance-type plants.

Grazing effects on plant community composition can also result from changes in seed production, establishment, survival of young plants and longevity of older plants (West et al. 1979). This applies to both grazed and ungrazed species. On a protected bluebunch wheatgrass and intermediate wheatgrass site there were 630 viable seed/m^2 compared to 12 viable seed/m^2 on a heavily grazed site (Hanson and Stoddart 1940). Hoof action by large herbivores has often been recommended to enhance establishment of seed. However, little data are available to support this hypothesis. In Nevada, trampling enhanced emergence of perennial grasses, but decreased the emergence of perennial forbs (Eckert et al. 1986a). In the southwest, seedbed disturbance by cattle appeared to enhance emergence of perennial grasses under conditions of moderate precipitation, but had little effect in wet years and is probably futile in dry years, depending on soils and species (Winkel and Roundy 1991). Reduced competition can also increase survival and growth of newly established plants. Neighboring plants accounted for 69% of the mortality and weight of cheatgrass (Mack and Harper 1977). The reduction of competition from neighbors enhanced seedling growth and survival of bitterbrush (*Purshia tridentata*) (Ferguson and Basile 1967) and big sagebrush (Romney et al. 1980, Pendery and Provenza 1987).

The effect of grazing on plant longevity is variable. Grazing has been reported to shorten (Hall et al. 1964, Barker and Lange 1970, Lange and Purdie 1976, Crisp and Lange 1976), have no effect (Wright 1972, Wright and Van Dyne 1976) and to increase (Canfield 1957, Williams 1970) individual plant longevity. The longevity of Great Basin species appears to vary with grazing history (Stewart et al. 1940) and plant species (West et al. 1979). In the sagebrush steppe, granite gilia (*Leptodactylon pungens*) lived longer on ungrazed plots while Indian ricegrass (*Oryzopsis hymenoides*) lived longer on grazed plots (West et al. 1979). Fall grazing on a threetip sagebrush shrub steppe increased the longevity of bluebunch wheatgrass, Indian ricegrass and needlegrass, while it decreased the longevity of three tipped sagebrush. Grazing generally has the greatest effect on mortality in one and two year old plants.

Livestock grazing the Intermountain Sagebrush Region may also alter nutrient availability thereby influencing plant community composition. Grazing can influence both the amount and rate of shoot and root turnover and litter accumulation. The feeding activities of grazers generally decreases the cycling time of nitrogen (Gist and Sferra 1978). The amount of nitrogen exported off site, lost via volatilization and redistributed on the site by livestock depends upon the intensity and season of use. Woodmansee (1978) estimated 50% of the N excreted by domestic herbivores is lost through volatilization. However, recent work concluded these losses were lower than originally expected, especially on heavily grazed grasslands (Schimel et al. 1986, Parton et al. 1988). Approximately 35% of the excreta was redistributed to 10% of the pasture. An additional amount of N and other nutrients are exported from the site via animal tissue. Woodmansee (1978) estimated that N input:output ratios are >1 on lightly grazed, approximately equal to 1 on moderately grazed and <1 on heavily grazed rangelands. Long term heavy grazing can gradually deplete soil nutrients. The greatest loss of nutrients may result from alteration of plant community structure which influences overland flow, erosion, infiltration rates and nutrient turnover rates.

Implications

Vegetation of the Intermountain Sagebrush Region has been in a state of flux for the past 10,000 years. Shifts in climate and fire return intervals have caused altitudinal changes in plant communities, increases and decreases in the abundance of salt desert species and the relative abundance of upland shrubs and grasses. Major holocene volcanic flows and ashfalls have also locally impacted plant and animal fauna. During the past 120 years, however, changes in plant composition across the sagebrush region have occurred at unprecedented rates. In addition to subtle changes in climate and large changes in fire return intervals, exotic Eurasian plant and animal species were introduced, large portions of the landscape were cultivated, water was diverted for agricultural and municipal use and CO_2 levels increased. When interpreting vegetation trends in the Intermountain Sagebrush Region it is often difficult or impossible to separate the effects of heavy livestock grazing from the myriad of interacting environmental parameters.

In general, heavy livestock grazing influences plant composition by; (1) increasing the dominance of unpalatable woody species and decreasing perennial forbs and grasses, or (2) increasing the dominance of introduced annuals. There is little argument that poor grazing practices were and in some areas still are, a primary cause of redirecting or accelerating succession towards less desirable new steady states. However, the practice of unwise grazing has not been the sole factor contributing to changes in plant composition. Grazing along with both natural and anthropogenic factors has had a cumulative influence on plant succession. Changes in plant composition would have occurred throughout this region without livestock grazing because of the introduction of alien plant species, cultivation, increased CO_2 levels and altered fire frequencies.

Evaluating the impacts of light to moderate grazing by domestic herbivores is even more difficult than with heavy grazing because; (1) a more subtle influence on ecosystem processes than heavy grazing, and (2) very limited long term research comparing various light to moderate grazing strategies with no grazing on both poor and good condition rangelands. How then can we address the questions: (1) What are the implications of light to moderate grazing on plant composition and ecosystem processes and functions? and (2) Is light to moderate grazing by domestic herbivores compatible with ecosystem processes in the Intermountain Sagebrush Region? The best documented long term evidence indicating that light to moderate livestock use can be compatible with the sagebrush ecosystem is from communities where plant species composition remained unchanged from adjacent ungrazed sites in good ecological condition (Mueggler 1950, Laycock 1967, Beedlow et al. 1988), or where native perennial grasses, forbs and palatable shrubs have reestablished under good grazing management (Sneva et al. 1984, Kindschy 1987). It appears that if plant communities have not crossed a threshold into another steady state that a return to good ecological condition can occur both under protection or light to moderate grazing by domestic herbivores. If a community has entered a new steady state, removal of livestock will most often not return this ecosystem to near pristine conditions.

Since Sampson's statement in 1914, pointing out the importance of information on the ecology of rangelands for proper management, our knowledge of range ecosystems has progressed well in some areas and little in others. Our understanding of individual plant mechanisms and process and to a lesser degree of ecosystem processes and function has progressed during the last 80 years. The science has progressed from looking at single factors to studying the interactions of multiple factors. However, our understanding of the long term effects of light to moderate livestock grazing on plant composition and ecosystem processes in the Intermountain Sagebrush Region has progressed little since the turn of the century. This points out the need for a long term research commitment evaluating the effects of best management practices on plant composition and ecosystem processes. It also points out the need and importance for designated research sites throughout the sagebrush region to conduct long term research.

It is obvious that we cannot protect our resources by exclusion from use. As long as humans have walked the surface of this earth we have used our resources and depended upon them for our survival. The efforts of scientists and resource managers must be focused on the proper management of resources for long term sustained use. An inadequate data base has tended to encourage the use of casual observations and untested theories in land management practices involving livestock grazing (Vallentine 1979). Development of proper use comes from an understanding of ecosystem processes and how certain variables, such as livestock grazing, interact with these processes. Archer and Smeins (1991) also point to the importance of identifying critical thresholds for plant communities and understanding succession, stability and resilience when developing management schemes for the landscape. Rather than managing for images of landscapes existing in the past, we should focus management and land restoration on a process level. For example, a high priority management objective on many of our uplands in the sagebrush region is to decrease overland flow and erosion while increasing the safe capture, storage and release of water. Another management objective may be to reduce the export of nutrients from a site. Leopold (1949) realized that our land use ethic, of looking at land as a commodity has gotten us into trouble. As a true conservationist he also knew we humans cannot separate ourselves from our environment, that we are a part of the ecosystem. However, he strongly emphasized the importance of treating our environment and our resources with respect and care. Although we are attempting to move in this direction, the challenge increases with continued increases in human population, demands for resources and the improvement of technology to exploit these resources.

Literature Cited

Aikens, C.M. 1986. Archaeology of Oregon. U.S. Dept. Inter. Bur. Land Manage.

Allen, M.F., E.B. Allen and N.E. West. 1987. Influence of parasitic and mutualistic fungi on *Artemisia tridentata* during high precipitation years. Bull. Torrey Bot. Club 114:272-279.

Allison, I.S. 1966. Fossil Lake, Oregon: its geology and fossil faunas. Oregon State University Press. Corvallis, OR.

Allison, I.S. 1982. Geology of pluvial lake Chewaucan Lake County, Oregon. Oregon State Univ. Press. Corvallis, OR.

Anderson, D.C. 1987. Below-ground herbivory in natural communities: a review emphasizing fossorial animals. Quart. Rev. Biol. 62:261-286.

Anderson, J.E. and K.E. Holte. 1981. Vegetation development over 25 years without grazing on sagebrush-dominated rangeland in southeastern Idaho. J. Range Manage. 34:25-29.

Anderson, J.E. and M.L. Shumar. 1986. Impacts of black-tailed jackrabbits at peak population densities on sagebrush-steppe vegetation. J. Range Manage. 39:152-156.

Antevs, E. 1938. Rainfall and tree growth in the Great Basin. Carnegie Instn. of Wash., Publ. 469, Amer. Geogr. Soc., Spec. Publ. 21.

Antevs, E. 1948. Climatic changes and pre-white man, p. 168-191. *In*: The Great Basin; with emphasis on glacial and postglacial times. Bull. Univ. Utah. Biol. Ser. 10(7).

Antevs, E. 1952. Cenozoic climates of the Great Basin: Geoloische Rundschau 40:94-108.

Archer, S. and L.L. Tiezen. 1986. Plant responses to defoliation: hierarchical considerations, p. 45-59. *In*: O. Gudmundsson (ed.), Grazing research at northern latitudes. Plenum Press, N.Y.

Archer, S., C. Scifres and C.R. Bassham. 1988. Autogenic succession in a subtropical savanna: conversion of grassland to thorn woodland. Ecol. Monogr. 58:111-127.

Archer, S. and F. Smeins. 1991. Ecosystem-level processes, p. 109-140. *In*: R.K. Heitschmidt and J.W. Stuth, (eds.), Grazing management: An ecological perspective. Timber Press, Portland OR.

Arnold, J.F., D.A. Jameson, and E.H. Reid. 1964. The pinyon-juniper type of Arizona: effects of grazing, fire and tree control. USDA Prod. Res. Rep. No. 84.

Barker, S. and R.T. Lange. 1970. Population ecology of *Atriplex* under sheep stocking, p. 105-120. *In*. R. Hones, (ed.), The biology of *Atriplex*. Commonwealth Sci. Indust. Res. Organ. Div. of Plant Indust. Canberra, Aust.

Barrett, S.W. and S.F. Arno. 1982. Indian fires as an ecological influence in the northern Rockies. J. For. 647-651.

Bedwell, S.F. 1973. Fort Rock Basin prehistory and environment. Univ. Oregon Books, Eugene, OR.

Beedlow, P.A., P. Van Voris and L.E. Rogers. 1988. Theoretical perspective on ecosystem disturbance and recovery, p. 257-269. *In*: W.H. Rickard, L.E. Rogers, B.E. Vaughan and S.F. Liebetrau (eds.), Shrub-steppe: balance and change in a semi-arid terrestrial ecosystem. Elsevier, N.Y.

Beiswenger, J.M. 1991. Late quaternary vegetational history of Grays Lake, Idaho. Ecol. Monogr. 6:165-182.

Billings, W.D. 1949. The shadscale vegetation zone of Nevada and eastern California in relation to climate and soils. Amer. Midl. Natur. 42:1225-1228.

Billings, W.D. 1990. *Bromus tectorum*, a biotic cause of ecosystem impoverishment in the Great Basin, p. 301-322. *In*: G.M. Woodwell (ed.), The earth in transition: patterns and processes of biotic impoverishment. Cambridge Univ. Press, N.Y.

Blackburn, W.H., and P.T. Tueller. 1970. Pinyon and juniper invasion in black sagebrush communities in east-central Nevada. Ecology 51:841-848.

Blackburn, W.H. 1975. Factors influencing infiltration rates and sediment production of semi-arid rangelands in Nevada. Water Resources Res. 11:929-937.

Blaisdell, J.P. and J.F. Pechanec. 1949. Effects of herbage removal at various dates on vigor of bluebunch wheatgrass and arrowleaf balsamroot. Ecology 30:298-305.

Blaisdell, J.P. 1953. Ecological effects on planned burning of sagebrush-grass range on the upper Snake River Plains. USDA Tech. Bull. 1075.

Branson, F.A. 1985. Vegetation changes on western rangeland. Range Monogr. No. 2. Soc. Range Manage., Denver, Co.

Brown, J.R. and S. Archer. 1989. Woody plant invasion of grasslands: establishment of honey mesquite (*Prosopis glandulosa* var. *glandulosa*) on sites differing in herbaceous biomass and grazing history. Oecologia 80:19-26.

Bryson, R.A. 1989. Late quaternary volcanic modulation of milankovith climate forming. Theor. Appl. Climat. 39:115-125.

Burkhardt, J.W. and E.W. Tisdale. 1976. Causes of juniper invasion in southwestern Idaho. Ecology 76:472-484.

Butler, B.R. 1976. The evolution of the modern sagebrush-grass steppe biome on the eastern Snake River Plain. Nevada Archeol. Surv. 6:4-39.

Byrne, R., C. Busby and R.F. Heizer. 1979. The altithermal revisited: Pollen evidence from the Leonard Rockshelter. J. Calif. and Great Basin Anthro. 1:280-294.

Caldwell, M.M. 1978. Physiology of sagebrush, p. 74-85. *In*: The sagebrush ecosystem: A symposium. Utah State Univ. Logan.

Caldwell, M.M., J.H. Richards, D.A. Johnson, R.S. Nowak, and R.S. Dzurec. 1981. Coping with herbivory: photosynthetic capacity and resource allocation in two semiarid *Agropyron* bunchgrasses. Oecologia 50:14-24.

Caldwell, M.M. 1985. Cold desert, p. 188-212. *In*: Physiological ecology of North American plant communities. B.F. Chabot and H.A. Mooney (eds.), Chapman and Hall, New York.

Caldwell, M.M., D.M. Eissenstat, J.H. Richards and M.F. Allen. 1985. Competition for phosphorus: differential uptake from dual-isotope interspace between shrub and grass. Science 229:384-386.

Campbell, G.S. and G.A. Harris. 1977. Water relations and water use patterns for *Artemisia tridentata* Nutt. in wet and dry years. Ecology 58:652-659.

Canfield, R.H. 1957. Reproduction and life span of some perennial grasses of southern Arizona. J. Range Manage. 10:199-203.

Christensen, E.M., and H.B. Johnson. 1964. Presettlement vegetation and vegetational change in three valleys in central Utah. Brigham Young Univ. Sci. Bull. 4:1-15.

Clark, R.C. 1932. Harney Basin Exploration, 1826-60. Oregon Hist. Quart. 33:101-114.

Cline, G.G. 1963. Exploring the Great Basin. Univ. Oklahoma Press. Norman, OK.

Conrad, C.E. and C.E. Poulton. 1966. Effects of wildfire on Idaho fescue and bluebunch wheatgrass. J. Range Manage. 19:138-141.

Cook, C.W. and R.D. Child. 1971. Recovery of desert plants in various states of vigor. J. Range Manage. 24:339-343.

Cooper, H.W. 1953. Amounts of big sagebrush in plant communities near Tensleep, Wyoming as affected by grazing treatment. Ecology 34:186-189.

Cottam, W.P. and G. Stewart. 1940. Plant succession as a result of grazing and of meadow desiccation by erosion since settlement in 1892. J. For. 38:613-626.

Cressman, L.S. 1977. Prehistory of the far west: homes of vanished peoples. Univ. Utah Press. Salt Lake City, UT.

Cressman, L.S. 1981. The sandal and the cave. Oregon State Univ. Press, Corvallis, OR.

Crawley, M.J. 1983. Herbivory: The dynamics of animal-plant interactions. Blackwell, Oxford.

Crisp, M.D. and R.T. Lange. 1976. Age structure, distribution and survival under grazing of the arid-zone shrub *Acacia burkitti*. Oikos 27:86-92.

Currie, P.O., and D.L. Goodwin. 1966. Consumption of forage by black-tailed jackrabbits on salt desert ranges of Utah. J. Wildl. Manage. 30:304-311.

Daddy, F., M.J. Trlica and C.D. Bonham. 1988. Vegetation and soil water differences among big sagebrush communities with different grazing histories. Southwestern Natur. 33:413-424.

Daubenmire, R. 1940. Plant succession due to overgrazing in the *Agropyron* bunchgrass prairie of southeastern Washington. Ecology 21:55-64.

Daubenmire, R. 1968. Ecology of fire in grasslands. Adv. Ecol. Res. 5:209-266.

Daubenmire, R. 1970. Steppe vegetation of Washington. Washington Agric. Exp. Sta. Tech. Bull. 62, Wash. State Univ., Pull.

Davis, J.O. 1982. Bits and pieces: the last 35,000 years in the Lahontan area. Soc. Amer. Arch. SAA Paper No. 2.

Davis, O.K., J.C. Sheppard and S. Robertson. 1986. Contrasting climatic histories for the Sake River Plain, Idaho. Quart. Res. 26:321-339.

Driscoll, R.S. 1964. A relict area in central Oregon juniper zone. Ecology 45:345-353.

Earle, C.J. and H.C. Fritts. 1986. Reconstructing river flow in the Sacramento Basin since 1560. Rep. Calif. Dept. Resources, Agreement no. DWR B-55395. Lab. of Tree-ring Res., Univ. Ariz., Tucson.

Easterling, W.E. 1990. Climate trends and prospects, p. 32-55. *In*: R.N. Sampson and D. Hair (eds.), Natural Resources for the 21st Century. Island Press. Washington D.C.

Eckert, R.E., Jr. and J.S. Spencer. 1986. Vegetation response on allotments grazed under rest-rotation management. J. Range Manage. 39:166-174.

Eckert, R.E., Jr., F.F. Peterson, M.S. Meurisse and J.L. Stephens. 1986a. Effects of soil-surface morphology on emergence and survival of seedling in big sagebrush communities. J. Range Manage. 39:414-420.

Eckert, R.E. Jr., F.F. Peterson and J.T. Belton. 1986b. Relation between ecological range condition and proportion of soil surface types. J. Range Manage 39:409-414.

Eckert, R.E. Jr. and J.S. Spencer. 1987. Growth and reproduction of grasses heavily grazed under rest-rotation management. J. Range Manage. 40:156-159.

Eddleman, L.E. 1987. Establishment and stand development of western juniper in central Oregon, p. 255-259. *In*: R.L. Everett (ed), Proc. Pinyon-Juniper Conf. USDA For. Ser. Intermountain Res. Sta. Gen. Tech. Rep. INT-215.

Eddleman, L.E. 1991. Professor Department of Rangeland Resources, Oregon State University. Corvallis, OR. personal communication.

Eissenstat, D.M. and M.M. Caldwell. 1987. Characteristics of successful competitors: an evaluation of potential growth rate in two cold desert tussock grasses. Oecologia 71:167-173.

Ellison, L. 1960. Influence of grazing on plant succession of rangelands. Bot. Rev. 26: 1-78.

EOARC. 1991. Eastern Oregon Agricultural Experiment Station, HC 71 4.51 Hwy 205, Burns, OR.

Escalante, Father. 1776. Sante Fe to Utah Lake and the Moqui villages in 1776, p. 487-495. *In*: Simpson, J.H. (ed.), Explorations across the Great Basin of the Territory of Utah. 1876. Engineering Dept, U.S. Army. Washington D.C. Printing Office.

Evans, R.A. 1961. Effects of difference densities of downy brome (*Bromus tectorum*) on growth and survival of crested wheatgrass (*Agropyron desertorum*) in the greenhouse. Weeds 9:216-223.

Evans, R.A., H.R. Holbo, R.E. Eckert, Jr., and J.A. Young. 1970. Functional environment of downy brome communities in relation to weed control and vegetation. Weed Sci. 18:154-162.

Ferguson, R.B. and J.V. Basile. 1967. Effect of seedling numbers on bitterbrush survival. J. Range Manage. 26:207-209.

Flaschka, I, C.W. Stockton, and W.R. Boggess. 1987. Climatic variation and surface water resources in the Great Basin. Water Res. Bull. 23:47-57.

Friedel, M.H. 1988. Range condition and the concept of thresholds, p. 1-3. *In*: Vol. I, Abstracts, 3rd Intern. Rangeland Congr. Range Manage. Soc. India.

Friedel, M.H. 1991. Range condition assessment and the concept of thresholds. A Viewpoint. J. Range Manage. 44:422-426.

Frison, R.H. 1978. Shoshonean antelope procurement, p. 254-256. *In*: Prehistoric hunters, of the high plains.

Fritts, H.C. and W. Xiangding. 1986. A comparison between response-function analysis and other regression techniques. Tree-ring Bull. 46:31-46.

Fryxell, R. 1965. Mazama and Glacier Peak volcanic ash layers: relative ages. Science 147:1288-1290.

Gabrielson, I.N. and S.G. Jewett. 1970. Birds of the Pacific Northwest. Dover Publ.

Gammon, R.H., E. Sundquist and P.J. Graser. 1985. History of carbon dioxide in the atmosphere, p. 25-62. **In**: J.R. Trabalk, (ed.), Atmospheric carbon dioxide and global carbon cycle. DOE/ER-0239. US Depart. Energy, Washington, D.C.

Ghil, M. and R. Vautgard. 1991. Interdecadal oscillations and the warming trend in global temperature time series. Nature 350:324-327.

Gist, C.S. and P.R. Sferra. 1978. Animal consumer roles in desert nitrogen cycles, p. 152-164. *In*: N.E. West and J.J. Skujins (eds.), Nitrogen in desert ecosystems. Dowden Hutchinson and Ross, Inc. Stroudsburg, PA.

Gibbens, R.P. and H.G. Fisser. 1975. Influence of grazing management systems on vegetation in the Red Desert region of Wyoming. Agric. Exp. Sta. Univ. Wyoming. Sci. Monogr. 29.

Graumlich, L. 1985. Long-term records of temperature and precipitation in the Pacific Northwest derived from tree rings. PhD. Thesis, Univ. of Washington, Seattle.

Grayson, D.K. 1987. The biogeographic history of small mammals in the Great Basin: observations on the last 20,000 years. J. Mammalogy 68:359-375.

Griffiths, D. 1902. Forage conditions on the northern border of the Great Basin. Bureau of Plant Industry. USDA, Bull 15.

Gross, J.E., L.C. Stoddart and F.H. Wagner. 1974. Demographic analysis of a northern Utah jackrabbit population. Wildl. Monogr. 40.

Gruel, G.E. 1985. Fire on the early western landscape: an annotated record of wildland fires 1771-1900. Northwest Sci. 59:97-107.

Hall, E.A.A., R.L. Specht and C.M. Eardley. 1964. Regeneration of the vegetation on Koonamore Vegetation Reserve. 1926-1962. Aust. J. Bot. 12:205-264.

Hanley, T.A. 1979. Application of an herbivore-plant model to rest-rotation grazing management on shrub-steppe ranges. J. Range Manage. 32:115-118.

Hansen, H.P. 1947a. Postglacial succession, climate and chronology in the Pacific Northwest. Am. Philos. Soc. Trans. Vol. 37.

Hansen, H.P. 1947b. Postglacial vegetation of the northern Great Basin. Amer. J. Bot. 34:164-171.

Hanson, W.R. and L.A. Stoddart. 1940. Effects of grazing upon bunch wheat grass. Amer. Soc. Agron. 32:278-289.

Harper, K.T. and J.R. Marble. 1988. A role for nonvascular plants in management of arid and semiarid rangelands, p. 135-169. *In*: P.T. Tueller (ed.), Application of Plant Sciences to Rangeland Management Inventory. Marinus Nijhoff/W. Junk, Amsterdam.

Harris, G.A 1967. Some competitive relationships between *Agropyron spicatum* and *Bromus tectorum*. Ecol. Monogr. 37:89-111.

Harris, G.A. and C.J. Goebel. 1976. Factors of plant competition in seeding Pacific Northwest bunchgrass ranges. Wash. State Univ. College of Agr. Res. Center. Bull. 820.

Hayes, W.P. 1927. Prairie insects. Ecology 8:238-250.

Heady, A.F. 1950. Studies on bluebunch wheatgrass in Montana and height-weight relationships of certain range grasses. Ecol. Monogr. 20:55-81.

Heady, A.F. 1975. Rangeland management. McGraw-Hill, N.Y.

Hewitt, G.B., E.W. Huddleston, R.J. Lavigne, D.N. Ueckert and J.G. Watts. 1974. Rangeland entomology. Soc. Range Manage. Denver, CO.

Holechek, J.L., and T. Stephenson. 1983. Comparison of big sagebrush vegetation in northcentral New Mexico under moderately grazed and grazing excluded conditions. J. Range Manage. 36:455-456.

Holechek, J.L., R.D. Pieper and C.H. Herbel. 1989. Range management principles and practices. Prentice Hall, Engelwood-Cliffs, N.J.

Houghton, J.G., C.M. Sakamoto and R.O. Gifford. 1975. Nevada's weather and climate. Nevada Bur. Mines and Geol. Spec. Publ. 2.

Houston, D. B. 1973. Wildfires in northern Yellowstone National Park. Ecology 54:1109-1117.

Hughes, L.E. 1980. Six grazing exclosures with a message. Rangelands 2:17-18.

Hugie, V.K. and H.B. Passey. 1963. Cicadas and their effect upon soil genesis in certain soils in southern Idaho, northern Utah, and northeastern Nevada. Soil Sci. Soc. Proc. 27:78-82.

Hull, A.C., Jr. 1963. Competition and water requirements of cheatgrass and wheatgrass in the greenhouse. J. Range Manage. 16:199-203.

Hull, A.C., Jr. 1976. Rangeland use and management in the Mormon west, *In:* Sym. on agriculture, food and man-century of progress. Brigham Young Univ., Provo, UT.

Hunter, R. 1990. Recent increases in *Bromus* populations on the Nevada Test Site, p. 22-25. *In*: McArthur et al. (eds.), Proc. - sym. on cheatgrass invasion, shrub die-off, and other aspects of shrub biology and management. USDA For. Ser., Gen. Tech. Rep. INT-276.

Hyder, D.N. and W.A. Sawyer. 1951. Rotation-deferred grazing as compared to season-long grazing on sagebrush-bunchgrass ranges in Oregon. J. Range Manage. 4:30-34.

Jackson, D. and M.L. Spence. 1970. The expeditions of John Charles Fremont: travels from 1838 to 1844. Univ. Illinois Press, Chicago.

Jaksic, F.M. and E.R. Fuentes. 1980. Why are native herbs in the Chilean matorral more abundant beneath bushes: microclimate or grazing? J. Ecol. 68:665-669.

Jameson, D.A. 1967. The relationship of tree overstory and herbaceous understory vegetation. J. Range Manage. 20:247-249.

Jellison, W.L., J.F. Bell, J.D. Vertrees, M.A. Holmes, C.L. Larson and C.R. Owens. 1958. Preliminary observations on diseases in the 1957-58 outbreak of *Microtis* in western United States. N. Am. Wildl. Conf. Trans. 23:137-145.

Johansen, J.R. 1984. Response of soil algae to a hundred-year storm in the Great Basin Desert, Utah. Phykos 23:51-54.

Johansen J.R. 1986. Importance of cryptogamic soil crusts to arid rangelands: Implications for short duration grazing, p. 127-136. *In*: J.A. Tiedeman (ed.), Short Duration Grazing: Proc. of a short course. Wash. State Univ. Pullman, WA.

Johansen, J.R. and W.R. Rayburn. 1989. Effects of rangefire on soil cryptogamic crusts, p. 161-164. *In*: D.M. Baugartner et al. (eds.), Prescribed fire in the intermountain region: Forest site preparation and range improvement. Wash. State Univ., Pullman, WA.

Johnsen, T.N. 1962. One-seed juniper invasion of northern Arizona grasslands. Ecol. Monogr. 32:187-207.

Johnson, C.W. M.R. Savabi and S.A. Loomis. 1984. Rangeland erosion measurements for the Universal Soil Loss Equation. Trans. ASAE 27:1313-1320.

Johnson, C.W., N.W. Gordon and C.L. Hanson. 1985. Northwest rangeland sediment yield analysis by MUSLE. Tans. ASAE 28:1889-1895.

Johnson, C.W. and N.D. Gordon. 1986. Runoff and erosion from rainfall simulator plots on sagebrush rangeland. Amer. Soc. Agric. Engin. Paper No. 86-2021.

Johnson, C.W. and W.H. Blackburn. 1987. Factors contributing to sagebrush rangeland soil loss. Am. Soc. Agric. Engin. Paper No. 87-2546.

Johnson, D.R. and J.M. Peek. 1984. The black-tailed jackrabbit in Idaho: Life history, populations dynamics and control. Univ. of Idaho, Coll. Agric. Coop. Ext. Serv. Bull. No. 637.

Johnson, E.A. and R.S. Larsen. 1991. Climatically induced change in fire frequency in the southern Canadian Rockies. Ecology 72:194-201.

Johnson, H.B. 1985. Consequences of species introduction and removals on ecosystem functions - implications for applied ecology, p. 27-56. *In*: E.S. Delfasse (ed.), Proc. IV Inter. Sym. Biol. Control Weeds, Vancouver, Canada.

Kamm, J.A. F.A. Sneva and L.M. Rittenhouse. 1978. Insect grazers on the cold desert biome, p. 479-483. *In:* D.N. Hyder (ed.), Proc. First International Rangeland Congress. Soc. Range Manage. Denver, CO.

Kapusta, G. and C.F. Strieker. 1975. Selective control of downy brome in alfalfa. Weed Sci. 23:202-216.

Keeling, C.D., R.B. Bacstow, A.F. Carter, S.C. Piper, T.P. Whorf, M. Heimann, W.G. Mook, and H. Roeloffzen. 1989. A three-dimensional model of CO_2 transport based on observed winds. I. Analysis of observational data. Amer. Geophy. Monogr. 55:165-234.

Kindschy, R.R. 1987. Sagehen exclosure: A history of bitterbrush reproduction. Rangelands 9:113-114.

Klebenow, D.A. 1985. Habitat management for sage grouse in Nevada. World Pheasant Assoc. J. 10:34-46.

Klemmedson, J.O. and J.G. Smith. 1964. Cheatgrass (*Bromus tectorum* L.). Bot. Rev. 30:226-262.

Kleiner, E.F. and K.T. Harper. 1972. Environment and community organization in grasslands of Canyonlands National Park. Ecology 53:299-309.

Klubek, B., P.J. Eberhart, and J. Skujins. 1978. Ammonia volatilization from Great Basin Desert soils, p. 107-129. *In*: N.E. West and J. Skujins (eds.), Nitrogen in desert ecosystems. US/IBP Synthesis Series Vol. 9. Dowden, Hutchinson and Ross, Stroudsberg, Penn.

Klubek, B. and J. Skujins. 1980. Heterotrophic N_2-fixation in arid soil crusts. Soil Biol. Biochem. 12:229-236.

Kuchler, A.W. 1970. Potential natural vegetation (map at scale of 1:7 500 000), p. 90-91. *In*: U.S. Geol. Surv, The National Atlas of the U.S.A. U.S. Gov. Printing Office, Washington D.C..

LaMarche, V.C., Jr. and H.A. Mooney. 1972. Recent climate change and development of the bristlecone pine, *P. logaeva* Bailey, Drummholz Zone, Mt. Washington, Nevada. Arctic Alpine Res. 4:61-72.

LaMarche, V.C., Jr. 1973. Holocene climatic variations inferred from treeline fluctuations in the White Mountains, California. Quater. Res. 3:632-660.

Lange, R. and R. Purdie. 1976. Western myall (*Acacia sowdenii*), its survival, prospects and management needs. Aust. Rangeland J. 1:64-69.

Laycock, W.A. 1967. How heavy grazing and protection affect sagebrush-grass ranges. J. Range Manage. 20:206-213.

Laycock, W.A. 1978. Factors affecting choice of management strategies within the sagebrush ecosystem, p. 230-236. *In*: The sagebrush ecosystem: A symposium. College of Natur. Resour., Utah State Univ., Logan, UT.

Laycock, W.A. 1991. Stable states and thresholds of range condition on North American rangelands: A viewpoint. J. Range Manage. 45:427-433.

Lederer, R.J. 1977. Winter territoriality and foraging behavior of the Townsend's Solitaire. Amer. Midl. Natur. 97:101-109.

Leopold, A. 1924. Grass, brush, timber and fire in southern Arizona. J. For. 22:1-10.

Leopold, A. 1949. Sand County Almanac. Ballentine Books, N.Y.

Mack, R.N. and J.L. Harper. 1977. Interference in dune annuals: spatial pattern and neighborhood effects. J. Ecol. 65:345-363.

Mack, R.N. 1981. The invasion of *Bromus tectorum* L. into western North America: an ecological chronicle. Agro-Ecosystems 7:145-165.

Mack, R.N., and J.N. Thompson. 1982. Evolution in steppe with few large, hoofed mammals. Amer. Natur. 119:757-773.

Mack, R.N. and D.A. Pyke. 1983. The demography of *Bromus tectorum*: variation in time and space. J. Ecol. 71:69-93.

Madsen, D.B. 1976. Pluvial-post pluvial vegetation changes in the southeastern Great Basin, p. 104-119. *In*: R. Elston and P. Headrick, (eds.), Holocene environmental change in the Great Basin. (Reno, Nevada: Nevada Arch. Survey, Res. Paper No. 6, University of Nevada, 1976)

Marble, J.R. and K.T. Harper. 1989. Effect of timing of grazing on soil-surface cryptogamic communities in a Great Basin low shrub desert: A preliminary report. Great Basin Natur. 49:104-107.

Martin, P.S. 1967. Prehistoric overkill, p. 75-120. *In*: P.S. Martin and H.E. Wright, Jr. (eds.), Pleistocene extinctions: The search for a cause. Vol. 6, Proc. VII Congress of the Inter. Assoc. for Quaternary Res. Yale Univ. Press.

McAdoo, J.K. and J.A. Young. 1980. Jackrabbits. Rangelands 2:135-138.

McArdle, R.E. et al., 1936. The whiteman's toll, p. 81-116. *In*: H.A. Wallace, Letter from the Secretary of Agriculture to the U.S. Senate - A Report on the Western Range - A Great But Neglected Resource. Senate Document 199.

Medina, E. and J.F. Silva. 1990. Savannas of northern South America: a steady state regulated by water-fire interactions on a background of low nutrient availability. J. Biogeog. 17:403-413.

Megee, C.R. 1938. Wild oats or downy brome, troublesome weed on sandy land. Michigan Agric. Exp. Station. Quarterly Bull. 20:153-156.

Mehringer, P.J., Jr. 1967. Pollen analysis of the Tule Springs area, Nevada. Nevada State Museum Anthropological Papers 13:129-200.

Mehringer, P.J., Jr. 1985. Late-quaternary pollen records from the interior Pacific Northwest and northern Great Basin of the United States, p. 167-189. *In*: Bryant, W.M., Jr. and R.G. Holloway (eds.), Pollen records of late-Quaternary North American sediments. Amer. Assoc. Stratigraphic Palynoligists.

Mehringer, P.J., Jr. 1986. Prehistoric environments, p. 31-50. *In*: Handbook of North American Indians Vol. 11, Great Basin. Smithsonian Inst. Washington D.C.

Mehringer, P.J., Jr. 1987. Late Holocene environments on the northern periphery of the Great Basin. Final Rep. Bur. Land Manage., Portland, OR.

Mehringer, P.J., Jr., and P.E. Wigand. 1987. Western juniper in the Holocene, p.13-16. *In:* R.L. Everett (ed.), Proc. Pinyon-Juniper Conf. USDA For. Ser. Gen. Tech. Rep. INT-215.

Mehringer, P.J., Jr. and P.E. Wigand. 1990. Comparison of late holocene environments from woodrat middens and pollen: Diamond Craters, Oregon, p. 294-325. *In*: J.L. Betancourt, T.R. Van Devender and P.S. Martin (eds.), Packrat Middens: The last 40,000 years of biotic change. Univ. Arizona Press, Tucson.

Milchunas, D.G., O.E. Sala, and W.K. Lauenroth. 1988. A generalized model of the effects of grazing by large herbivores on grassland community structure. Amer. Natur. 132:87-106.

Miller, C.D. 1973. Chronology of neoglacial deposits in the northern Sawatch Range, Colorado. Arctic and Alpine Res. 5:385-400.

Miller, R.F. and L.I. Shultz. 1987. Development and longevity of ephemeral and perennial leaves on *Artemisia tridentata* subsp *wyomingensis*. Great Basin Natur. 47:227-230.

Miller, R.F. 1988. Comparison of water use by *Artemisia tridentata* spp. *wyomingensis* and *Chrysothamnus viscidiflorus* spp. *viscidiflorus*. J. Range Manage. 40:58-62.

Miller, R.F., M.R. Haferkamp and R.F. Angell. 1990. Clipping date effects on soil water and regrowth in crested wheatgrass. J. Range Manage. 43:253-257.

Miller, R.F., P.S. Doescher and J. Wang. 1991. Response of *Artemisia tridentata* spp. *wyomingensis* and *Stipa thurberiana* to nitrogen amendments. Amer. Midl. Natur. 125:104-113.

Morrison, R.B. 1964. Lake Lahontan: Geology of Southern Carson Desert, Nevada. Geol. Surv. Profess. Paper 401. U.S. Gov. Printing Office, Washington D.C.

Mueggler, W.F. 1950. Effects of spring and fall grazing by sheep on vegetation of the Upper Snake River Plains. J. Range Manage. 3:308-315.

Mueggler, W.F. 1967a. Voles damage big sagebrush in southwestern Montana. J. Range Manage. 20:88-90.

Mueggler, W.F. 1967b. Response of mountain grassland vegetation to clipping in southwestern Montana. Ecology 48:942-949.

Mueggler, W.F. 1970. Influence of competition on the response of Idaho fescue to clipping. USDA For. Serv. Res. Paper INT-73.

Mueggler, W.F. 1972. Influence of competition on the response of bluebunch wheatgrass to clipping. J. Range Manage. 25:88-92.

Mueller, R.J. and J.H. Richards. 1986. Morphological analysis of tillering in *Agropyron spicatum* and *Agropyron desertorum*. Ann. Bot. 58:911-921.

Neilson, R.P. 1986. High resolution climatic analysis and southwest biogeography. Science 232:27-34.

Neilson, R.P. 1987. On the interface between current ecological studies and the paleobotany of pinyon-juniper woodlands, p. 93-98. *In*: R. Everett (ed.), Proc. Pinyon-Juniper Conf. USDA For. Ser. Inter. Mountain Sta., Tech. Rep INT-215.

Nicholls, N. 1988. El Nino-southern oscillation and rainfall variability. J. Climate 1:418-421.

Niering, W.A., and R.H. Goodwin. 1974. Creation of relatively stable shrublands with herbicides: arresting "succession" on rights-of-way and pastureland. Ecology 55:784-795.

Noy-Meir, I. 1985. Desert ecosystem structure and function, p. 93-104. *In*: M. Evenari, I Noy-Meir and D.W. Goodall (eds.), Ecosystems of the world: hot deserts and arid shrublands. Elsevier, N.Y.

Nowak, R.S. and M.M. Caldwell. 1984. A test of compensatory photosynthesis in the field: implications for herbivory tolerance. Oecologia 61:311-318.

Oliphant, J.O. 1968. On the cattle ranges of the Oregon country. Univ. Wash. Press. Seattle, WA.

Parmenter, R.R. and J.A. MacMahon. 1983. Factors determining the abundance and distribution of rodents in a shrub-steppe ecosystem: the role of shrubs. Oecologia 59:145-156.

Parton, W.J., A.R. Mosier and D.S. Schimel. 1988. Rates and pathways of nitrous oxide production in a shortgrass steppe. Biogeochemistry 6:45-58.

Passey, H.B., W.K Hugie, E.W. Williams and D.E. Ball. 1982. Relationships between soil, plant community, and climate on rangelands of the intermountain west. USDA Soil Cons. Ser. Tech Bull. No. 1669.

Pearson, G.A. 1942. Herbaceous vegetation as a factor in natural regeneration of ponderosa pine in the southwest. Ecol. Monogr. 12:315-338.

Pechanec, J.F., G.D. Pickford, and G. Stewart. 1937. Effects of the 1934 drought on native vegetation of the Upper Snake River Plains, Idaho. Ecology 18:490-505.

Pendery, B. and F. Provenza. 1987. Interplanting crested wheatgrass with shrubs and alfalfa: effects of competition and preferential clipping. J. Range Manage. 40:514-520.

Pielou, E.C. 1991. After the ice age: The return of life to glaciated North America. Univ. Chicago Press.

Pipe, A.E. and L.E. Shubert. 1983. The use of algae as indicators of soil fertility, p. 213-233. *In*: L.E. Shubert (ed.), Algae as ecological indicators. Academic Press, London.

Piper, S.E. 1909. The Nevada mouse plague of 1907-1908. USDA Farmers Bull. 352:1-23.

Poddar, S. and R.J. Lederer. 1982. Juniper berries as an exclusive winter forage for Townsend's Solitaires. Amer. Midl. Natur. 108:34-40.

Pyne, S.J. 1988. Fire in America: a cultural history of wildland and rural fire. Princeton Univ. Press.

Reveal, J.L. 1979. Biogeography of the intermountain region: a speculative appraisal. Mentzelia No. 4.

Rich, E.E., A.M. Johnson and B.R. Baker (eds.). 1950. Peter Skene Ogden's Snake Country Journals: 1824-25 and 1825-26. The Hudson Bay Society, London.

Richards, J.H. 1984. Root growth response to defoliation in two *Agropyron* bunchgrasses: field observation with an improved root periscope. Oecologia 64:21-25.

Richards, J.H. and M.M. Caldwell. 1985. Soluble carbohydrates, concurrent photosynthesis and efficiency in regrowth following defoliation: a field study with *Agropyron* species. J. Appl. Ecol. 22:907-920.

Richards, J.H., R.J. Mueller and J.J. Mott. 1988. Tillering in tussock grasses in relation to defoliation and apical bud removal. Ann. Bot. 62:173-179.

Rickard, W.H. and R.H. Sauer. 1982. Self-revegetation of disturbed ground in the deserts of Nevada and Washington. Northwest Sci. 56:41-47.

Rickard, W.H. and B.E. Vaughan. 1988. Plant communities: characteristics and responses, p. 109-181. *In:* W.H. Rickard, L.E. Rogers, B.E. Vaughan and S.F. Liebetrau (eds.), Shrub-steppe: balance and change in a semi-arid terrestrial ecosystem. Elsevier, N.Y.

Ritchie, G.A. 1974. Atmospheric and soil water influences on the plant water balance. Agric. Metero. 14:183-198.

Robertson, J.H. 1971. Changes on a grass-shrub range in Nevada ungrazed for 30 years. J. Range Manage. 24:397-400.

Rogers, G.F. 1982. Then and Now: a photographic history of vegetation change in the central Great Basin desert. University of Utah Press, Salt Lake City.

Rogers, L.E., R.E. Fitzner, L.L. Caldwell and B.E. Vaughan. 1988. Terrestrial animal habitats and population responses, p. 181-256. *In*: W.H. Rickard, L.E. Rogers, B.E. Vaughan and S.F. Liebetrau (eds.), Shrub-steppe: balance and change in a semi-arid terrestrial ecosystem. Elsevier, N.Y.

Romney, E.M., A. Wallace and R.B. Hunter. 1980. The pulse hypothesis in the establishment of *Artemisia* seedlings at Pahute Mesa, Nevada. Great Basin Natur. 4:28-33.

Ropelewski, C.F. and M.S. Halpert. 1986. North American precipitation and temperature patterns associated with the El Nino/Southern Oscillation (ENSO). Monthly Weather Rev. 114:2352-2362.

Rosentreter, R. 1986. Compositional patterns within a rabbitbrush (*Chrysothamnus*) community of the Idaho Snake River Plain, p. 273-277. *In*: Proc. of a symp. on the Biol. *Artemisia* and *Chrysothamnus*. USDA For. Ser. Gen. Tech. Rep. No. INT-200.

Russell, I.C. 1885. Geological history of Lake Lahontan, a Quaternary lake of northwestern Nevada. U.S. Geol. Surv. Mon. 11.

Rychert, R.C. and J. Skujins. 1974. Nitrogen fixation by blue-green algae-lichen crusts in the Great Basin Desert. Soil Sci. Soc. Amer. Proc. 38:768-771.

Rychert, R.C., J. Skujins, D. Sorenson and D. Porcella. 1978. Nitrogen fixation by lichens and free living microorganisms in deserts, p. 20-30. *In*: N.E. West and J. Skujins (eds.), Nitrogen in desert ecosystems. Dowden, Hutchinson and Ross, Stroudsburg, PA.

Rydrych, D.J. 1974. Competition between winter wheat and downy brome. Weed. Sci. 22:211-214.

St. Clair, L.L., B.L. Webb, J.R. Johansen and G.T. Nebeker. 1984. Cryptogamic soil crusts: enhancement of seedling establishment in disturbed and undisturbed areas. Reclam. Reveg. Res. 3:129-136.

Sampson, A.W. 1914. Natural revegetation of rangelands based upon growth requirements and life history of the vegetation. J. Agr. Res. 3:93-147.

Sanders, K.D. and A.A. Voth. 1983. Ecological changes of grazed and ungrazed plant communities, p. 176-179. *In*: Managing intermountain rangelands-improvement of range and wildlife habitats. USDA, For. Serv. Gen. Tech. Rep. INT-157.

Savage, M. and T.W. Swetman. 1990. Early 19th century fire decline following sheep pasturing in a Navajo ponderosa pine forest. Ecology 71:2374-2378.

Scharff, J. 1991. US Fish and Wildlife Refuge Manager, Malheur National Wildlife Refuge, retired. Burns, OR. personal communication.

Schlatterer, E.F. and E.W. Tisdale. 1969. Effects of litter of *Artemisia, Chrysothamnus* and *Tortula* on germination and growth of three perennial grasses. Ecology 50:869-873.

Schimel, D.S., W.J. Parton, P.J. Adamsen, R.G. Woodmansee, R.L. Senft and M.A. Stillwell. 1986. The role of cattle in the volatile loss of nitrogen from shortgrass steppe. Biogeochemistry 2:39-52.

Sheldon, J.K. and L.E. Rogers. 1978. Grasshopper food habits within a shrub-steppe community. Oecologia 32:85-92.

Simpson, J.H. 1876. Explorations across the Great Basin of the Territory of Utah. Washington, Engineering Dept, U.S. Army. Gov. Printing Office, Washington D.C.

Skujins, J. and P. Trujillo Y Fulgham. 1978. Nitrification in Great Basin desert soils, p. 60-74. *In*: N.E. West and J. Skujins (eds.), Nitrogen in desert ecosystems. Dowden, Hutchinson and Ross, Stroudsburg, Penn.

Skujins, J. and B. Klubek. 1978a. Nitrogen fixation and cycling by blue-green algae-lichen crusts in arid rangeland soils. Ecol. Bull. 26:164-171.

Skujins, J. and B. Klubek. 1978b. Nitrogen fixation and denitrification in arid soil cryptogamic crust microenvironments, p. 543-552. *In*: W. Krumbein (ed.), Environmental biogeochemistry and geomicrobiology. Vol. 2: The terrestrial environment. Ann Arbor, Sci., Ann Arbor, Mich.

Slayter, R.O. 1970. Comparative photosynthesis, growth, and transpiration of 2 species of *Atriplex*. Planta 93:175-189.

Smith, C.C. 1940. The effects of overgrazing and erosion upon biota of the mixed-grass prairie of Oklahoma. Ecology 21:381-297.

Smith, S.D., B.R. Strain and T.D. Sharkey. 1987. Effects of CO_2 enrichment on four Great Basin grasses. Functional Ecol. 1:139-143.

Sneva, F.A., R.L. Rittenhouse, P.T. Tueller, and P. Reece. 1984. Changes in protected and grazed sagebrush-grass in eastern Oregon, 1937 to 1974. Agric. Exp. Sta. Oregon State Univ. Sta. Bull. 663.

Sparks, S.R. N.E. West and E.B. Allen. 1990. Changes in vegetation and land use at two townships in Skull Valley, western Utah, p. 26-36. *In:* Proc. Cheatgrass Invasion Shrub Die-off, and other aspects of shrub biology and management. USDA For. Ser. Inter. Res. Sta. Gen. Tech. Rep. INT-276.

Stebbins, G.L. 1981. Coevolution of grasses and herbivores. Ann. Missouri Bot. Garden 68:75-86.

Stewart, G., W.P Cottam, and S.S. Hutchings. 1940. Influence of unrestricted grazing on northern salt desert plant associations in western Utah. J. Agric. Res. 60:289-316.

Stewart, O.C. 1951. Burning and natural vegetation in the United States. Geogr. Rev. 41:317-320.

Stewart, O.C. 1963. Barriers to understanding the influence of use of fire by aborigines on vegetation. Tall Timbers Fire Ecol. Conf. 2:117-126.

Stoddart, L.A. 1941. The Palouse grassland association in northern Utah. Ecology 22:158-163.

Stoddart, L.A. 1946. Some physical and chemical responses of *Agropyron spicatum* to herbage removal at various seasons. Utah Agr. Exp. Sta. Bull. 324.

Stoddart, L.A. and A.D. Smith. 1955. Range management. McGraw-Hill Book Co..

Strong, E. 1969. Stone age in the Great Basin. Binford and Mort Publ. Portland, OR.

Stuart, R. 1812. Discovery of the Oregon trail. Scibner's Sons. New York.

Sturges, D.L. 1977. Soil water withdrawal and root characteristics of big sagebrush (*Artemisia tridentata*) in the United States. Amer. Midl. Natur. 98:257-274.

Sutherland, J.P. 1974. Multiple stable points in natural communities. Amer. Natur. 108:859-873.

Svejcar, T. and R. Tausch. 1991. Anaho Island, Nevada: a relict area dominated by annual invader species. Rangelands 13:233-236.

Svejcar, T. 1990. Root length, leaf area, and biomass of crested wheatgrass and cheatgrass seedlings. J. Range Manage. 43:446-448.

Thompson, D. 1916. David Thompson's narrative. J.B. Tyrrel, (ed.), The Champlain Society, Toronto, Ontario, Canada.

Thompson, R.S. and E.M. Hattori. 1983. Packrat (*Neotoma*) middens from Gatecliff Shelter and Holocene migrations of woodland plants, p. 167-171. *In*: D.H. Thomas (ed.), The archeology of Monitor Valley, 2. Gatecliff Shelter. Anthropological Papers Amer. Museum of Natural History 59, No. 1.

Thompson, R.S. 1990. Late Quaternary vegetation and climate in the Great Basin, p. 200-239. *In*. J.L. Betancourt, T.R. Van Devender and P.S. Martin (eds.), Packrat Middens: The last 40,000 years of biotic change. Univ. Arizona Press, Tucson.

Thwaites, R.G. (ed.). 1959. Original journals of the Lewis and Clark expedition. 1804-1806. Vol. 2,3, and 5. Antiquarian Press, N.Y.

Tisdale, E.W., M. Hironaka and F.A. Fosberg. 1965. An area of pristine vegetation in Craters of the Moon National Monument, Idaho. Ecology 46:349-352.

Trlica, M.J. and C.W. Cook. 1971. Defoliation effects on carbohydrate reserves of desert species. J. Range Manage. 24:418-424.

Trlica, M.J., M. Buwai and J.W. Menke. 1977. Effects of rest following defoliations on the recovery of several range species. J. Range Manage. 30:21-27.

Upadhyaya, M.K., R. Turkington and D. McIlvride. 1986. The biology of Canadian weeds. 75. *Bromus tectorum* L.. Can. J. Plant Sci. 66:689-709.

Uresk, D.W. 1978. Diets of the black-tailed hare in steppe vegetation. J. Range Manage. 31:439-442.

Urness, P. 1989. Why did bison fail west of the Rockies? Utah Sci. 50:175-179.

Vale, T.R. 1975. Presettlement vegetation in the sagebrush-grass area of the intermountain west. J. Range Manage 28:32-36.

Vallentine, J.F. 1979. Grazing systems as a management tool, p. 214-219. *In*: The sagebrush ecosystem: A symposium. Utah State Univ. Logan.

Van Hulst, R. 1979. On the dynamics of vegetation: Markov chains as models of succession. Vegetatio 40:3-14.

Van Vuren, D. 1987. Bison west of the Rocky Mountains: A review of the theories and an explanation. Northwest Sci. 61:65-69.

Walker, B.H., D. Ludwig, C.S. Holling, and R.M. Peterman. 1981. Stability of semi-arid savanna grazing systems. J. Ecol. 69:473-499.

Webb, R.H., J.W. Steiger and E.B. Newman. 1988. The response of vegetation to disturbance in Death Valley National Monument, California. US Geol. Surv. Bull. 1793.

West, N.E. and P.T. Tueller. 1972. Special approaches to studies of competition and succession in shrub communities, p. 165-171. *In*: Wildland Shrubs. Their Biology and Utilization. Interm. For. Range Sta. USDA. Gen. Tech Rep. INT-1.

West, N.E., K.H. Rea and R.J. Tausch. 1975. Basic synecological relationships in juniper-pinyon woodlands, p. 41-53. *In:* The pinyon juniper ecosystem: A symp. Utah Agric. Exper. Sta. Logan.

West, N.E., K.H. Rea, and R.O. Harniss. 1979. Plant demographic studies in sagebrush-grass communities of southeastern Idaho. Ecology 60:376-388.

West, N.E. 1983. Western intermountain sagebrush steppe, p. 351-397. *In*: N.E. West (ed.), Ecosystem of the World 5: Temperate deserts and semi-deserts. Elsevier Scientific Publ. Co. N.Y.

West, N.E., F.D. Provenza, P.S. Johnson, and K.M. Owens. 1984. Vegetation change after 13 years of livestock grazing exclusion on sagebrush semi-arid-desert in West Central Utah. J. Range Mange. 37:262-264.

West, N.E. and M.A. Hassan. 1985. Recovery of sagebrush-grass vegetation following wildfire. J. Range Manage. 38:131-134.

West, N.E. 1987. Human impacts on vegetation of the semiarid temperate portion of North America. *In*: Symp. on degradation of land cover in arid areas of temperate zones. XIV International Bot. Congr. West Berlin, Germany.

West, N.E. 1988. Intermountain deserts, shrub steppes, and woodlands, p. 209-230. *In*: M.B. Barbour and W.D. Billings (eds.), North American Terrestrial Vegetation. Cambridge Univ. Press.

West, N.E. 1990. Structure and function of microphytic soil crusts in wildland ecosystems to arid to semi-arid regions. Advan. Ecol. Res. 20:179-223.

Westman, W.E. 1990. Managing for biodiversity. Bioscience 40:26-33.

Westoby, M. B. Walker and I. Noy-Meir. 1989. Opportunistic management for rangelands not at equilibrium. J. Range Manage 42:266-274.

Whisenant, S.G. 1990. Changing fire frequencies on Idaho's Snake River Plains: ecological and management implications, p. 4-10. *In:* Proc. Cheatgrass Invasion Shrub Die-off, and other aspects of shrub biology and management. USDA For. Ser. Inter. Res. Sta. Gen. Tech. Rep. INT-276.

Wigand, P.E. 1985. Diamond Pond, Harney County, Oregon: Man and marsh in the eastern Oregon desert. Ph.D. diss., Washington State Univ. Pullman.

Wigand, P.E. and P.F. Mehringer, Jr. 1985. Pollen and seed analysis. Anthrop. Papers Amer. Museum of Natural Hist. 61:108-124.

Wigand, P.E. 1987. Diamond Pond, Harney County, Oregon: vegetation history and water table in the eastern Oregon desert. Great Basin Nat. 47:427-458.

Williams, G., D.E. Miller and D.H. Miller. 1971. Peter Skene Ogden's Snake Country journals: 1827-28 and 1828-29. Hudson's Bay Record Society. London.

Williams, O.B. 1970. Population dynamics of two perennial grasses in Australian semi-arid grassland. J. Ecol. 58:869-876.

Wilson, A.M., G.A. Harris and D.H. Gates. 1966. Cumulative effects on clipping on yield of bluebunch wheatgrass. J. Range Manage. 19:90-91.

Winward, A.H. 1991. A renewed commitment to management of sagebrush grasslands, p. 2-7. *In*: Research in Rangeland Management. Oregon State Univ. Agric. Exper. Sta. Spec. Rep. 880.

Winkel, V.K. and B.A Roundy. 1991. Effects of trampling and mechanical seedbed preparation on grass seedling emergence. J. Range Manage. 44:176-180.

Woodmansee, R.G. 1978. Additions and losses of nitrogen in grassland ecosystems. Bio-Science 28:448-453.

Wraith, J.M., D.A. Johnson, R.J. Hanks and D.V. Sisson. 1987. Soil and plant water relations in a crested wheatgrass pasture: response to spring grazing by cattle. Oecologia 73:573-578.

Wright, H.A. and J.O. Klemmedson. 1965. Effects of fire on bunchgrasses of the sagebrush-grass region in southern Idaho. Ecology 46:680-688.

Wright, H.A. and A.W. Bailey. 1982. Fire Ecology: United States and southern Canada. John Wiley and Sons, N.Y.

Wright, R.G. 1972. Computer processing of charge quadrate maps and their use in plant demographic studies. J. Range Mange. 25:476-478.

Wright, R.G. and G.M. Van Dyne. 1976. Environmental factors influencing semidesert grassland perennial grass demography. Southwestern Natur. 21:259-274.

Young, J.A., R.D. Evans, and R.C. Eckert, Jr. 1969. Population dynamics of downy brome. Weed Sci. 17:20-26.

Young, J.A. and R.A. Evans. 1973. Downy brome - intruder in the plant succession of big sagebrush communities in the Great Basin. J. Range Manage. 26:410-415.

Young, J.A. and R.A. Evans. 1974. Population dynamics of green rabbitbrush in disturbed big sagebrush communities. J. Range Manage. 27:127-132.

Young, J.A., R.A. Evans. 1976. Response of weed populations to human manipulations of the natural environment. Weed Sci. 24:186-190.

Young, J.A. R.A. Evans, and P.T. Tueller. 1976. Great Basin plant communities-pristine and grazed, p. 187-215. *In*: R. Elston and P. Headrick (eds.), Holocene environmental change in the Great Basin. (Reno, Nevada: Nevada Arch. Survey, Res. Paper No. 6, University of Nevada, 1976)

Young, J.A. and R.A. Evans. 1978. Population dynamics after wildfires in sagebrush grasslands. J. Range Manage. 31:283-289.

Young, J.A., and R.A. Evans. 1981. Demography and fire history of a western juniper stand. J. Range Manage. 34:501-506.

Young, J.A. and R.A. Evans. 1985. Demography of *Bromus tectorum* in *Artemisia* communities. *In*: J. White (ed.), Population Structure of Vegetation. Dr. W. Junk Publishers, Dordrecht.

Young, J.A. and B.A. Sparks. 1985. Cattle in the cold desert. Utah State Univ. Press, Logan.

Young, J.A. 1989. Intermountain shrub steppe plant communities pristine and grazed, p.3-14. *In*: Proc. Western Raptors Sym. Workshop, Inst. Wildl. Res. Sci. Tech.

Young, J.A., R.D. Evans, R.E. Eckert, Jr., and B.L. Kay. 1987. Cheatgrass. Rangelands 9:266-270.

Young, J.A. 1991. Cheatgrass, p. 408-417. *In*: L.F. James et al. (eds.), Noxious range weeds. Westview Press. San Francisco.

Young, J.A. and F. Tipton. 1990. Invasion of cheatgrass into arid environments of the Lahontan basin, p. 37-40. *In*: Proc. Cheatgrass Invasion Shrub Die-off, and other aspects of shrub biology and management. USDA For. Ser.

Young, R.P. and R.F. Miller. 1985. Response of *Sitanion hystrix* (Nutt.)J.G. to prescribed burning. Amer. Midl. Natur. 113:182-187.

Physiological Responses of Individual Plants to Grazing: Current Status and Ecological Significance

D. D. Briske and J. H. Richards

Abstract

Individual plant responses to herbivory were initially investigated to establish prudent guidelines for livestock grazing on pastures and rangelands. A great deal has been learned concerning the effects of defoliation on the morphology, physiology and productivity of important forage and browse species. However, our perception and understanding of individual plant responses to herbivory have been substantially altered by research findings developed during the past 15 years. Information within this contemporary research perspective can be generalized into three broad categories. First, several traditional assumptions regarding plant responses to defoliation have proven to be over simplifications. For example, questions have arisen concerning the value of carbohydrate reserves as an indicator of plant growth following defoliation and the assumption that tillering is exclusively regulated by auxin produced in the apical meristem. Second, several novel hypotheses have been proposed to explain the potentially beneficial effects of herbivory. Increased photosynthetic rates of residual foliage, accelerated growth rates, modified allocation patterns and enhanced nutrient absorption have been observed in various plant species following defoliation. Third, the ability of biotic interactions and abiotic constraints to accentuate herbivore-mediated plant responses has only recently been fully appreciated. For example, herbivore-induced modifications of competitive interactions may potentially constrain plant growth to a greater extent than the direct effects of biomass removal. Evidence from both the traditional and contemporary literature has been surveyed to assess the current status of our knowledge and evaluate the ecological significance of individual plant responses to grazing by large herbivores.

Introduction

Domestic herbivores were initially introduced into North America in the early 1500's and their density and distribution increased steadily with colonization by the Spanish in the south and southwest, the English on the eastern seaboard, and the French in the midwest and Louisiana (Sampson 1952). Domestic herbivore populations increased to a maximum during the inflationary period following the Civil War with approximately 26 million cattle and 20 million sheep in the 17 western states in 1890 (Young, this volume). Escalating animal numbers and intensive grazing of western rangelands during the late 1800's contributed to a growing public concern for conservation of lands in the public domain. Investigations of plant responses to defoliation were soon initiated to establish a scientific basis for the conservation and management of rangeland vegetation in the western United States.

A great deal has been learned concerning the effects of grazing on the morphology, physiology, and productivity of important forage and browse plants during the 20th century. Ellison (1960) and Jameson (1963) provide excellent summaries of research addressing individual plant and community responses to grazing during the first half of the century. Examples of some of the prominent concepts developed during this period include: 1) identification of critical frequencies and intensities of defoliation to ensure plant survival and maintain productivity, 2) documentation of a proportional relationship between root growth and the severity of shoot defoliation, 3) appreciation of the significance of shoot morphology for grazing resistance, 4) development of the concept of carbohydrate reserves to explain plant survival and growth following defoliation, and 5) recognition of the interactive effects of apical dominance and defoliation on tiller initiation. However, not all of the traditional concepts have proven to be compatible with subsequent research findings.

The objective of this chapter is to provide both a synopsis of the current status of our knowledge and a discussion of the ecological significance of the physiological and morphological responses of individual plants to grazing. We have approached this objective by comparing the traditional literature with the more contemporary literature developed since 1975. Although all major growth forms have been referenced, grasses have been given the greatest emphasis. A brief overview of the hierarchical organization of grassland vegetation is presented to emphasize the significance of individual plant responses to community ecology and rangeland management.

Ecological Scales

Grassland communities can be viewed at various scales reflecting their hierarchical organization. Individual grass plants consist of an assemblage of tillers, grass populations are comprised of individual plants with varying tiller numbers per plant, and grassland communities are further composed of an aggregation of species populations variously arranged in space and time (Briske 1991). Each of the scales is associated with overlapping temporal responses and spatial dimensions. For example, the physiological and developmental processes of individual plants react in much shorter time frames and at smaller spatial dimensions than do population or community dynamics.

Recognition of the linkage between processes at various ecological scales determines that insight into mechanisms occurring at higher scales (e.g., population and community) requires that processes at lower scales (e.g., plant and tiller) be investigated (Brown and Allen 1989, Allen and Hoekstra 1990). Although proportionate responses may occur among ecological scales to affect ecosystem structure and function, additional complexity is frequently encountered by the occurrence of disproportionate responses between or among scales. Consequently, it is essential that several ecological scales be considered simultaneously when evaluating vegetation responses to herbivory to avoid incomplete or erroneous conclusions (Brown and Allen 1989).

Plant Responses To Herbivory

Carbohydrate reserves

The major premise for monitoring carbohydrate reserves was to provide an index of potential regrowth based upon the assumption that these carbohydrates provided the major source of carbon for leaf regrowth. The depletion of carbohydrate reserves in response to defoliation, environmental stresses, or as a result of normal phenological processes, was thought to be the major cause of subsequent growth reductions. In extreme cases, plant mortality could also result from a substantial reduction in carbohydrate reserves (Weinmann 1948). Carbohydrate reserves, referred to as total nonstructural carbohydrates (TNC) or total available carbohydrates (TAC), are a product of photosynthesis in excess of requirements for growth and maintenance (White 1973). These labile carbon compounds are utilized for plant growth and maintenance when photosynthetic capacity is limited as evidenced by the reduction in carbohydrate reserves following defoliation, shading or other manipulations that limit photosynthesis, in a large number of grass species (Deregibus et al. 1982) (Fig. 1). The traditional interpretation assumed that carbohydrates stored in roots were remobilized to support shoot growth following defoliation.

However, a substantial amount of information has been compiled during the last three decades indicating that the contribution of carbohydrate reserves to leaf regrowth of perennial grasses may be much smaller than previously assumed (May 1960, Davidson and Milthorpe 1965, Ryle and Powell 1975, Atkinson and Farrar 1983, Richards and Caldwell 1985, Richards 1986). Carbohydrate pools (tissue mass x TNC concentration) within root systems are primarily used for root growth and respiration and are not usually remobilized for subsequent use in shoots following defoliation. TNC pools in root systems are dispersed among the numerous fine roots and are insufficient to maintain pre-defoliation rates of root growth and function following defoliation (Richards 1984). Therefore, a decrease in TNC pools in root systems following defoliation results from the combined effects of a reduction of current photosynthates allocated to the root system and continued TNC utilization in root respiration, rather than remobilization and allocation to the shoot system (Fig. 1). This interpretation is supported by a decrease in both TNC pools and TNC concentrations of grasses even in the absence of shoot regrowth following defoliation (Richards and Caldwell 1985).

A consistent, positive relationship between the size of carbohydrate pools and plant regrowth has not been definitively established (Busso et al. 1990). In seasonal comparisons, an inverse relationship is frequently observed with carbohydrate pools increasing during periods of minimal growth and then decreasing during rapid growth (Davidson and Milthorpe 1966, Caldwell 1984). The potential

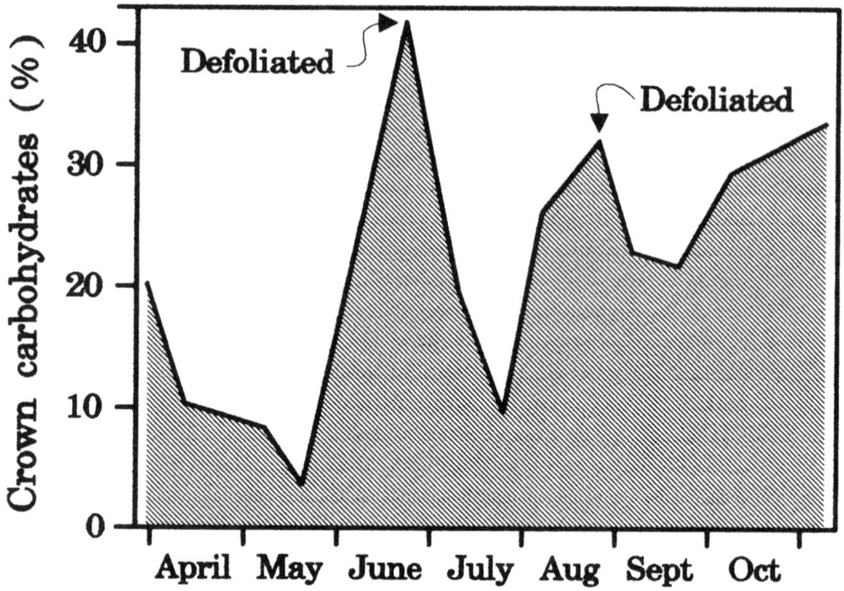

Figure 1. Carbohydrate concentrations in the crown of timothy (*Phleum pratense* L.) in response to defoliation in June and August during a single season. Rapid reductions in crown and root carbohydrate concentrations are traditionally interpreted as the remobilization of carbohydrates to support shoot growth following defoliation. The contemporary interpretation indicates that carbohydrate concentrations in the crown and roots decline in response to the combined effects of reduced carbon allocation from the shoot and continued carbohydrate utilization in root respiration following defoliation (from Smith 1972).

for plant regrowth increased only when TNC concentrations in roots and crowns of crested wheatgrass [*Agropyron desertorum* (Fisch. ex Link) Schult.] and bluebunch wheatgrass [*Pseudoroegneria spicata* (Pursh) A. Löve ssp. *spicata*; syn: *A. spicatum* (Pursh) Scribn. and Smith] increased to exceptionally high levels in response to an experimentally imposed drought (Fig. 2). Under all other conditions in a 2-year field study, crown and root TNC concentrations or pools were poor indicators of regrowth potential in early spring when these two grasses exhibit their greatest growth potential (Busso et al. 1990). In addition, neither carbohydrate concentrations nor pools account for the wide variation in grazing tolerance between crested wheatgrass and bluebunch wheatgrass (Richards and Caldwell 1985). Carbohydrate pools within the crowns, including sheath bases, of the two wheatgrass species were equivalent to carbohydrates produced in only 3 days of photosynthesis (Fig. 3). Consequently, plant growth was more dependent upon current photosynthesis than stored carbohydrates within 3 days following severe defoliation. Similarly, leaf elongation or tillering were not proportionately related to carbohydrate concentrations in leaves, crowns or roots of tall fescue (*Festuca arundinacea* Schreb.), orchardgrass (*Dactylis glomerata* L.) or reed canarygrass (*Phalaris arundinacea* L.) (Sambo 1983, Zarrough et al. 1984, Volenec and Nelson 1984). These findings support the view of May (1960), who indicated that use of the term carbohydrate "reserve" evokes a false conception of their contribution to growth following defoliation (Deregibus et al. 1982).

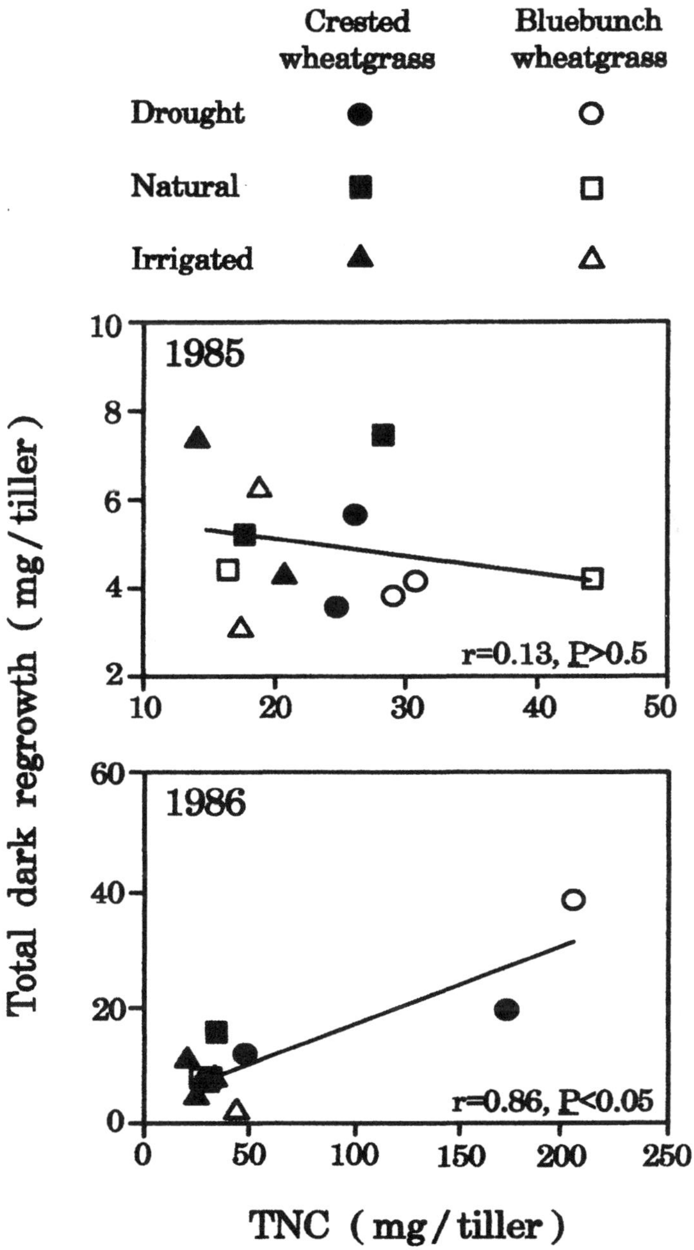

Figure 2. Relationship between root and crown TNC (total nonstructural carbohydrates) pools and regrowth of crested and bluebunch wheatgrasses produced in the dark under irrigated, natural rainfed, and drought conditions. Regrowth produced in the dark quantifies the contribution of carbohydrate reserves to regrowth in the absence of current photosynthesis. A significant, positive relationship between TNC pools and regrowth was only observed in 1986 when severely water stressed plants had unusually large TNC pools and produced large amounts of regrowth. Under all other growth conditions in both years the relationship was not significant (from Busso et al. 1990).

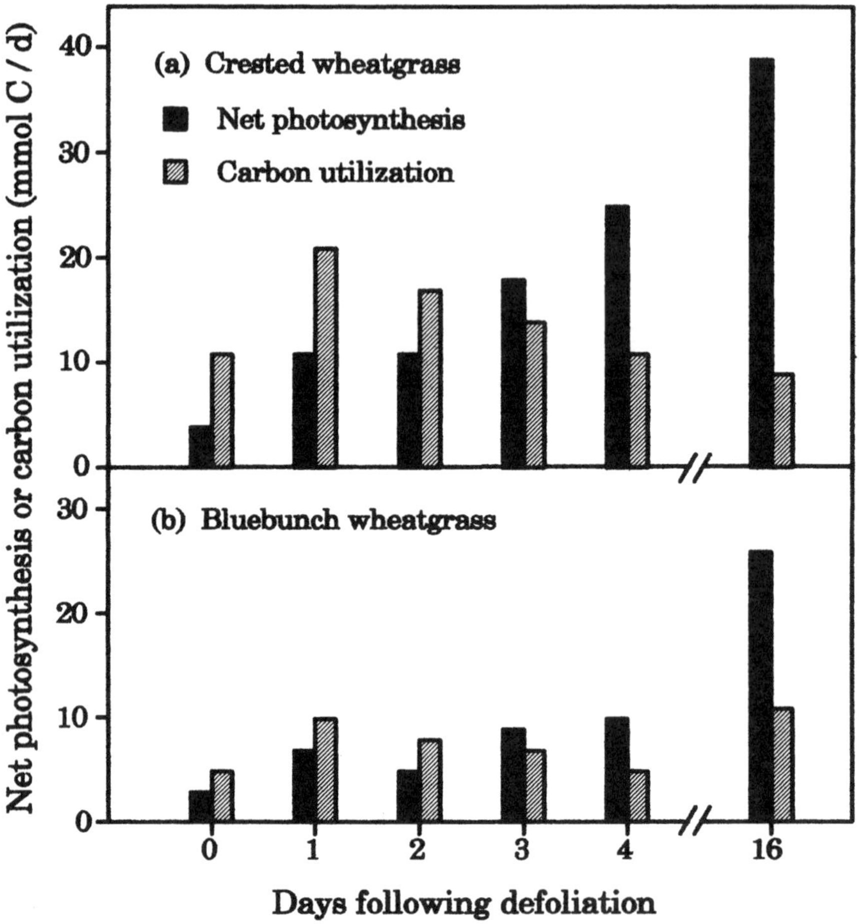

Figure 3. Daily net photosynthetic carbon uptake and carbon utilization by shoots and crowns of (a) crested wheatgrass and (b) bluebunch wheatgrass plants on the day of defoliation (day zero) and 1, 2, 3, 4, and 16 days following an early spring defoliation. The contribution of current photosynthesis exceeds the requirement of shoot regrowth and shoot and crown respiration within 3 days of severe defoliation in both species (from Richards and Caldwell 1985).

Several other considerations further limit the utility of carbohydrate pools as indicators of regrowth potential. First, TNC mass does not always account for the total reductions in crown weight or the amount of regrowth produced in the dark following defoliation indicating that compounds other than nonstructural carbohydrates (e.g., proteins, hemicellulose and organic acids), which are not commonly evaluated, may also function as carbon or energy sources during regrowth (Davidson and Milthorpe 1965, 1966, Richards and Caldwell 1985). Second, the carbohydrate concentration or pool-size necessary to ensure plant survival and maintain maximum leaf replacement potential has not been established for individual species or species groups. Carbohydrate concentrations of 1-6% have been suggested as minimal concentrations in grasses, but these estimates are far from conclusive (Caldwell 1984). Considering the dynamics of carbohydrate allocation, accumulation, and utilization it is unlikely such a simple estimate would have much predictive value. Third, total carbohydrate pools must be quantified by determining both carbohydrate concentrations and weight of the associated plant organ(s) (Richards and Caldwell 1985). TNC concentration may decrease while total TNC pool size increases during intervals of rapid plant growth (Caldwell et al. 1981) Fourth, inconsistencies associated with analytical techniques for carbohydrate extraction have also curtailed research progress. TNC estimates derived by acid extraction may be two to three times greater than for boiling water extraction because a greater amount of structural carbohydrates are digested (Smith 1972, 1981, Richards and Caldwell 1985).

Carbohydrate pools play an important role in initiating plant growth when photosynthetic capacity is severely limited. However, the limited amount of carbohydrates stored in tiller bases (crowns and sheaths), the inaccessibility of root carbohydrates to support shoot growth, and the poor correlation between shoot growth and carbohydrate concentrations or pools limits their use as an effective index of shoot regrowth in perennial grasses and potentially other growth forms as well. The traditional interpretation of carbohydrate reserves appears to have been founded on an over-simplified, static conceptualization of the carbon economy of plants. The amount of residual photosynthetic area or the availability of active meristems following defoliation may be of equal or greater consequence in determining potential leaf regrowth of grasses.

Apical dominance

Apical dominance in grasses was initially described by Leopold (1949) as the production of auxin within the apical meristem which suppressed the growth of axillary buds. This interpretation was based on the earlier work of Thimann and Skoog (1933) who had established that auxin controlled lateral bud growth in bean (*Vicia Faba* L.). Leopold (1949) destroyed apical meristems in twenty plants of each of two annual grasses, teosinte (*Euchlaena mexicana* Schrad.) and barley (*Hordeum vulgare* L.), and observed tillering to increase while a synthetic auxin (naphthalene acetic acid) applied to the location of the excised apical meristems reversed this effect. This single investigation has largely shaped our perception of how the tillering process is regulated in grasses, but has been criticized as being less than definitive from an experimental perspective (Bunting and Drennan 1966, Williams and Langer 1975).

This theory of apical dominance, referred to as the direct theory of auxin inhibition, encountered several technical difficulties and was soon abandoned for a theory proposing a more indirect effect of auxin. Major inconsistencies associated with the direct inhibition theory included: 1) experimental auxin concentrations several times greater than endogenous concentrations were required to inhibit bud growth, 2) auxin concentrations frequently increased rather than decreased in lateral buds shortly following plant decapitation, and 3) unidirectional auxin transport from the apical meristem toward the stem base limited the potential for entry into lateral buds (Gregory and Veale 1957, Martin 1987). The alternative indirect theory proposed that auxin inhibited lateral bud growth by either the production of a secondary inhibitor or by the regulation of resource allocation to regions other than lateral buds. Unfortunately, the direct inhibition theory continues as the prevailing view of apical dominance in rangeland ecology and management even though insight into the physiological process has advanced substantially.

Continued adherence to the direct inhibition theory of apical dominance poses numerous inconsistencies of particular relevance to rangeland ecology and management (Murphy and Briske 1992). For example, tiller recruitment has been observed to occur in response to grazing even though apical meristems of little bluestem (*Schizachyrium scoparium* var. *frequens* Hubb.) tillers were insufficiently elevated to be removed by livestock (Butler and Briske 1988). Conversely, removal of apical meristems from tillers of crested wheatgrass and bluebunch wheatgrass did not always accelerate tiller recruitment (Olson and Richards 1988a, Richards et al. 1988). In contrast to the responses of temperate species, axillary bud growth was stimulated by both apical meristem and leaf removal in three tropical grasses. These conflicting observations attest to the complexity of apical dominance in perennial grasses and the limited interpretative ability of the direct inhibition theory.

Rigid adherence to the direct inhibition theory of apical dominance may have biased our interpretation of tillering following plant defoliation. Both Ellison (1960) and Jameson (1963), following a thorough evaluation of the early literature, concluded that defoliation does not consistently increase tillering in perennial grasses. Furthermore, much of the recent literature substantiates their conclusions (Murphy and Briske 1992). For example, four consecutive years of biweekly defoliation to stubble heights of 5, 10, and 15 cm decreased tiller densities proportionately in pinegrass (*Calamagrostis rebescens* Buckl.) (Stout et al. 1981) (Fig. 4). Two seasons of livestock grazing did not increase, and in the case of intensive grazing, decreased the total number of tillers recruited from crested wheatgrass (Olson and Richards 1988b) and little bluestem (Butler and Briske 1988) in comparison with ungrazed plants. The complexity of the physiological mechanisms regulating axillary bud growth (e.g., apical dominance, resource availability, and phytochrome reactions) and the large number of potentially intervening factors (e.g., environmental variables, species-specific responses, stage of phenological development, frequency and intensity of defoliation) also minimizes the likelihood of consistent tillering responses to defoliation.

Defoliation may more appropriately be viewed as a means of altering the timing or seasonality of tiller recruitment rather than increasing total tiller recruitment over the long-term in many range grasses (Murphy and Briske 1992). Tillers are frequently initiated within 2-3 weeks of defoliation (Olson and Richards 1988a, Richards et al. 1988), and are much more obvious following partial removal of the canopy. However, this short-term "flush" of tiller initiation following defoliation may reduce maximum tiller recruitment in defoliated plants relative to periods of maximum recruitment in undefoliated plants (Butler and Briske 1988). In addition, increased tiller recruitment following spring grazing of crested wheatgrass did not contribute to tiller replacement the following season because of greater tiller mortality in grazed plants than in ungrazed plants during the winter (Olson and Richards 1988b). Therefore, in spite of a brief period of increased recruitment immediately following defoliation, defoliated plants may not produce a greater number of tillers than undefoliated plants when evaluated over one or more growing seasons.

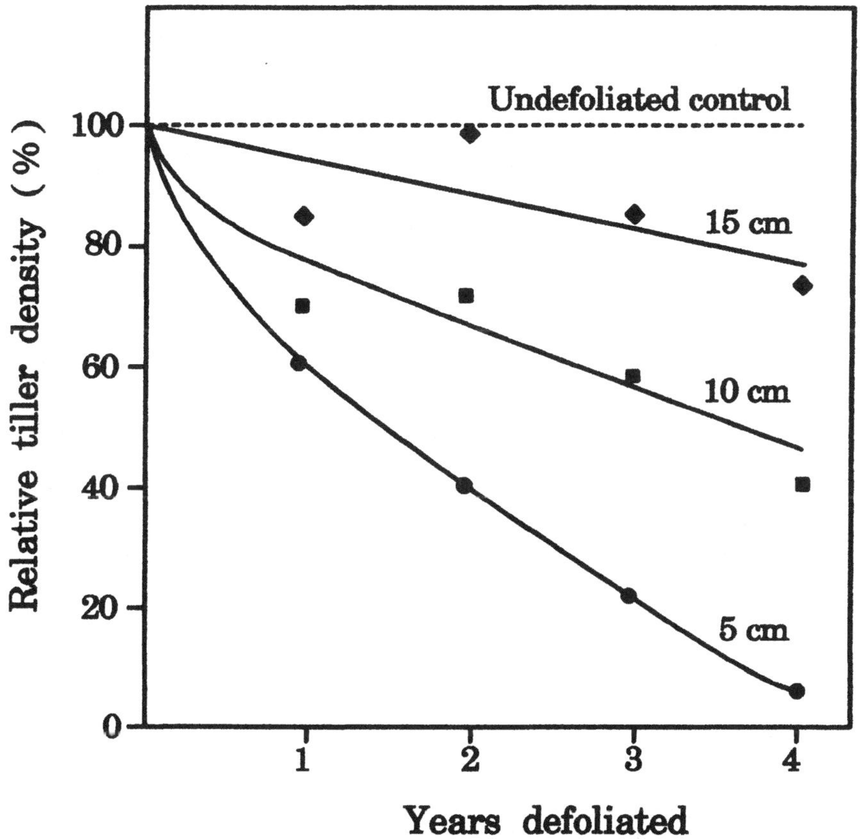

Figure 4. Relative density of pinegrass tillers in response to biweekly clipping to stubble heights of 5, 10, and 15 cm for four successive years in comparison with tiller densities of undefoliated plants. These data demonstrate that tillering was not promoted by defoliation at any time in this species population grown in the field (from Stout et al. 1981).

The current theory of apical dominance proposes that auxin produced in the apical meristem or young leaf primordia provides the principle correlative signal for bud inhibition and that a second hormone, cytokinin, is also involved (Phillips 1975, Hillman 1984, Cline 1991). Auxin appears to catabolize or prevent cytokinin transport from the root system thereby preventing bud growth. Two additional hormones, abscisic acid (Nojima et al. 1989) and ethylene (Harrison and Kaufman 1982), both of which are plant growth inhibitors, have also been associated with apical dominance in grasses, but their specific functions have not yet been established. Additional insight into the physiological mechanism(s) of apical dominance and inconsistent tillering responses following defoliation indicate that a more comprehensive model is required to interpret and explain the timing and magnitude of tiller initiation in grasses.

Radiation quality has been implicated as an environmental signal regulating tiller initiation in grasses (Kasperbauer and Karlen 1986, Casal et al. 1987, Simon and Lemaire 1987). Red radiation is attenuated more rapidly than far-red radiation as it passes through a plant canopy. A reduction in the red:far-red ratio associated with increasing canopy development may signal the diminishing availability of resources and suppress additional tiller initiation (Deregibus et al. 1985, Casal et. al. 1987). Alternatively, partial canopy removal by grazing may increase the red:far-red ratio and promote tillering in the absence of disturbance to the apical meristem. The role of radiation quality as an environmental signal requires documentation in a greater number of species under field conditions prior to its acceptance as a process common to grasses. A cause-effect relationship has not been established between the red:far-red ratio and the physiological mechanisms of apical dominance.

Compensatory processes

Range plant responses to defoliation were initially investigated to minimize the detrimental consequences of livestock grazing. This perspective on plant response to grazing is inherent in much of the early scientific literature. However, an alternative perspective indicating that an optimal level of grazing may increase plant growth (i.e., overcompensation) above that of ungrazed plants was proposed by several ecologists in the mid-1970's (Dyer 1975, McNaughton 1976, 1979). This perspective stimulated a great deal of research exploring the beneficial aspects of grazing on plant growth and function. Several novel research hypotheses were proposed and evaluated. The status of the most significant of these hypotheses is summarized below.

Photosynthesis

Grazing has long been known to dramatically affect the photosynthetic capacity of plants by reducing total photosynthetic area. However, it has more recently been appreciated that the relative proportion of foliage elements (i.e., blade, sheath, culm), age of individual foliage elements and photosynthetic rate of foliage elements are also important determinants of photosynthetic capacity following defoliation. Removal of young foliage elements, which attain maximum photosynthetic capacity at about the time of complete expansion, reduces canopy photosynthesis to a much greater extent than equivalent removal of older foliage (e.g. Ludlow and Charles-Edwards 1980, Gold and Caldwell 1989, 1990). Photosynthetic rates among foliage elements of comparable age often vary with blades frequently possessing greater rates than sheaths or culms (Caldwell et al. 1981, Nowak and Caldwell 1984, Ludlow and Mott 1988). Therefore, the photosynthetic capacity of defoliated plants may differ from that of nondefoliated plants because of a modification in the proportion and age structure of foliage elements, in addition to a reduction in total photosynthetic area and altered microenvironmental variables within the canopy.

Greater photosynthetic rates of leaves on defoliated plants than for comparably aged leaves on undefoliated plants is referred to as compensatory photosynthesis (Detling et al. 1979b, Nowak and Caldwell 1984) (Fig. 5). This response occurs over a several day period in both mature and expanding leaves remaining on the defoliated plant as well as leaves initiated following defoliation. Compensatory photosynthesis has been observed in many species. A few examples among grasses include crested wheatgrass and bluebunch wheatgrass (Nowak and Caldwell 1984),

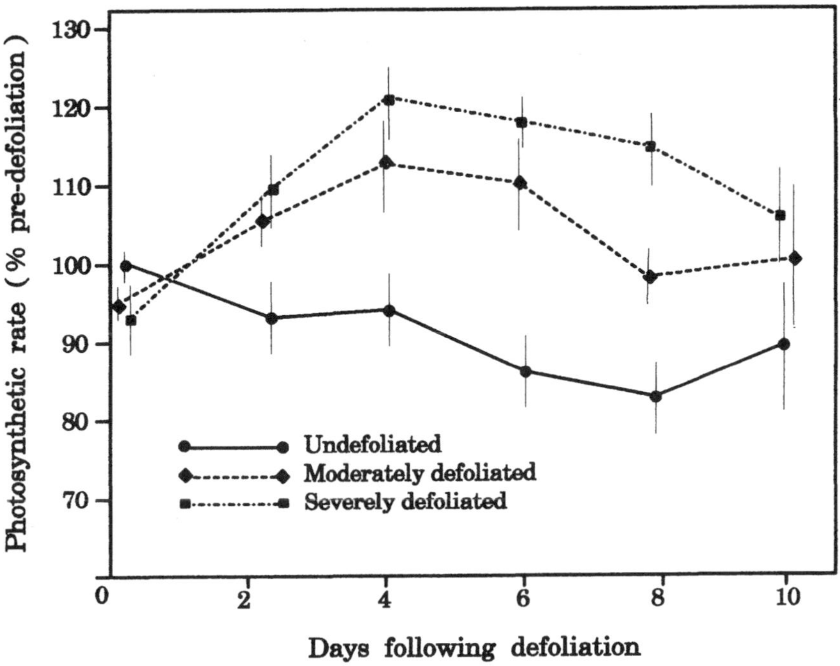

Figure 5. Photosynthetic rates (expressed as a percentage of the pre-defoliation rate) of expanded leaves on undefoliated tillers of western wheatgrass plants from which 50 (moderately defoliated) or 75% (severely defoliated) of all tillers had been defoliated at ground level. Photosynthetic rates decreased slightly immediately following defoliation, but increased to a maximum 4 days following defoliation while photosynthetic rates of leaves on undefoliated plants declined with age during the 10 day period (from Painter and Detling 1981).

western wheatgrass [*Pascopyrum smithii* (Rydb.) A. Löve, syn: *Agropyron smithii* (Rydb.] (Painter and Detling 1981), sheep fescue (*Festuca ovina* L.) (Atkinson 1986), Italian ryegrass (*Lolium multiflorum* Lam.) (Gifford and Marshall 1973), and (*Themeda triandra* Forsk.) (Wallace et al. 1984), as well as numerous herbaceous and woody dicotyledonous species. However, photosynthetic rates less than or equal to those on undefoliated plants have also been reported for some species following defoliation (Ryle and Powell 1975, Detling et al. 1979a, Ferree and Hall 1981, Atkinson 1986). Contrasting reports of reduced photosynthetic rates and those reporting compensatory photosynthesis have not been adequately explained in most cases. In addition, repeated severe defoliation can reduce photosynthetic rates, even in very grazing-tolerant species (Wallace et al. 1985, Hodgkinson et al. 1989).

Compensatory photosynthesis results from rejuvenation or inhibition of the normal decline in photosynthetic capacity as leaves age and senesce. Rejuvenation of net photosynthetic rate in leaves on defoliated plants may be partial, complete or, in some cases, result in photosynthetic rates that exceed those of newly developed leaves on undefoliated plants (Hodgkinson 1974, Heichel and Turner 1983). The ability of mature leaves to rejuvenate their photosynthetic capacity declines as they age and older leaves also require longer than young leaves to achieve maximum recovery (Hodgkinson 1974, Nowak and Caldwell 1984). Leaf photosynthetic rates generally increase less than 50% of pre-defoliation rates. However, relative to the initially low rates of net photosynthesis in older leaves, proportional increases may be greater in old than in young leaves following defoliation. Leaves on defoliated plants often possess greater photosynthetic rates than comparable leaves on undefoliated plants even if they are not rejuvenated because the normal decline in photosynthetic capacity of aging leaves is inhibited (Gifford and Marshall 1973).

Compensatory photosynthesis appears to mechanistically result from a greater or more efficient enzymatic capacity to assimilate CO_2, rather than an enhancement of stomatal CO_2 conductance (Gifford and Marshall 1973). Leaves displaying compensatory photosynthesis generally have increased chlorophyll, nitrogen and soluble protein contents (Satoh et al. 1977, Nowak and Caldwell 1984, Yamashita and Fujino 1986). Ribulose-1, 5-bisphosphate carboxylase/oxygenase (RuBPCO) activity or amount, capacity for RuBPCO regeneration, and electron transport capacity have also been documented to increase within several days following defoliation (Neales et al. 1971, Jenkins and Woolhouse 1971, von Caemmerer and Farquhar 1984).

The mechanisms regulating compensatory photosynthesis have not yet been definitively established. Compensatory photosynthesis can be induced by increases in irradiance on residual photosynthetic tissues, modification of physiological plant function induced by defoliation or by a combination of these processes. However, it has been well documented that compensatory photosynthesis can occur independently of changes in the light environment associated with plant defoliation (Hodgkinson et al. 1972, Hodgkinson 1974, Woledge 1977). Increased sink strength produced by the rapid growth of defoliated shoots and the increased transport of hormones or other substances to shoots from roots represent the two prominent physiological modifications. Compensatory photosynthesis is apparently induced by phytohormones (potentially cytokinins) transported from roots to leaves on defoliated plants, rather than in response to direct sink:source interactions (Wareing et al. 1968, Carmi and Koller 1979). The absence of an *immediate* increase in photosynthesis following an increase in the sink:source ratio following defoliation further supports the conclusion that compensatory photosynthesis is not the direct result of increased sink strength (Geiger 1976, von Caemmerer and Farquhar 1984).

Although the physiological process of compensatory photosynthesis does occur in a large number of plants, its significance to grazing tolerance and plant productivity must be assessed on a case-by-case basis because of interspecific variation and the influence of a large number of plant attributes and environmental variables. For example, compensatory photosynthesis occurs in both crested wheatgrass and bluebunch wheatgrass, but is most evident in the oldest leaves which comprise a small percentage of the total photosynthetic area (Nowak and Caldwell 1984). The total amount of compensatory photosynthesis resulting from this portion of the canopy is insufficient to explain the large difference in grazing tolerance between these two species and even very grazing tolerant species fail to exhibit compensatory photosynthesis under all conditions. Total photosynthetic area, the occurrence of favorable environmental conditions, and age of photosynthetic tissues may have an equal or greater effect on the total carbon gain capacity of plants than does photosynthetic rate.

Allocation priorities

Defoliation consistently modifies allocation priorities within plants in addition to affecting total carbon gain. Increased carbon export from remaining photosynthetically active leaves and increased carbon allocation to growing shoots minimizes the reduction in the absolute amount of carbon available to shoots following defoliation (Marshall and Sagar 1965, Bassmann and Dickmann 1985). For example, a 51% reduction in whole plant photosynthesis on the day of defoliation reduced carbon allocation to shoots by only 40% in barley (Ryle and Powell 1975). The relative amount of carbon allocated to defoliated shoots continued to increase during the next 7 days. Increased carbon allocation to shoots occurred concomitantly with a comparable reduction in allocation to roots, which received much less carbon than in undefoliated plants (Ryle and Powell 1975) (Fig. 6).

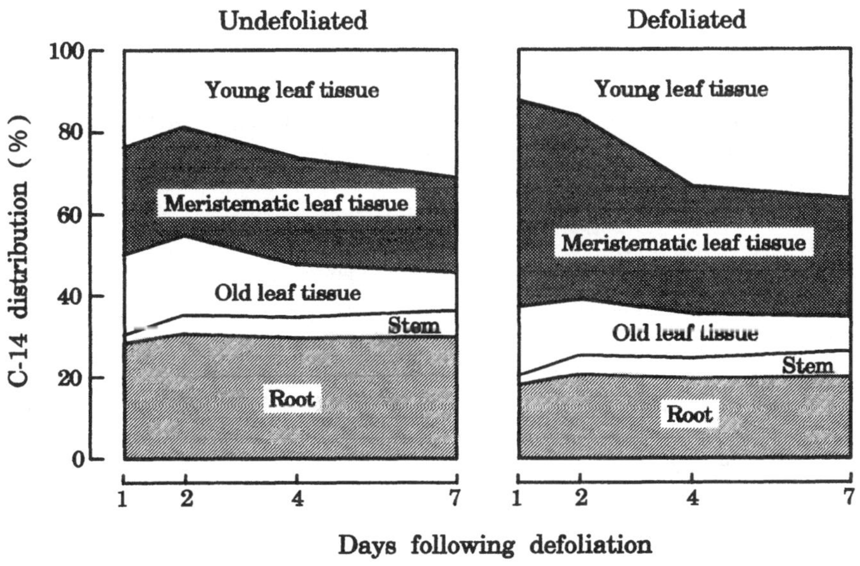

Figure 6. Relative allocation of ^{14}C-labeled photosynthate to various tissues of defoliated and undefoliated barley plants. Plants were labeled for 25 minutes and harvested at 1, 2, 4, and 7 days following defoliation. Defoliated plants were cut at the ligule of the 3rd leaf at the beginning of day 1. Carbon allocation to young leaf tissue including the apical meristem, leaf primordia, and expanding leaves increased while carbon allocation to roots decreased at day 4 and 7 following defoliation (from Ryle and Powell 1975).

Defoliation may also modify allocation priorities within the shoot system. A greater proportion of available carbon is allocated to leaf blades than leaf sheaths following defoliation of blue grama [*Bouteloua gracilis* (H.B.K.) Lag. ex Steud.] (Detling et al. 1979b) and western wheatgrass (Painter et al. 1989). Increased blade:sheath ratios may function to increase the total photosynthetic capacity of plants immediately following defoliation. Leaf blades often have a greater photosynthetic capacity than leaf sheaths (Caldwell et al. 1981, Ludlow and Mott 1988). Therefore, carbon allocation to blades, rather than sheaths, would potentially provide the greatest photosynthetic gain. However, this response requires evaluation in a greater number of species, consideration of water and nitrogen-use efficiencies, and determination of blade and sheath longevities before it is assumed to be a widely distributed tolerance mechanism (Jaramillo and Detling 1988, Polley and Detling 1988).

Inter-tiller resource (C and N) allocation following defoliation provides a potential mechanism of herbivory tolerance by facilitating tiller survival and rapid reestablishment of photosynthetic surfaces. Rates of resource import from attached, undefoliated tillers increase rapidly following partial tiller defoliation with multiple defoliations of an individual tiller further increasing the rate of resource import (Gifford and Marshall 1973, Welker et al. 1985). However, increased rates of resource import may not necessarily represent an absolute increase in resource accumulation because defoliation reduces the total resource requirement by decreasing shoot mass and subsequently sink size (Welker et al. 1987). Resource import from attached undefoliated tillers again decreases to a low level with the reestablishment of photosynthetic surfaces within several days of a single defoliation (Gifford and Marshall 1973, Welker et al. 1985, 1987). For example, growth of tall fescue plants increased progressively following defoliation when either a greater percentage of tillers remained undefoliated or when the defoliation intensity was reduced (Matches 1966). A portion of this growth response was attributable to increased rates of ^{14}C allocation from undefoliated to defoliated tillers (Watson and Ward 1970). Similar conclusions have been drawn from investigations with several rhizomatous graminoids (Mattheis et al. 1976, Archer and Tieszen 1986, Jonsdottir and Callaghan 1989).

Short-term modifications in resource allocation patterns following defoliation are controlled by the relative sink strengths of the respective tissues involved, either through source-sink feedback or hormonal signals (e.g. Bucher et al. 1987, Bassmann and Dickmann 1985, Geiger and Fondy 1985). Apical meristems and associated young leaves are sufficiently strong sinks to deprive other sinks, such as roots or young tillers, of resources until the amount of source tissue increases to exceed the demands of these sinks. If defoliation removes the predominant shoot sinks (apical and intercalary meristems), available resources are allocated to weaker sinks including roots, or storage in sheaths and stems (Bucher et al. 1987). Long-term changes in allocation patterns are mediated by hormonal signals and resulting changes in developmental morphology and sink activities (Wareing et al. 1968, Geiger 1976, Geiger and Fondy 1985).

Flexible carbon allocation patterns may be causally associated with grazing tolerance in grasses. As an example, defoliated crested wheatgrass plants reallocate carbon to reestablishment of photosynthetic tissue while temporarily reducing allocation to roots (Caldwell et al. 1981, Richards 1984). Reallocated carbon is converted into leaf area following rapid activation and growth of axillary buds (Mueller and Richards 1986). Conversely, bluebunch wheatgrass continues to allocate a large proportion of carbon to the root system following defoliation, thus reestablishing its canopy much less rapidly than crested wheatgrass. These data indicate that a short-term reduction in carbon allocation to root growth minimizes the detrimental effects of defoliation on leaf production and the long-term competitive ability of grazed plants. Short-term effects of defoliation on competitive ability are discussed below.

Nutrient absorption

Root respiration and nutrient acquisition are rapidly reduced following shoot defoliation (Fig. 7). Root respiration begins to decline within hours after defoliation and it is substantially reduced within 24 hours (Davidson and Milthorpe 1966, Clement et al. 1978). Concomitant with a defoliation-induced reduction in root respiration, nutrient absorption is also strongly inhibited. Nitrate absorption decreased within 30 minutes following 70% shoot removal (dry weight basis) in perennial ryegrass (*Lolium perenne* L.), and decreased to less than 40% of the pre-defoliation rate within 2 hours (Clement et al. 1978). Nitrate uptake continued to decline over the next 4-12 h and remained minimal for 2 or 7 days before recovery began under high and low light conditions, respectively. Nitrate absorption did not increase until plants reestablished a positive carbon balance (Clement et al. 1978).

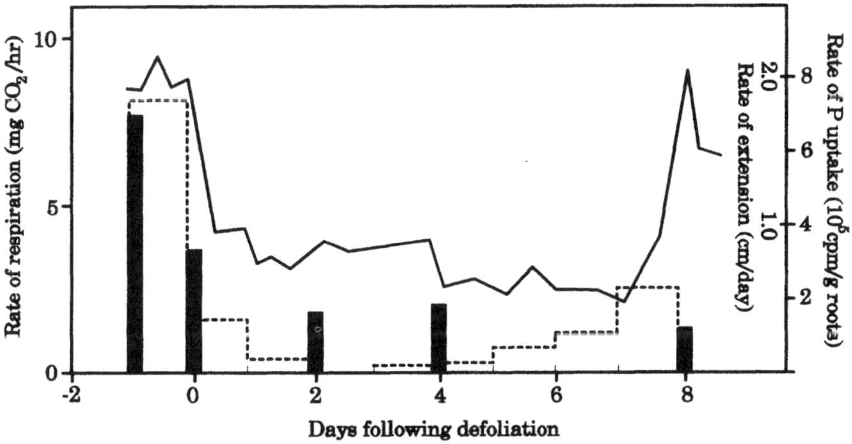

Figure 7. Rates of respiration (solid line), root length extension (broken line), and relative uptake of ^{32}P (solid bars) in orchardgrass plants prior to and for 8-days following defoliation to a height of 2.5 cm on day 0. Reductions in root growth and function result from a decrease in total carbon assimilation and a reduction in the relative proportion of carbon allocated to roots following defoliation. Plants were grown in solution culture to allow continuous physiological measurements (from Davidson and Milthorpe 1966).

Rapid reductions in root respiration and nutrient absorption following defoliation are positively related to the intensity of plant defoliation (Davidson and Milthorpe 1966, Thorgeirsson 1988). Similarly to defoliation, shading can also produce large decreases in root respiration and nutrient absorption (Massimino et al. 1981, Aslam and Huffaker 1982, Bloom and Caldwell 1988). Defoliation-induced reductions in root respiration and nutrient absorption result from a decrease in total carbon assimilation and a decrease in the relative proportion of carbon allocated to roots following defoliation.

Root function in plants from infertile habitats apparently differs from that of the more productive temperate species considered previously. Root growth, respiration and nutrient absorption are often maintained, or even increased, following defoliation (Chapin and Slack 1979). In addition to these short term responses, long-term increases in root respiration and nutrient absorption have been found in both tropical and tundra graminoids growing under nutrient-limited conditions

(Chapin and Slack 1979, Ruess et al. 1983). Energy requirements for root respiration and nutrient absorption were most likely met by carbohydrates translocated from remaining shoot tissues, or following rapid reestablishment of leaf area, because TNC concentrations in the roots were comparable to those measured in temperate perennial grasses (Davidson and Milthorpe 1966, Chapin and Slack 1979, Bloom and Caldwell 1988).

Accelerated rates of nutrient absorption have also been documented in several temperate, perennial grasses including western wheatgrass (Polley and Detling 1988) and blue grama (Jaramillo and Detling 1988) following defoliation. Mean nitrogen accumulation rate increased 200 and 122% in western wheatgrass and blue grama, respectively, even though defoliation reduced root growth in both species. A greater relative growth rate of shoots in defoliated plants may have increased shoot sinks thereby stimulating nitrogen accumulation (Polley and Detling 1988). In addition, growth may have been limited by nutrients other than carbon, in a manner comparable to that of plants in infertile habitats, thereby sustaining carbon allocation belowground to maintain root respiration and nutrient absorption.

Increased rates of root respiration and nutrient absorption in defoliated plants may function as compensatory mechanisms to offset nutrient losses to herbivores (Ruess 1984, Jaramillo and Detling 1988, Polley and Detling 1988). Increased rates of nutrient absorption may potentially restore the root:shoot equilibrium to defoliated plants and enhance growth and competitive ability. This compensatory mechanism may be especially effective in systems where grazing increases nitrogen availability by accelerating the rate of nutrient cycling (Polley and Detling 1988, Hik and Jefferies 1990). However, species-specific responses, the influence of environmental variables, and physiological mechanisms must be more clearly understood before compensatory nutrient absorption can be assumed to operate in a majority of grass species.

Shoot growth

Defoliation frequently decreases plant production because leaf removal reduces the potential for total carbon gain (e.g. Belsky 1986). However, the occurrence of compensatory physiological processes raises the possibility that growth of defoliated plants may approach or exceed that of undefoliated plants. The term "compensatory growth" has been used to describe plant growth responses ranging from a partial replacement of removed biomass to production exceeding that of ungrazed plants. Specific terms to clarify these growth responses include overcompensation which refers to instances where the cumulative total weight of defoliated plants, including biomass removed by defoliation, exceeds the cumulative total weight of undefoliated plants (Belsky 1986). Partial compensation defines instances where more growth is produced by defoliated plants than would be expected if no increase in growth rate had occurred. The grazing optimization hypothesis is based on the occurrence of compensatory growth and states that primary production increases above that of ungrazed plants with an increase in grazing intensity to an optimal level and then primary production decreases with greater grazing intensity (Dyer 1975, McNaughton 1976, 1979) (Fig. 8). Unfortunately, the general applicability of the grazing optimization hypothesis to vegetation management has not yet been resolved.

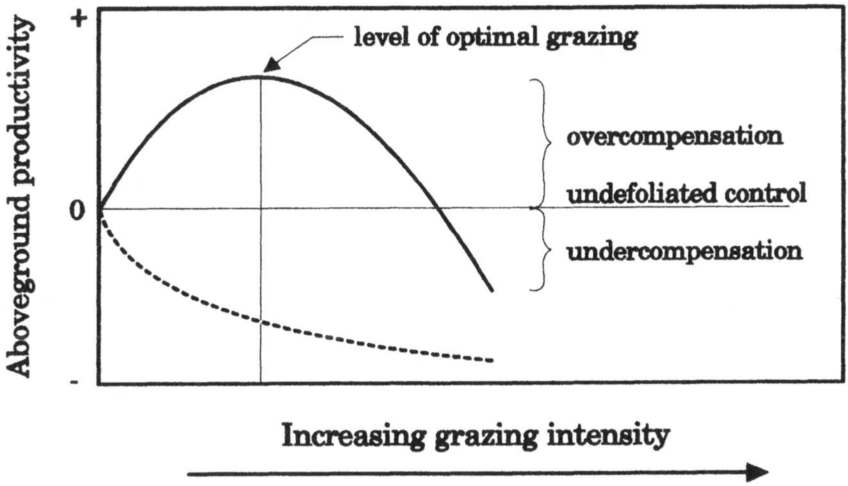

Figure 8. Illustration of the grazing optimization hypothesis proposing that aboveground primary productivity increases, above that of ungrazed plants, as grazing intensity increases to an optimal level and then decreases at greater grazing intensities (solid line). Values above those of undefoliated (control) plants indicate an increase in productivity (overcompensation) while values below those of undefoliated plants indicate a decrease in productivity (undercompensation) in response to grazing (from Belsky 1986). The alternative hypothesis proposes that grazing progressively suppresses aboveground productivity with increasing intensity (dashed line). A large number of potential production responses exist between these two extremes.

Much of the evidence in support of the grazing optimization hypothesis has been derived from containerized plant investigations conducted under highly experimental conditions (Belsky 1986). Overcompensation has been documented for *Kyllinga nervosa* Steud., an African C_4 sedge (McNaughton et al. 1983, Wallace et al. 1985), *Themeda triandra* (Oesterheld and McNaughton 1988) and *Sporobolus kentrophyllus* K. Shcum. (Georgiadis et al. 1989). Although these data substantiate the occurrence of overcompensation, they also demonstrate that specific growth conditions are required to induce this response. Plants with relatively low growth rates (i.e., unfavorable conditions for growth) that are provided with relatively long intervals between defoliation events possess the greatest potential for compensatory growth (Hilbert et al. 1981, Georgiadis et al. 1989).

Examples of overcompensation from field investigations are less numerous and are frequently less definitive than those from containerized plant investigations. Overcompensation by individual plants has been observed in scarlet gilia [*Ipomopsis aggregata* (Pursh) V. Grant], a herbaceous dicot from the Intermountain West (Paige and Whitham 1987), bitterbrush [*Purshia tridentata* (Pursh) DC.], a defoliation tolerant Intermountain shrub (Bilbrough and Richards, 1993), and crested wheatgrass, an Eurasian perennial grass (Olson and Richards 1988a). Examples of compensatory growth within plant communities include *Andropogon greenwayi* Napper grasslands in the Serengeti Plains of east-central Africa (McNaughton 1979), intertidal graminoid communities near Hudson Bay (Cargill and Jefferies 1984a and b), and a mixed prairie in north-central Texas (Heitschmidt et al. 1982). A greater diversity of communities require investigation to more clearly define the significance of compensatory growth to the production dynamics of grazed systems.

Numerous mechanisms have been proposed to explain the occurrence of compensatory growth. Intrinsic mechanisms, which are associated with herbivore-induced physiological processes, have received the greatest attention (McNaughton 1979, 1983) while extrinsic mechanisms, those involving herbivore-induced environmental modification, have been less thoroughly investigated, but may be of equal or greater importance. For example, accelerated rates of nutrient cycling appear to play an essential role in the occurrence of compensatory growth within plant communities (Ruess and McNaughton 1987, Hik and Jefferies 1990) thus verifying the assumption of Ellison (1960) that if vegetation benefits from herbivory it will occur as a result of ecosystem processes rather than individual plant function. Herbivory increases the rate of mineralization thereby increasing the rate at which nutrients become available for reabsorption within the system (Woodmansee et al. 1981, Floate 1981). However, intrinsic mechanisms must be the predominant explanation for compensatory growth in individual plant investigations, especially those in which defoliation is imposed by clipping.

Both intrinsic and extrinsic mechanisms presumably contribute to compensatory growth, but their relative contribution varies with species, defoliation regime, associated environmental variables and the ecological scale investigated. Compensatory growth in *Ipomopsis arizonica* (Greene) Wherry decreased in response to increasing interspecific competition, decreasing nutrient availability, and grazing during the later portion of the growing season (Maschinski and Whitham 1989). These data indicate that the potential occurrence of compensatory growth is not a constant species-specific response.

Currently, insufficient data are available to support the conclusion that overcompensation is a widely occurring phenomena in grassland ecosystems. Belsky (1986) reviewed 48 reports in the literature referencing aboveground production in response to grazing and found that 34 documented a decrease in production, 5 reported no change and 9 reported an increase in production. However, it is possible that plants increase their growth rate following defoliation, but only partially compensate for the total amount of biomass removed (Hilbert et al. 1981, Belsky 1986). Partial compensation implies that plant productivity is not suppressed in direct proportion to frequency and intensity of defoliation imposed (McNaughton 1983, 1985). Additional quantification of the magnitude, frequency and underlying mechanisms of compensatory growth is required to definitively assess its ecological significance in grazed systems.

Biotic interactions

Defoliation x competition interactions

Plants in field settings do not respond to grazing as isolated individuals, but against a background of intra- and inter-specific competition. The frequency and intensity of grazing between and among species affects these competitive interactions and directly influences growth following defoliation (Mueggler 1972, Banyikwa 1988). The rapid and substantial affect of defoliation on resource competition has clearly been illustrated in a perennial grass-shrub association. Big sagebrush [*Artemisia tridentata* ssp. *vaseyana* (Rydb.) Beetle] acquired 6 times as much labeled phosphorus from interspaces shared with defoliated (85% canopy removal) bluebunch wheatgrass or crested wheatgrass neighbors as it did from interspaces shared with undefoliated grass plants (Caldwell et al. 1987) (Fig. 9). Significant increases in phosphorus absorption by big sagebrush occurred within 2 weeks of grass defoliation indicating that belowground competition is rapidly modified by grazing.

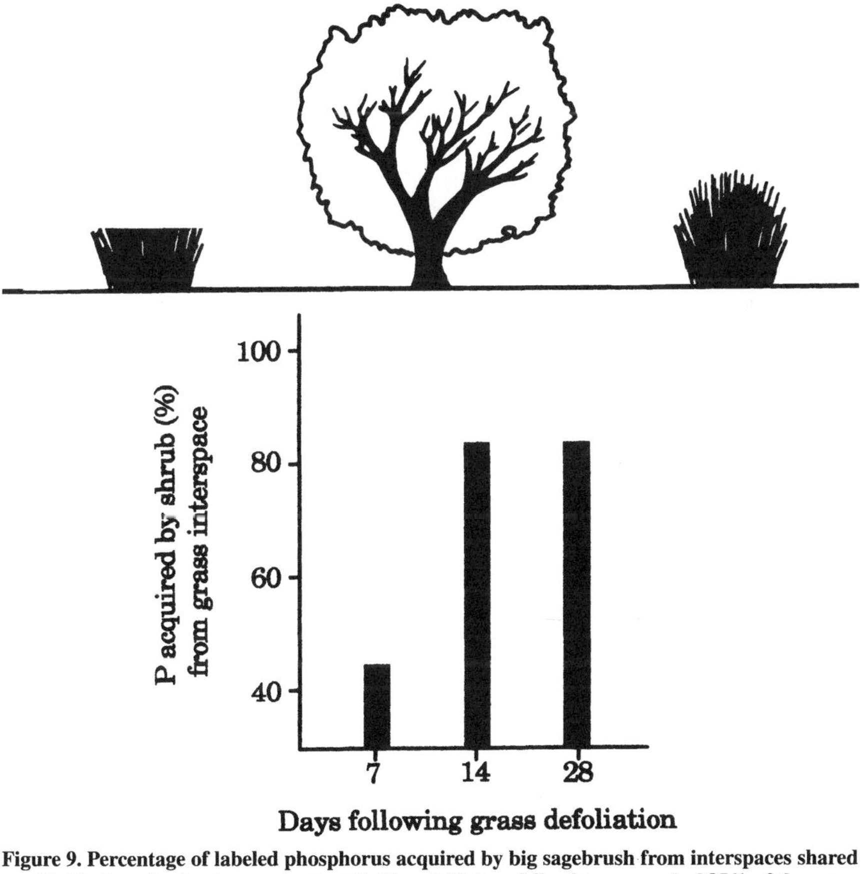

Figure 9. Percentage of labeled phosphorus acquired by big sagebrush from interspaces shared with bluebunch wheatgrass plants at 7, 14 and 28 days following removal of 85% of the grass canopy. Big sagebrush acquired as much as 6 times the amount of phosphorus from interspaces shared with defoliated bunchgrass competitors as it did from interspaces shared with undefoliated bunchgrass plants (from Caldwell et al. 1987).

The first, and most direct, mechanism by which herbivores alter competitive interactions involves selective plant utilization based on the relative display of avoidance mechanisms among species (Briske 1991). Species grazed severely are placed at a disadvantage when competing with associated species grazed less severely. Production of bluebunch wheatgrass plants subjected to 50% canopy removal just prior to culm elongation was equivalent to the production of undefoliated plants growing with full competition when competition was reduced by clipping the associated vegetation within a 90 cm radius to ground level (Mueggler

1972). When competition from associated vegetation was completely removed by tilling within a 90 cm radius, defoliated plants produced three times the biomass of undefoliated plants growing with full competition. These data clearly demonstrate that the ability of plants to respond to defoliation is not only determined by an inherent suite of morphological and physiological characteristics, but also by competition from associated species (Caldwell 1984). The inherent competitive ability of a species within the community is an important determinant of its persistence in a grazed system.

The second mechanism by which grazing may alter competitive interactions among plant species involves the differential ability of species to regrow following a similar intensity of defoliation based upon the possession of various tolerance mechanisms (Briske 1991). Species rapidly replacing photosynthetic surfaces gain a competitive advantage over associated species that grow more slowly following defoliation. Crested wheatgrass exhibits a 2-3 fold greater rate of leaf replacement than does bluebunch wheatgrass following an equivalent intensity of defoliation (Caldwell et al. 1981, Richards 1984, Richards and Caldwell 1985). This response is partially a function of the ability of crested wheatgrass to rapidly initiate a greater number of tillers and to allocate carbon to reestablish photosynthetic surfaces while temporarily decreasing allocation to the root system. Inequitable responses to defoliation between these two species can be attributed to the differential expression of tolerance mechanisms and inherent competitive ability because the similarity in plant chemistry and architecture minimizes the influence of avoidance mechanisms (Caldwell 1984).

Species grazed less severely (i.e., avoidance mechanisms), capable of regrowing more rapidly following defoliation (i.e., tolerance mechanisms), or possessing a combination of these two resistance components realize a competitive advantage within the community. These species, through the possession of a greater canopy area, are able to intercept greater amounts of solar energy, and assimilate greater amounts of carbon, further enhancing their competitive ability. Consequently, grazing resistant species are able to preempt resources that may have been utilized by associated grazing sensitive species prior to grazing (Mueggler 1972, Caldwell et al. 1987, Banyikwa 1988).

Caldwell (1984) has cautioned that many growth responses attributed to grazing resistance may actually be a consequence of the competition experienced by plants following defoliation. Although, inherent physiological and morphological mechanisms regulate plant growth following defoliation, differential resource acquisition among species may be of equal or greater significance (Mueggler 1972, Banyikwa 1988). Grazing resistance and competitive ability need to be more clearly distinguished to accurately evaluate the mechanisms of individual plant response to defoliation and to extrapolate these data to the population and community levels of grassland organization.

Grazing morphs

Long-term herbivory of several perennial grass populations is known to have selected against genotypes possessing an erect canopy architecture. Genotypic selection has been induced by several species of herbivores in approximately equal proportion to the severity and duration of herbivory (Scott and Whalley 1984). The plant genotypes remaining in the population were characterized by a large number of small tillers with reduced leaf numbers and blade areas (Alexander and Thompson 1982, Detling and Painter 1983, Jaramillo and Detling 1988) (Fig. 10). Although the concept of biotic selection by herbivores was introduced early in the 20th century (Gregor and Sansome 1926, Kemp 1937), its potential ecological significance has only recently been appreciated.

Figure 10. Individual western wheatgrass plants collected from a grazing exclosure (left) and a prairie dog colony (right) at Wind Cave National Park, South Dakota. Both plants were propagated from rhizome segments collected three years earlier and maintained in a greenhouse without defoliation. Both plants were defoliated to a height of 2 cm 9 weeks prior to the time the photo was taken (from Detling and Painter 1983).

These decumbent genotypes or "grazing morphs" possess greater grazing avoidance because a greater amount of meristematic and photosynthetic tissue is protected from removal by grazing animals to facilitate growth following defoliation (Carman and Briske 1985, Briske and Anderson 1992) (Fig. 10). It is generally assumed that grazing-induced selection primarily affects architectural attributes of plants, rather than physiological processes, including photosynthetic rate, transpiration rate and water-use efficiency (Detling and Painter 1983, Jaramillo and Detling 1988). However, root systems of grazing morphs in western wheatgrass populations had a 3-fold greater rate of nitrogen accumulation (Polley and Detling 1988) and grazing morphs of both western wheatgrass and blue grama were more productive and had less tiller mortality following defoliation than did populations with no history of grazing (Painter et al. 1989). Consequently, it must be assumed that both components of grazing resistance, avoidance and tolerance, can be affected by grazing-induced selection.

Grazing morphs are frequently less competitive than the more erect genotypes in a species population in the absence of herbivory (Eagles 1983, Painter et al. 1989). Contrastingly, genotypes characterized by fewer, larger tillers with greater leaf area are more competitive in environments with dense canopies (Grime 1977, Alexander and Thompson 1982). Correspondingly, intra-specific competition is greatest within populations composed of erect genotypes (Painter et al. 1989), but grazing is much more detrimental to the competitive ability of erect genotypes in comparison with grazing morphs because a greater portion of the canopy is removed.

The process by which grazing-induced selection occurs within species populations has not been definitively established. Although phenotypic changes occur in grazed plants, changes in plant stature and associated morphological characteristics are very likely genetically based (McNeilly 1984, Scott and Whalley 1984, Carman and Briske 1985, Painter et al. 1989). The observation that grazing morphs of western wheatgrass retained their decumbent stature in a controlled environment without defoliation for a period comparable to three growing seasons supports this viewpoint (Polley and Detling 1988). However, the involvement of developmental plasticity in the occurrence of grazing morphs can not be entirely ruled out. Several investigators have observed a loss or reduction in architectural variation between plants with contrasting grazing histories following the cessation of herbivory (Peterson 1962, Quinn and Miller 1967, Painter et al. 1989).

Although grazing-induced genotypic selection increases grazing avoidance, it also has the potential to simultaneously decrease productivity (Coppock et al. 1983, Detling 1988). Therefore, grasslands consisting of high proportions of grazing morphs within various species populations should be less productive than ungrazed or leniently grazed grasslands dominated by more erect morphs. However, grazing morphs may possess tolerance mechanisms which partially compensate for their diminished architectural stature and biomass. For example, grazing morphs have been documented to possess greater leaf N concentrations (Detling and Painter 1983), increased rates of nutrient absorption per unit root mass (Ruess et al. 1983, Polley and Detling 1988), and more rapid rates of tiller recruitment (Carman and Briske 1985, Briske and Anderson 1992). These compensatory mechanisms may potentially increase plant productivity and partially compensate for the reduced aboveground biomass associated with grazing morphs (Coppock et al. 1983, Detling 1988). However, our understanding of the extent to which species populations have been influenced by genotypic selection from domestic herbivores on a regional scale and the associated ecological consequences of this selection process is extremely limited.

Summary And Conclusions

An evaluation of both the traditional and contemporary literature addressing individual plant responses to defoliation indicates that the widely held concepts of carbohydrate reserves and apical dominance are in need of revision. Physiological concepts were applied to rangeland ecology and management early in the century to interpret and anticipate the responses of individual plants to grazing, even though a thorough understanding of whole plant function had not yet been achieved. Consequently, concept reevaluation and refinement are a natural and necessary consequence associated with the progressive development and application of scientific information to vegetation management.

The concept of reserve carbohydrates is valid from a physiologically perspective, but was over extended as a managerial criteria. Research information developed during the past two decades indicates that carbohydrate concentrations or pools are not effective indicators of plant survival and growth following defoliation because they are much more limited than previously assumed and the total pool size of TNC has not been directly correlated with plant growth. Similarly, the direct inhibition hypothesis of apical dominance has been adopted as the sole mechanism explaining tiller development in grasses. However, physiological insight into apical dominance has advanced substantially and the traditional interpretation does not adequately explain many documented tillering responses of direct relevance to rangeland ecology and management. Even though the contemporary concept of apical dominance is not sufficiently understood to determine the specific environmental variables and physiological processes regulating tillering, the concept should no longer be viewed as the direct inhibition of lateral buds by a single growth hormone produced in the apical meristem. These observations substantiate the need to periodically reevaluate the physiological basis for vegetation management within the context of contemporary physiological information and documented plant responses.

The perspective advocating that grazing may potentially benefit plant growth and function has substantially altered our approach to plant-animal interactions research. Several physiological and morphological mechanisms potentially capable of increasing plant growth following defoliation have been identified. Compensatory photosynthesis, resource allocation, nutrient absorption and shoot growth have all been documented in a variety of species by independent investigators. However, the occurrence, magnitude and significance of these compensatory processes to individual plant and community productivity is much less clearly understood. These compensatory mechanisms may frequently prevent plant growth from being suppressed in direct proportion to the frequency and intensity of defoliation, but only infrequently increase total growth beyond that of undefoliated plants. Our understanding of the mechanisms and magnitude of compensatory growth is incomplete suggesting that the application of this concept to vegetation management is currently unwarranted.

The biotic processes of competition and grazing induced genotypic selection have long been recognized as components of grazed systems. However, the ecological significance of these processes to plant growth has only recently been fully appreciated. Competitive ability may constrain growth to an equal or greater extent than the direct effects of defoliation. However, the relative contribution of these two processes are largely unknown, undoubtedly vary in response to plant and environmental variables, and require additional investigation. Similarly, it is well documented that both native and domestic herbivores can induce genotypic selection within species populations within a relatively short period, but the cumulative influence that grazing has had on the genotypic composition and potential productivity of rangeland plants remains largely unexplored.

The contemporary perspective on individual plant responses to grazing indicate that individual plant research is in itself insufficient to develop a complete information base for vegetation management. Individual plant research is an integral component of any investigation designed to evaluate vegetation responses to grazing, because a mechanistic understanding of community and landscape responses requires that plant and population responses must also be at least partially understood (Brown and Allen 1989, Allen and Hoekstra 1990). However, individual plant research must eventually be conducted and interpreted within the context of associated abiotic and biotic environmental variables because of the ability of these variables to constrain or modify individual plant responses. Individual plant research designed to incorporate the effects of biotic and abiotic variables provides an opportunity to determine the relative contribution of these variables on plant growth following defoliation. Individual plant data acquired in this manner will be much more appropriate for extrapolation to the plant community where vegetation management is most frequently conducted.

Literature Cited

Alexander, K.I. and K. Thompson. 1982. The effect of clipping frequency on the competitive interaction between two perennial grasses. Oecologia 53:251-254.

Allen, T.F.H. and T.W. Hoekstra. 1990. The confusion between scale-defined levels and conventional levels of organization in ecology. J. Veg. Sci. 1:5-12.

Archer, S.R. and L.L. Tieszen. 1986. Plant response to defoliation: hierarchical considerations. p. 45-59: *In*: O. Gudmundsson (ed.) Grazing research at northern latitudes. Plenum Publ. Corp.

Aslam, M. and R.C. Huffaker. 1982. In Vivo nitrate reduction in roots and shoots of barley(*Hordeum vulgare* L.) seedlings in light and darkness. Plant Physiol. 70:1009-1013.

Atkinson, C.J. 1986. The effect of clipping on net photosynthesis and dark respiration rates of plants from an upland grassland, with reference to carbon partitioning in *Festuca ovina*. Ann. Bot. 58:61-72.

Atkinson, C.J. and J.F. Farrar. 1983. Allocation of photosynthetically-fixed carbon in *Festuca ovina* L. and *Nardus stricta* L. New Phytol. 95:519-531.

Banyikwa, F.F. 1988. The growth response of two East African perennial grasses to defoliation, nitrogen fertilizer and competition. Oikos 51:25-30.

Bassman, J.H. and D.I. Dickmann. 1985. Effects of defoliation in the developing leaf zone on young populus x-euramericana plants. II. Distribution of 14-C-photosynthate after defoliation. For. Sci. 31:358-366.

Belsky, A.J. 1986. Does herbivory benefit plants? A review of the evidence. Am. Nat. 127:870-892.

Bilbrough, C.J. and J.H. Richards. 1993. Growth of sagebrush and bitterbrush following simulated winter browsing: mechanisms of tolerance. Ecology 74:481-492.

Bloom, A.J. and R.M. Caldwell. 1988. Root excision decreases nutrient absorption and gas fluxes. Plant Physiol. 87:794-796.

Briske, D.D. and V.J. Anderson. 1992. Competitive ability of the bunchgrass *Schizachyrium scoparium* as affected by grazing history and defoliation. Vegetatio 103:41-49.

Briske, D.D. 1991. Developmental morphology and physiology of grasses. p. 85-108. *In*: R.K. Heitschmidt and J.W. Stuth (eds.) Grazing management: an ecological perspective. Timber Press, Portland, Oregon.

Brown, B.J. and T.F.H. Allen. 1989. The importance of scale in evaluating herbivory impacts. Oikos 54:189-194.

Bucher, H.P., F. Machler, and J. Noesberger. 1987. Sink control of assimilate partitioning in meadow fescue (*Festuca pratensis* Huds.). J. Plant Physiol. 129:469-477.

Bunting, A.H. and D.S.H. Drennan. 1966. Some aspects of the morphology and physiology of cereals in the vegetative phase. *In*: F.L. Milthorpe and I.D. Ivins (eds.) The growth of cereals and grasses. Butterworths Sci. Publs., London.

Busso, C.A., J.H. Richards and N.J. Chatterton. 1990. Nonstructural carbohydrates and spring regrowth of two cool-season grasses: Interaction of drought and clipping. J. Range Manage. 43:336-343.

Butler, J.L., and D.D. Briske. 1988. Population structure and tiller demography of the bunchgrass *Schizachyrium scoparium* in response to herbivory. Oikos 51:306-312.

Caldwell, M.M. 1984. Plant requirements for prudent grazing. p. 117-152: *In*: Developing strategies for rangeland management. Westview Press, Boulder, Colorado.

Caldwell, M.M., J.H. Richards, D.A. Johnson, R.S. Nowak, and R.S. Dzurec. 1981. Coping with herbivory: photosynthetic capacity and resource allocation in two semiarid *Agropyron* bunchgrasses. Oecologia 50:14-24.

Caldwell, M.M., J.H. Richards, J.H. Manwaring, and D.M. Eissenstat. 1987. Rapid shifts in phosphate acquisition show direct competition between neighbouring plants. Nature. 327:615-616.

Cargill, S.M. and R.L. Jefferies. 1984a. Nutrient limitation of primary production in a sub-arctic salt marsh. J. Appl. Ecol. 21:657-668.

Cargill, S.M. and R.L. Jefferies. 1984b. The effects of grazing by lesser snow geese on the vegetation of a sub-arctic salt marsh. J. Appl. Ecol. 21:669-686.

Carman, J.G. and D.D. Briske. 1985. Morphologic and allozymic variation between long-term grazed and non-grazed populations of the bunchgrass *Schizachyrium scoparium* var. *frequens*. Oecologia 66:332-337.

Carmi, A. and D. Koller. 1979. Regulation of photosynthetic activity in the primary leaves of bean (*Phaseolus vulgaris* L.) by materials moving in the water-conducting system. Plant Physiol. 64:285-288

Casal, J.J., R.A. Sanchez, and V.A. Deregibus. 1987. Tillering responses of *Lolium multiflorum* plants to changes of red/far-red ratio typical of sparse canopies. J. Exp. Bot. 38:1432-1439.

Chapin III, S.F. and M. Slack. 1979. Effect of defoliation upon root growth, phosphate absorption and respiration in nutrient-limited tundra graminoids. Oecologia 42:67-79.

Clement, C.R., M.J. Hopper, L.H.P. Jones, and E.L. Leafe. 1978. The uptake of nitrate by *Lolium perenne* from flowing nutrient solution. II. Effects of light, defoliation, and relationship to CO_2 flux. J. Exp. Bot. 29:1173-1183.

Cline, M.G. 1991. Apical dominance. Bot. Rev. 57:318-358.

Coppock, D.L., J.K. Detling, J.E. Ellis, and M.I. Dyer. 1983. Plant-herbivore interactions in a North American mixed-grass prairie. I. Effects of black-tailed prairie dogs on intraseasonal aboveground plant biomass and nutrient dynamics and plant species diversity. Oecologia 56:1-9.

Davidson, J.L. and F.L. Milthorpe. 1966. Leaf growth in *Dactylis glomerata* following defoliation. Ann. Bot. 30:173-184.

Davidson J.L. and F.L. Milthorpe. 1965. Carbohydrate reserves in the regrowth of cocksfoot (*Dactylis glomerata* L.) J. Br. Grassld. Soc. 20:15-18.

Deregibus, V.A., M.J. Trlica, and D.A. Jameson. 1982. Organic reserves in herbage plants: Their relationship to grassland management. p. 315-344. *In*: M.J. Recheigl (ed.), Handbook of agricultural productivity. Vol. I. Plant productivity. CRC Press, Inc.

Deregibus, V.A., R.A. Sanchez, J.J. Casal, and M.J. Trlica. 1985. Tillering responses to enrichment of red light beneath the canopy in a humid natural grassland. J. Appl. Ecol. 22:199-206.

Detling, J.K. 1988. Grasslands and savannas: Regulation of energy flow and nutrient cycling by herbivores. *In*: L.R. Pomeroy and J.J. Alberts (eds.) Concepts of ecosystem ecology—A comparative view. Springer-Verlag New York, Inc.

Detling, J.K. and E.L. Painter. 1983. Defoliation responses of western wheatgrass populations with diverse histories of prairie dog grazing. Oecologia 57:65-71.

Detling, J.K., M.I. Dyer, and D.T. Winn. 1979a. Effect of simulated grasshopper grazing on CO_2 exchange rates of western wheatgrass leaves. J. Econ. Ent. 72:403-406.

Detling, J.K., M.I. Dyer, and D.T. Winn. 1979b. Net photosynthesis, root respiration, and regrowth of *Bouteloua gracilis* following simulated grazing. Oecologia 41:127-134.

Dyer, M.I. 1975. The effects of red winged blackbirds (*Agelaius phoeniceus* L.) on biomass production of corn grains (*Zea mays* L.) J. Appl. Ecol. 12:719-726.

Eagles, C.F. 1983. Relationship between competitive ability and yielding ability in mixtures and monocultures of populations of *Dactylis glomerata* L. Grass For. Sci. 38:21-24.

Ellison, L. 1960. Influence of grazing on plant succession of rangelands. Bot. Rev. 26:1-78.

Ferree, D.C. and F.R. Hall. 1981. Influence of physical stress on photosynthesis and transpiration of apple leaves. J. Amer. Soc. Hort. Sci. 106:348-351.

Floate, M.J.S. 1981. Effects of grazing by large herbivores on nitrogen cycling in agricultural ecosystems. *In*: F.E. Clark and T. Rosswall (eds.) Terrestrial nitrogen cycles. Ecol. Bull. (Stockholm) 33:585-601.

Geiger, D.R. 1976. Effects of translocation and assimilate demand on photosynthesis. Can. J. Bot. 54:2337-2345.

Gieger, D.R. and B.R. Fondy. 1985. Responses of export and partitioning to internal and external factors. p. 177-194: *In*: B. Jeffcoat, A.F. Hawkins and A.D. Stead (eds.) Regulation of sources and sinks in crop plants. British Plant Growth Regulator Group Monograph 12, Bristol, UK.

Georgiadis, N.J., R.W. Ruess, S.J. McNaughton, and D. Western. 1989. Ecological conditions that determine when grazing stimulates grass production. Oecologia 81:316-322.

Gifford, R.M. and C. Marshall. 1973. Photosynthesis and assimilate distribution in *Lolium multiflorum* Lam. following differential tiller defoliation. Aust. J. Biol. Sci. 26:517-526.

Gold, W.G. and M.M. Caldwell. 1989. The effects of the spatial pattern of defoliation on regrowth of a tussock grass. II. Canopy gas exchange. Oecologia 81:437-442.

Gold, W.G. and M.M. Caldwell. 1990. The effects of the spatial pattern of defoliation on regrowth of a tussock grass. III. Photosynthesis, canopy structure and light interception. Oecologia 82:12-17.

Gregor, J.W. and F.W. Sansome. 1926. Experiments on the genetics of wild populations. I. Grasses. J. Genet. 17:349-364.

Gregory, F.G. and J.A. Veale. 1957. A reassessment of the problem of apical dominance. Symp. Soc. Expt. Biol. 11:1-20.

Grime, P.J. 1977. Plant strategies and vegetation processes. John Wiley and Sons, Inc., New York.

Harrison, M.A. and P.B. Kaufman. 1982. Does ethylene play a role in the release of lateral buds (tillers) from apical dominance in oats? Plant Physiol. 70:811-814.

Heichel, G.H. and N.C. Turner. 1983. CO_2 assimilation of primary and regrowth foliage of red maple (*Acer rubrum* L.) and red oak (*Quercus rubra* L.): response to defoliation. Oecologia 57:14-19.

Heitschmidt, R.K., D.L. Price, R.A. Gordon, and J.R. Frasure. 1982. Short duration grazing at the Texas Experimental Ranch: effects on above ground primary production and seasonal growth dynamics. J. Range Manage. 35:367-372.

Hik D.S. and R.L. Jefferies. 1990. Increases in the net above-ground primary production of a salt-marsh forage grass: a test of the predictions of the herbivore-optimization model. J. Ecol. 78:180-195.

Hilbert, D.W., D.M. Swift, J.K Detling, and M.I. Dyer. 1981. Relative growth rates and the grazing optimization hypothesis. Oecologia 51:14-18.

Hillman, J.R. 1984. Apical dominance. p. 127-148: *In*: M.B. Wilkins (ed.) Advanced plant physiology. Pitman Publ. London.

Hodgkinson, K.C. 1974. Influence of partial defoliation on photosynthesis, photorespiration and transpiration by lucerne leaves of different ages. Aust. J. Plant Physiol. 1:561-578.

Hodgkinson, K.C., M.M. Ludlow, J.J. Mott, and Z. Baruch. 1989. Comparative responses of the savanna grasses *Cenchrus ciliaris* and *Themeda triandra* to defoliation. Oecologia 79:45-52.

Hodgkinson, K.C., N.G. Smith, and G.E. Miles. 1972. The photosynthetic capacity of stubble leaves and their contribution to growth of the lucerne plant after high level cutting. Aust. J. Agri. Res. 23:225-238.

Jameson, D.A. 1963. Responses of individual plants to harvesting. Bot. Rev. 29:532-594.

Jaramillo, V.J. and J.K. Detling. 1988. Grazing history, defoliation, and competition: effects on shortgrass production and nitrogen accumulation. Ecology 69:1599-1608.

Jenkins, G.I. and H.W. Woolhouse. 1971. Photosynthetic electron transport during senescence of the primary leaves of *Phaseolus vulgaris* L. I. Non-cyclic electron transport. J. Exp. Bot. 32:467-478.

Jonsdottir, I.S. and T.V. Callaghan. 1989. Localized defoliation stress and the movement of ^{14}C-photoassimilates between tillers of *Carex bigelowii*. Oikos 54:211-219.

Kasperbauer, M.J. and D.L. Karlen. 1986. Light-mediated bioregulation of tillering and photosynthate partitioning in wheat. Physiol. Plant. 66:159-163.

Kemp, W.B. 1937. Natural selection within plant species as exemplified in a permanent pasture. J. Hered. 28:329-333.

Leopold, A.C. 1949. The control of tillering in grasses by auxin. Amer. J. Bot. 36:437-440.

Ludlow, M.M. and D.A. Charles-Edwards. 1980. Analysis of regrowth of a tropical grass/legume sward subjected to different frequencies and intensities of defoliation. Aust. J. Agric. Res. 31:673-692.

Ludlow, M.M. and J.J. Mott. 1988. Photosynthetic capacities of lamina and sheaths. pp. 47 *In*: CSIRO Division of tropical crops and pastures annual report 1987-88, CSIRO, Brisbane, Australia.

Marshall, C. and G.R. Sagar. 1965. The influence of defoliation on the distribution of assimilates in *Lolium multiflorum* Lam. Ann. Bot. 29:365-370.

Martin, G.C. 1987. Apical dominance. HortScience 22:824-833.

Maschinski, J. and T. G. Whitham. 1989. The continuum of plant responses to herbivory: The influence of plant association, nutrient availability, and timing. Am. Nat. 134:1-19.

Massimino, D., M. Andre, C. Richaud, J. Massimino and J. Vivoli. 1981. The effect of a day at low irradiance of a maize crop. I. Root respiration and uptake of N, P and K. Physiol. Plant. 51:150-155.

Matches, A.G. 1966. Influence of intact tillers and height of stubble on growth responses of tall fescue (*Festuca arundinacea* Schreb.). Crop Sci. 6:484-487.

Mattheis P.J., L.L. Tieszen, and M.C. Lewis. 1976. Responses of *Dupontia fischeri* to simulated lemming grazing in an Alaskan arctic tundra. Ann. Bot. 40:179- 197

May, L.H. 1960. The utilization of carbohydrate reserves in pasture plants after defoliation. Herb. Abst. 30:239-245.

McNaughton, S.J. 1985. Ecology of grazing ecosystem: the Serengeti. Ecol. Monogr. 55:259-294.

McNaughton, S.J., L.L. Wallace, and M.B. Coughenour. 1983. Plant adaptation in an ecosystem context: effects of defoliation, nitrogen, and water on growth of an African C_4 sedge. Ecology 64:307-318.

McNaughton, S.J. 1983. Compensatory plant growth as a response to herbivory. Oikos 40:329-336.

McNaughton, S.J. 1979. Grazing as an optimization process: grass-ungulate relationships in the Serengeti. Am. Nat. 113:691-703.

McNaughton, S.J. 1976. Serengeti migratory wildebeest: facilitation of energy flow by grazing. Science 191:92-94.

McNeilly, T. 1984. Ecotypic differentiation in *Poa annua*: within population variation in response to competition and cutting. New Phytol. 96:307-316.

Mueggler, W.F. 1972. Influence of competition on the response of bluebunch wheatgrass to clipping. J. Range Manage. 25:88-92.

Mueller, R.J. and J.H. Richards. 1986. Morphological analysis of tillering in *Agropyron spicatum* and *Agropyron desertorum*. Ann. Bot. 58:911-921.

Murphy, J.S. and D.D. Briske. 1992. Regulation of tillering by apical dominance: Chronology, interpretive value, and current perspectives. J. Range Manage. 45: 419-430.

Neales, T.F., K.J. Treharne, and P.F. Wareing. 1971. A relationship between net photosynthesis, diffusive resistance, and carboxylating enzyme activity in bean leaves. p. 89-97. *In*: M.D. Hatch, C.B. Osmond, and R.O. Slatyer (eds.) Photosynthesis and photorespiration. New York:Wiley-Interscience.

Nojima, H., S. Fuketa, A. Isoda, and Y. Takasaki. 1989. Levels of abscisic acid in lateral buds during regrowth of *Sorghum bicolor* M. p. 505-506: *In*: Proc. XVI Int. Grassld. Cong., Nice, France.

Nowak, R.S. and M.M. Caldwell. 1984. A test of compensatory photosynthesis in the field: implications for herbivory tolerance. Oecologia 61:311-318.

Oesterheld M. and S.J. McNaughton. 1988. Intraspecific variation in the response of *Themeda triandra* to defoliation: the effect of time of recovery and growth rates on compensatory growth. Oecologia 77:181-186.

Olson, B.E. and J.H. Richards. 1988a. Tussock regrowth after grazing: intercalary meristem and axillary bud activity of *Agropyron desertorum*. Oikos 51:374-382.

Olson, B.E. and J.H. Richards. 1988b. Annual replacement of the tillers of *Agropyron desertorum* following grazing. Oecologia 76:1-6.

Paige, K.N. and T.G. Whitham. 1987. Overcompensation in response to mammalian herbivory: the advantage of being eaten. Am. Nat. 129:407-416.

Painter, E.L. and J.K Detling. 1981. Effects of defoliation on net photosynthesis and regrowth of western wheatgrass. J. Range Manage. 34:68-71.

Painter, E.L., J.K. Detling, and D.A. Steingraeber. 1989. Grazing history, defoliation, and frequency-dependent competition: effects on two North American grasses. Amer. J. Bot. 76:1368-1379.

Peterson, R.A. 1962. Factors affecting resistance to heavy grazing in needle-and-thread grass. J. Range Manage. 15:183-189.

Phillips, I.D.J. 1975. Apical dominance. Ann. Rev. Plant Physiol. 26:341-367.

Polley, H.W. and J.K. Detling. 1988. Herbivory tolerance of *Agropyron smithii* populations with different grazing histories. Oecologia 77:261-267.

Quinn, J.A. and R.V. Miller. 1967. A biotic selection study utilizing *Muhlenbergia montana*. Bull Torrey Bot. Club. 94:423-432.

Richards, J.H., R.J. Mueller, and J.J. Mott. 1988. Tillering in tussock grasses in relation to defoliation and apical bud removal. Ann. Bot. 62:173-179.

Richards, J.H. 1986. Plant response to grazing: the role of photosynthetic capacity and stored carbon reserves. p. 428-430. *In*: P.J. Joss, P.W. Lynch, and O.B. Williams (eds.) Rangelands: a resource under siege. Cambridge Univ. Press.

Richards, J.H. and M.M. Caldwell. 1985. Soluble carbohydrates, concurrent photosynthesis and efficiency in regrowth following defoliation: A field study with *Agropyron* species. J. Appl. Ecol. 22:907-920.

Richards, J.H. 1984. Root growth response to defoliation in two *Agropyron* bunchgrasses: field observations with an improved root periscope. Oecologia 64:21-25.

Ruess, R.W. 1984. Nutrient movement and grazing: experimental effects of clipping and nitrogen source on nutrient uptake in Kyllinga nervosa. Oikos 43:183-188.

Ruess, R.W. and S.J. McNaughton. 1987. Grazing and the dynamics of nutrient and energy regulated microbial processes in the Serengeti grasslands. Oikos 49:101-110.

Ruess, R.W., S.J. McNaughton, and M.B. Coughenour. 1983. The effects of clipping, nitrogen source and nitrogen concentration on the growth responses and nitrogen uptake of an east african sedge. Oecologia 59:253-261.

Ryle, G.J. and C.E. Powell. 1975. Defoliation and regrowth in the graminaceous plant: the role of current assimilate. Ann. Bot. 39:297-310.

Sambo, E.Y. 1983. Leaf extension rate in temperate pasture grasses in relation to assimilate pool in the extension zone. J. Exp. Bot. 34:1281-1290.

Sampson, A.W. 1952. Range management principles and practices. John Wiley and Sons, Inc., New York.

Satoh, M., P.E. Kriedemann, and B.R. Loveys. 1977. Changes in photosynthetic activity and related processes following decapitation in mulberry trees. Physiol. Plant. 41:203-210.

Scott, A.W. and R.D.B. Whalley. 1984. The influence of intensive sheep grazing on genotypic differentiation in *Danthonia linkii. D. richardsonii* and *D. racemosa* on the New England Tablelands. Aust. J. Ecol. 9:419-429.

Simon, J.C. and G. Lemaire. 1987. Tillering and leaf area index in grasses in the vegetative phase. Grass Forage Sci. 72:373-380.

Smith, D. 1972. Carbohydrate reserves of grasses. *In*: V.B. Youngner and C.M. McKell (eds.) The biology and utilization of grasses. Academic Press, New York.

Smith, D. 1981. Removing and analyzing total nonstructural carbohydrates from plant tissue. University of Wisconsin, College of Agriculture and Life Sciences Publication, R-2107. University of Wisconsin, Madison.

Stout, D.G., J. Hall, B. Brooke, and A. McLean. 1981. Influence of successive years of simulated grazing (clipping) on pinegrass growth. Can. J. Plant Sci. 61:939-943.

Thimann, K.V. and F. Skoog. 1933. Studies on the growth hormone of plants. 3. The inhibiting action of the growth substance on bud development. Proc. Nat. Acad. Sci. USA 19:714-716.

Thorgeirsson, H. 1988. The modeling and measurement of respiratory carbon use and net carbon gain of two *Agropyron* bunchgrasses. Ph.D. dissertation, Utah State University, Logan, Utah.

Volenec, J.J., and C.J. Nelson. 1984. Carbohydrate metabolism in leaf meristems of tall fescue. II. Relationship to leaf elongation rates modified by nitrogen fertilization. Plant Physiol. 74:595-600.

von Caemmerer, S. and G.D. Farquhar. 1984. Effects of partial defoliation, changes in irradiance during growth, short-term water stress and growth at enhanced p(CO_2) on the photosynthetic capacity of leaves of *Phaseolus vulgaris* L. Planta 160:320-329.

Wallace, L.L., S.J. McNaughton, and M.B. Coughenour. 1985. Effects of clipping and four levels of nitrogen on the gas exchange, growth, and productivity of two east African graminoids. Amer. J. Bot. 72:222-230.

Wallace, L.L., S.J. McNaughton, and M.B. Coughenour. 1984. Compensatory photosynthetic responses of three African graminoids to different fertilization, watering, and clipping regimes. Bot. Gaz. 145:151-156.

Wareing, P.F., M.M. Khalifa, and K.J. Treharne. 1968. Rate-limiting processes in photosynthesis at saturating light intensities. Nature 220:453-457.

Watson, V.H. and C.Y. Ward. 1970. Influence of intact tillers and height of cut on regrowth and carbohydrate reserves of Dallisgrass (*Paspalum dilatatum* Poir.) Crop Sci. 10:474-476.

Weinmann, H. 1948. Underground development and reserves of grasses. J. Brit. Grassld. Soc. 3:115-140.

Welker, J.M., D.D. Briske, and R.W. Weaver. 1987. Nitrogen-15 partitioning within a three generation tiller sequence of the bunchgrass *Schizachyrium scoparium*: response to selective defoliation. Oecologia 24:330-334.

Welker, J.M., E.J. Rykiel, Jr., D.D. Briske, and J.D. Goeschl. 1985. Carbon import among vegetative tillers within two bunchgrasses: assessment with carbon-11 labeling. Oecologia 67:209-212.

White, L.M. 1973. Carbohydrate reserves of grasses: A review. J. Range Manage. 26:13-18.

Williams, R.F. and R.H.M. Langer. 1975. Growth and development of the wheat tiller. II. The dynamics of tiller growth. Aust. J. Bot. 23:745-759.

Woledge, J. 1977. The effects of shading and cutting treatments on the photosynthetic rates of ryegrass leaves. Ann. Bot. 41:1279-1286.

Woodmansee, R.G., I. Vallis, and J.J. Mott. 1981. Grassland nitrogen. *In*: F.E. Clark and T. Rosswall (eds.) Terrestrial nitrogon cycles. Ecol. Bull. 33:443-462.

Yamashita, T. and A. Fujino. 1986. Effects of pruning of young and old shoots on ribulose bisphosphate carboxylase and other constituents in leaves of the mulberry tree (*Morus alba* L.). J. Exp. Bot. 37:1836-1841.

Zarrough, K.M., C.J. Nelson, and D.A. Sleper. 1984. Interrelationships between rates of leaf appearance and tillering in selected tall fescue populations. Crop Sci. 24:565-569.

Ecological Implications of Livestock Grazing

Rex D. Pieper

Abstract

Herbivory is an important and nearly universal ecosystem process. In the western U. S. domestic livestock were introduced into some rangeland ecosystems where large ungulates were not abundant before settlement by European man. Grazing by introduced livestock probably has had much larger impacts on vegetation of the Intermountain Region, Great Basin, and the Southwest, than the Great Plains where livestock replaced bison as the dominant herbivore. Destructive livestock grazing in the West occurred in the late 1800s and early 1900s, but was largely arrested after WWII when more intensive management was applied to most rangelands. Unfortunately, excessive grazing still occurs in some riparian and upland situations. Much early range management was based on the ideas of plant succession developed by F. E. Clements. However, recently some of these ideas have been modified and are being incorporated into dynamic models. Removing livestock from rangelands is unlikely to return these ecosystems to their pristine conditions. Many other changes including climatic shifts, increases in woody plant species, reduction of fire frequency and intensity, introduction of alien plant species, and other human activities have occurred, resulting in undetermined impacts on the native vegetation. Livestock grazing has played a role in reducing the amount of fuel for wildfires, altering nutrient distribution, acting to create patchiness at landscape levels in the environment for many animal species, and disrupting cryptogamic crusts. Nevertheless, domestic livestock grazing at conservative levels appears to be sustainable, even on sensitive western rangelands.

Introduction

Herbivory is an important and nearly universal ecosystem process. Herbivores are the link between primary producers (green plants) and carnivores. They must obtain all their nutritional requirements from the plants they consume and, in turn, have a direct impact on these plants. Herbivores also serve as the main food sources for carnivores. Consequently, some workers view herbivores as regulators, occupying a central position in ecosystems (Dyer et al. 1982). The importance of the herbivory process is evidenced by the increased interest in research concerning herbivory of all kinds. For example, as early as 1964, the British Ecological Society published a symposium on grazing, and nearly half the volume was devoted to grazing in marine environments (Crisp 1964). Crawley (1983) has written a book on herbivory with little mention of livestock grazing.

Those interested in livestock grazing should recognize there are now many more players in the herbivory game. These range from insect herbivory to herbivory in the oceans just mentioned. This expanded thrust is beneficial for those of us in the livestock grazing area because it brings new and innovative approaches to our understanding of the grazing process. We need to take advantage of knowledge derived from these other approaches to improve our understanding of the role of livestock as herbivores in range ecosystems.

Livestock are often viewed as if they are not a part of rangeland ecosystems, but rather some extrinsic factor. While it is true that livestock are different from native herbivores, they still are herbivores and can serve as integral components of rangeland ecosystems. Some of the questions considered in this paper concern the role of livestock in the deterioration of these ecosystems, and the possibility of their restoration through manipulation of livestock grazing. The main objective of this paper is to consider the ecological role of domestic livestock in rangeland ecosystems of the western United States.

Rangeland Deterioration

The common perception that more livestock on an area than there is forage to support them leads to range deterioration is the heart of the ecological argument favoring removing livestock from public rangelands. Many ecologists and plant physiologists agree that herbivory can alter plant communities (Caldwell 1984, Pieper 1984). The problem becomes one of determining how much change can be accepted before deterioration is judged to have occurred. However, this degree of change is often tempered by one's background and outlook. To some, one cow pie per section may be too many while others may view a rangeland grazed uniformly by cattle as the optimum condition. Most would agree that too many livestock grazing on rangeland for too long a period is not desirable.

Vegetational changes resulting from excessive herbivory have been depicted as a series of changes as shown in Figure 1. This diagram illustrates that excessive grazing pressure stresses the most palatable plants by removing too much photosynthetic tissue. The competitive advantage is then gained by less palatable species that increases their influence in the plant community. Often the less palatable species is shrubby, such as honey mesquite (*Prosopis glandulosa)* in the Southwest and big sagebrush (*Artemisia tridentata)* in the Intermountain Region and the northern Great Basin. As plant cover is decreased through heavy grazing, soil loss is accelerated and the changes result in a downward spiral (Strange 1980). The degree of grazing pressure needed to initiate and continue these kinds of change differs among different situations. Similar kinds of change were depicted by a geographer (Vale 1982) (Fig. 2.). This model depicts two conditions: pre-grazed vegetation and grazed vegetation with intermediate stages resulting from cyclical changes. If livestock grazing is removed, a new equilibrium is formed which may not be similar to either the pre-grazed or the grazed vegetation. The diagrams pre-

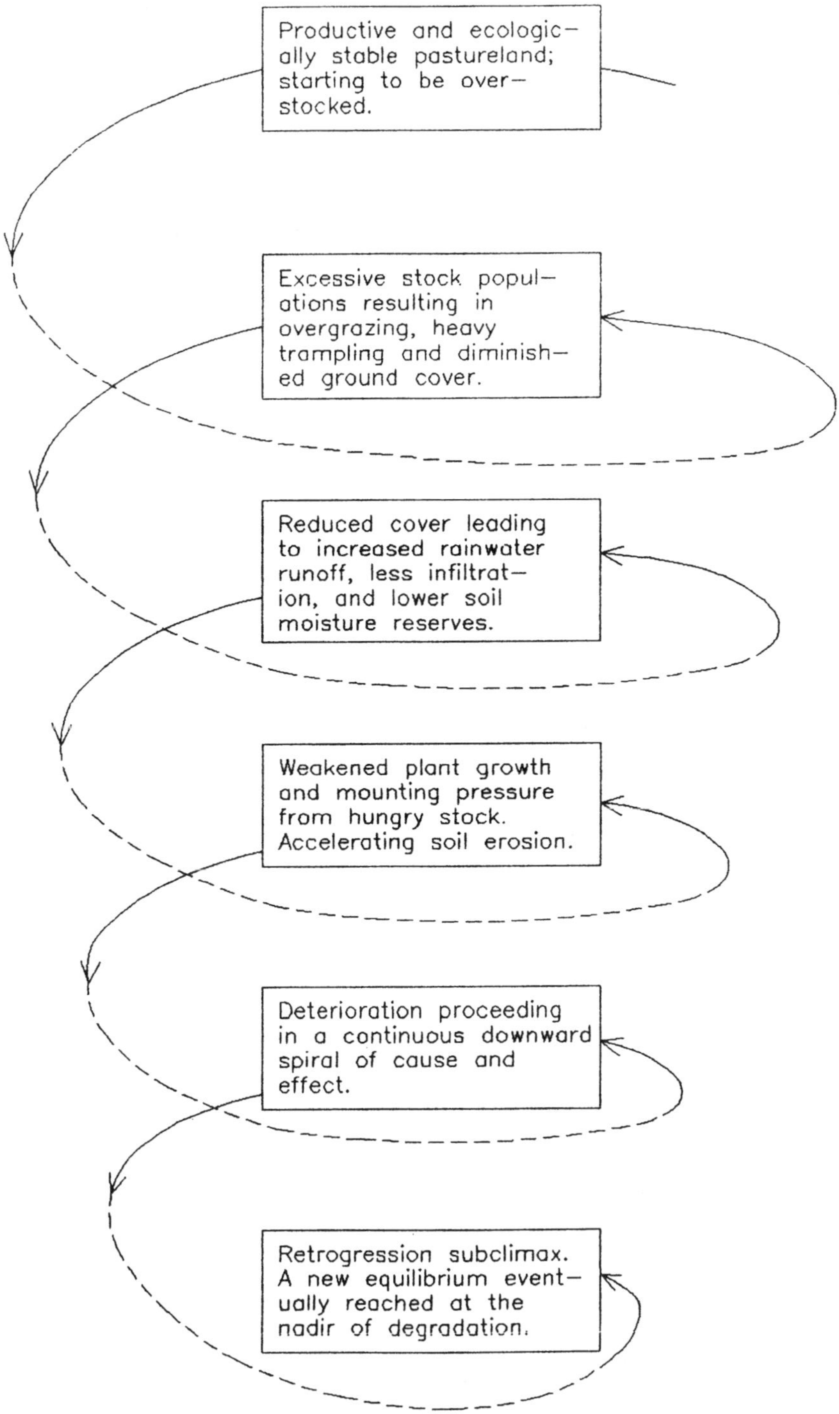

Figure 1. Retrogressive spiral on rangeland under heavy grazing pressure (from Strange 1980).

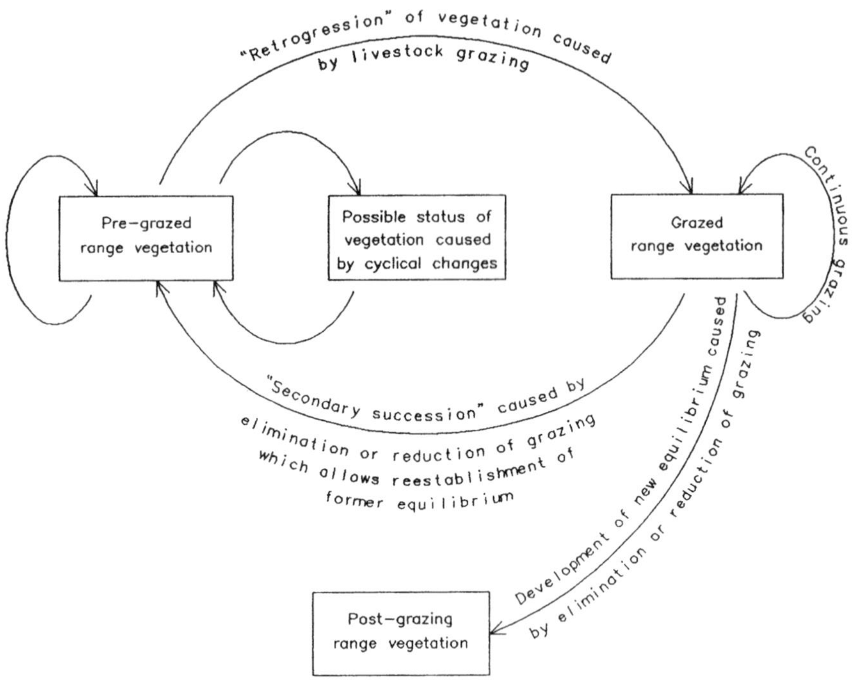

Figure 2. Generalized vegetation changes associated with livestock grazing (from Vale 1982).

sented by Vale are similar to those presented by Heady (1973). A more detailed diagram including soil development also was presented by Heady (1975). Such patterns of deterioration under heavy livestock grazing were recognized early during the time of unrestricted livestock grazing in the West (Colville 1898, Kennedy and Doten 1901, Jardine and Anderson 1919, Nelson 1898, Smith 1895, 1899, Thornber 1910, Wooton 1908).

Excessive livestock grazing also has been implicated in global desertification problems. Overgrazing is listed as the principal mechanism in many of the steps leading to desertification of different areas in Chile in South America (Mabbutt and Floret 1980). Many other examples were presented concerning the detrimental role of grazing in the desertification process in this UNESCO publication. Pulyarkin and Rakitnikov (1986) discussed problems of livestock distribution, mediated by stock water location and different plant communities, as they relate to desertification in the USSR. However, recent satellite data suggest the area of the Sahara Desert in northern Africa has varied considerably from 1980 through 1990 (Tucker et al. 1991). The greatest area of the Sahara occurred in 1984 which coincided with the severe drought in the Sahel (Sollod 1990).

Historically, there has been a natural tendency for producers to stock at a rate that leads gradually to some degree of rangeland deterioration (Pieper and Heitschmidt 1988). Therefore, it is not surprising that a common perception is livestock grazing is not compatible with many other uses of western rangeland. Perceived higher monetary returns have been one rationale for heavy stocking rates.

Vegetational Response To Grazing

Plant responses to defoliation are complex. Heady (1975) listed intensity, frequency and season as the main factors involved in defoliation impacts on plants. Each of these factors produce a gradient of plant responses, and when coupled with different plant susceptibilities to defoliation, constitute a myriad of possible combinations. Workers first correlated grazing resistance to carbohydrate reserve levels of plants at various phenological stages (McCarty and Price 1942, Sampson and McCarty 1930). Additional research identified several types of annual carbohydrate cycles and related them to grazing resistance (Buwai and Trlica 1977, Menke and Trlica 1981). However, other studies indicated carbohydrate level was only one of several operating factors (Jameson 1963, Caldwell 1984). Caldwell (1984) considered carbohydrate levels as a small buffer rather than a major factor in grazing resistance. Morphological characteristics of plants also are important in determining plant responses to grazing (Branson 1953, Hyder 1972, 1974, Heady 1975, Dahl and Hyder 1977 and Caldwell 1984, Sosebee and Wan 1988, Hart and Norton 1988). Many subsequent studies have been conducted to elucidate these complex interactions and are reviewed by Briske (this volume).

Translation of defoliation effects upon individual plants to the community level also involves the competitive relationships among the species involved. Crawley (1983) outlined a simple 2×2 matrix to illustrate the interaction of competitive ability and resistance of individual species to defoliation. If some plant species have low grazing resistance and others high resistance and if plants may have low or high competitive ability, four outcomes are possible (Fig. 3). In one case, if a highly competitive plant is also highly palatable and susceptible to grazing, it might

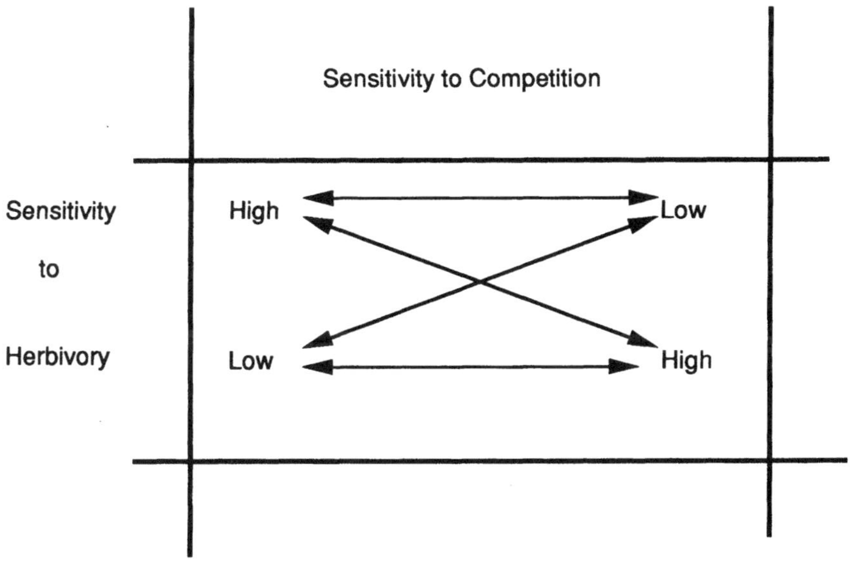

Figure 3. Matrix showing possible outcomes of interactions among plants based on sensitivity to herbivory and competition.

be grazed to the extent that it is dominated by a less competitive and less palatable species. If the less competitive plant is also highly palatable, it is at a disadvantage from both standpoints and its role in the community will likely be reduced relatively quickly. In some cases, the herbivore may switch from one species to another (Pieper et al. 1959) and have small net effect on community composition. If herbivores exhibit little preference among species, competitive ability alone may determine the final plant community composition. Reaction of individual species to defoliation also depends on relative resistance to grazing. Silvertown (1982) has discussed competitive relationships in stable and unstable environments. These considerations are relevant to equilibrial dynamics discussed later in this paper.

A good example of these interactions involves the perennial grass species bluebunch wheatgrass (*Agropyron spicatum*). This species was once widespread throughout the sagebrush-grass and Palouse region of the Intermountain area (Branson 1985, West 1979, 1988). Caldwell et al. (1981) elucidated one of the reasons why bluebunch wheatgrass is more susceptible to herbivory than crested wheatgrass (*Agropyron desertorum*). Crested wheatgrass has the capacity to reconstitute its canopy much more quickly than bluebunch wheatgrass following defoliation. Mueggler (1972) demonstrated that bluebunch wheatgrass was also highly susceptible to competition from other herbaceous species in southern Montana. Heady (1984) summarized the response of bluebunch wheatgrass to various grazing treatments under poor and excellent range condition (Table 1). This table indicates that early grazing is detrimental, early rest from grazing beneficial and late rest neutral to bluebunch wheatgrass.

Under partial competition, unclipped plants produced twice as much herbage and three times as many reproductive shoots as unclipped plants subjected to full competition (Mueggler 1972). Clipping and competition together reduced substantially many characteristics, when compared to plants without clipping or competitive stress. Because grazing resistance, competitive ability, and palatability are all continuous variables, the possible combinations of reactions to different herbivores in enormous. These interactions constitute internal (intrinsic) processes operating within ecosystems.

Grazing Optimization

If as Caldwell (1984) states, "removal of foliage from a plant must to some degree reduce the potential of plants to compete and retain their status in the community", grazing might logically be considered as reducing primary productivity of grazed communities as outlined earlier. Yet, many workers have seen plant productivity stimulated by defoliation or grazing (Heady 1975, McNaughton 1979). Later publications expanded on the concept and refined it (McNaughton 1979, 1983, 1984, 1985,). Other workers have also contributed to these ideas (Dyer et al. 1982, Hilbert et al. 1981, Paige and Whitham 1987, Williamson et al. 1989). Basically, the proposal presented by McNaughton (1983) indicated that plants vary in their response to herbivory (Fig 4). Some plants are detrimentally impacted by even low levels of grazing [e. g. so-called *ice cream* plants of Stoddart and Smith (1943)]. These are represented by line O-A in Figure 4. Other plants may not be

Table 1. Response of bluebunch wheatgrass (*Agropyron spicatum*) to different periods of grazing, deferment and rest (from Heady 1984)

Treatment	Range Condition	Plant Production	Vigor	Seed Production	Seedling Establishment
Seasonlong	Poor	–[1]		–	
Grazing	Excellent	0	0	–	0
Late	Poor	0	+	+	+
Grazing	Excellent	0	+	+	–
Early	Poor	–	–	–	–
Grazing	Excellent	0	0	0	–
Short Term	Poor	–	0	–	0
Grazing	Excellent	–	0	–	–
Seasonlong	Poor	+	+	+	+
Nongrazing	Excellent	–	+	+	–
Late	Poor	0	0	0	0
Nongrazing	Excellent	0	0	0	0
Early	Poor	+	+	+	+
Nongrazing	Excellent	0	+	+	–
Short	Poor	0	0	0	0
Nongrazing	Excellent	0	0	0	0
Nonuse	Poor	+	+	+	+
Grazing	Excellent	–	+	+	–

[1] "–" indicates harmful effects, "+" beneficial effects, "0" no effects and a blank space means no information available.

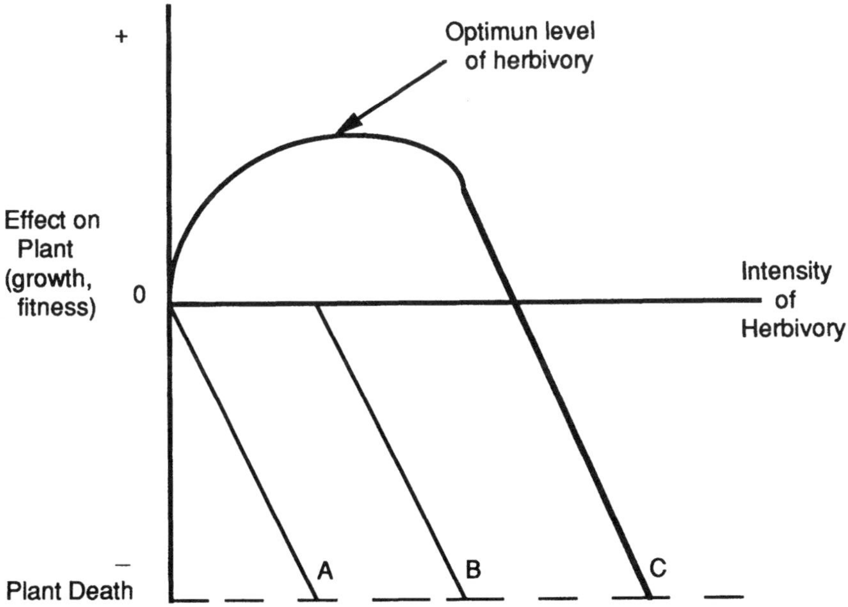

Figure 4. Grazing optimization curve hypothesized by McNaughton (1983).

detrimentally impacted until defoliation reaches a certain point (line O-B in Fig. 4). Still other plants can be stimulated until a certain level of defoliation is reached (i. e., the optimum point) and then detrimentally impacted by increased defoliation until plant death occurs (line O-C in Fig. 4). Some plants may fall into intermediate classes rather than one of these three, shown by the later curves presented by McNaughton (1983). McNaughton developed his hypothesis from work in the highly productive Serengeti Plains in East Africa. Evidence from other areas is limited. It seems likely that stimulation by grazing is related to regrowth potential of the plants and the amount and timing of precipitation. Consequently, if grazing optimization does occur, it is more plausible in mesic rangelands and less likely to occur in arid and semi-arid ecosystems (Pieper and Heitschmidt 1988).

Not all workers agree with the grazing optimization hypothesis. Notably, Belsky has been an outspoken critic of many of these ideas. She wrote in 1986:

> In an early review of the effects of grazing on rangeland species and communities, Ellison (1960) concluded that there was no evidence to support the claims that grazing benefits plants. In the last quarter century, there have been more reports claiming the benefits of grazing to plants. However, other than for plants grown under artificial conditions, no new evidence has been reported that renders Ellison's conclusion any less accurate today.

Central to the grazing optimization theory is the idea of coevolution of grass plants and herbivores that feed upon them. These ideas are not new, but have been reemphasized recently. Many believe grasses have been shaped by grazers, but recently, several workers have suggested that grasses depend on, or could not exist in the absence of, grazers (Aarssen and Irwin 1991, Owen 1980, 1990, McNaughton 1991, Owen and Weigert 1976, Paige and Whitham 1987, Platou and Tueller 1985). Maschinski and Whitham (1989) considered reactions of plants consumed by herbivores to represent a continuum from those detrimentally impacted to those benefited by herbivory. These ideas have important implications for livestock grazing in the western United States.

Livestock Grazing In The West

This section will focus on three areas: the Southwest, the Great Plains, and the Intermountain Area/Great Basin. These areas have contrasting histories that can shed some light on the impact of livestock grazing. In the Great Plains, bison was the dominant large herbivore before the herds were depleted by hunting. Seton (1927) estimated as many as 40 to 60 million bison grazed in the Great Plains at one time. Other estimates are even higher (Arthun and Holechek 1982, Blaylock 1982). There is some disagreement concerning the role and impact of bison in the Great Plains, but Larson (1940) contended that, at least in the more xeric portions of the Great Plains, bison were largely responsible for maintaining short-grass vegetation rather than mixed-prairie vegetation. Bison were highly mobile and undoubtedly could move long distances for water and green forage. However, England and DeVos (1969) presented several historical accounts of heavy grazing by bison, and Kirsch and Druse (1972) also reported accounts of heavy bison grazing, at least in localized areas. Conservative estimates of bison numbers would indicate heavy stocking and if those mentioned by Blaylock (15 to 20 bison per acre!) (1982) or Hart and Norton (1988) (3 per hectare near Missoula, Montana) occurred even on isolated tracts for limited time, the impacts must have been substantial.

Situations in the Great Plains contrast sharply with those in the Southwest and the Great Basin/Intermountain Area since the Pleistocene. In recent geological time, native large herbivores were scarce since large extinctions occurred during the Pleistocene (Martin and Wright 1967, Platou and Tueller 1985, Mack and Thompson 1982). Martin (1968) especially alluded to empty niches of large browsing mammals in the Southwest since the Pleistocene. Janzen (1986) also described the lack of large mammal components for the Chihuahuan Desert. Undoubtedly, there were some pronghorn antelope, bison, mule deer and elk inhabiting these ranges, but they probably had minimal impact on other ecosystem components.

The stage was set in the 1500s when domestic livestock were introduced into the Southwest (Schickedanz 1980). At first, these livestock had mostly local impacts around settlements and Spanish missions. By 1580, some sheep had been acquired by Native Americans and the impacts of livestock had become more pervasive. However, the large influx of domestic livestock into the West did not occur until settlement of these areas after the Civil War (Galbraith and Anderson 1971, Krueger 1988, Schickedanz 1980, Peterson 1985). As the West was settled, increases in livestock numbers were fueled by the demand of military forces, miners and eastern cities that could acquire western grown beef transported by railroad. Considerable range deterioration occurred during this time (Krueger 1988, Holechek et al. 1989, Holechek 1981). Accurate data on livestock numbers are difficult to obtain for these early periods, but sheep numbers in Utah increased from 1 million head in 1885 to 1.5 million head in 1890 and to 3.8 million head by 1900 (Peterson 1985). In 1860, there were 65,000 people and about 200,000 cattle in the state of Oregon (Galbraith and Anderson 1971). Cattle numbers in the 17 western states increased from about 4.6 million in 1870 to 35-40 million in 1884 (Holechek et al. 1989). In Arizona, cattle numbers reached a peak in 1891 (Humphrey 1987). Many cattle were trailed to the Northwest and Great Basin/Intermountain Area from Texas (Galbraith and Anderson 1971, Peterson 1985). Early estimates of stocking rates that these arid and semiarid ranges could support were greatly exaggerated. Early stockman were familiar with productive pastures of the humid East and South and did not recognize western limitations. Considerable degradation occurred before the fragility of western rangelands was realized. In many cases, stocking rates remained excessive until well into the 1910-20 period (Branson 1985, Wagner 1978).

In the Southwest, Great Basin and Intermountain Regions, livestock were introduced into ecosystems that had not supported large numbers of ungulates since the Pleistocene (Martin 1968, Mack and Thompson 1982, Young and Evans 1985). Consequently, the introduction of domestic livestock into these rangelands had relatively larger impacts than when cattle replaced bison in the Great Plains. However, replacing bison with cattle was not without some impacts. Bison were primarily grazers (Kautz and Van Dyne 1978). Peden et al. (1974) and Van Dyne et al. (1980) reported as much as 90% of bison diet consisted of grass. Cattle, in contrast, consume more browse and forbs. Kautz and Van Dyne (1978) reported cattle diets comprised 16% browse and 26 % forbs based on 52 studies on short grass vegetation. Bison diets were more digestible than cattle diets suggesting that bison were more adapted to native forage than cattle (Hawley et al. 1981). In addition, cattle grazing was restricted by fencing while bison were able to move about unrestricted.

Vegetational changes in the Great Basin and Intermountain Region have been pronounced during the past 150 years. Although the original extent of sagebrush-grass range has probably not expanded much during this period, increases in sagebrush density and decreases in perennial understory grass cover and density have undoubtedly occurred (Branson 1985, Young et al. 1976, Young and Evans 1979). Decrease in the amount of perennial grasses was accompanied by concomitant increases in introduced annuals, especially cheatgrass (*Bromus tectorum*) and Russian thistle (*Salsola australis*) (Klemmedson and Smith 1964, Fleming 1942, Young et al. 1972, Young and Tipton 1990). On some ranges, cheatgrass has been replaced by medusahead rye (*Taeniatherum asperum*) (Branson 1985, Turner et al. 1963).

The precise role of grazing by domestic livestock in these vegetational changes can probably never be determined completely. Heavy grazing coupled with fire suppression, introduction of alien species, and homesteading have all interacted to influence vegetation in the Intermountain West. In addition, the final outcome resulting from the change in herbivory type and level in these ecosystems has not yet been determined. Young and Evans (1985) termed ranching in the Great Basin as the "Grand Experiment."

The dominant change in southwestern vegetation has been an increase in both density and extent of woody species, mainly mesquite (*Prosopis* spp.) and creosotebush (*Larrea tridentata*) (Branson 1985, Branscomb 1958, Buffington and Herbel 1965, Donart 1984, Humphrey 1987, Hastings and Turner 1965, Smeins 1983). These changes have often been at the expense of herbaceous species that were valuable for livestock forage. At first, many ecologists thought that the increase in shrubs was directly related to livestock grazing pressure, but in some cases, the increase in shrub density and extent seems to be independent of livestock grazing intensity. On the other hand, black grama (*Bouteloua eriopoda*) appears to be fairly susceptible to cattle grazing (Nelson 1934, Paulsen and Ares 1962, Valentine 1970). The role of livestock grazing in these vegetational changes will be considered in more detail in a later section. Hart and Norton (1988) have reviewed the impact of domestic livestock grazing in many other parts of the world.

Stability And Recovery

Early attempts to address problems of rangeland degradation were based on plant successional theories proposed by Clements (1907, 1916, 1936) and Sampson (1917, 1919). Clements' idea of plant succession, developed in the Great Plains, was essentially a linear model applied over large landscape units. He described succession as a six-step process (Barbour et al. 1987):

1. Nudation — The exposure of a new surface in primary succession or the clearing away of previous vegetation in secondary succession.

2. Migration — Of seeds, spores or vegetative propagules from adjacent areas, though in secondary succession many of these are already present in the soil.
3. Ecesis — The germination, early growth, and establishment of plants.
4. Competition — Among the established plants.
5. Reaction — The autogenic effects of plants on the habitat.
6. Stabilization — The climax

This type of succession has also been called autogenic succession (Tansley 1935, Egler 1954, McCormick 1968). Two features in this model are that the climax is relatively stable (i. e., the climax species are capable of replacing themselves and the overall nature of the vegetation does not change drastically) and the stages in succession are predictable, given a certain environment. Clements envisioned these changes occurring over large landscapes during the geological changes involving peneplanation. He developed terminology to describe any deviation from the perceived climax. Clements viewed the climate as having dominant influence over the climax vegetation and referred to climatic climax. Competition and reaction were key processes in this model that purportedly led to equilibrial conditions, at least in the climax. This view held sway for half a century and had profound influence over the newly developing discipline of range management (Tobey 1981).

Not every ecologist embraced Clement's ideas. Among his more vocal critics were Henry Gleason who espoused an "individualistic" concept of the plant association (1917, 1926). Gleason argued that chance played a major role in the development of plant communities, and such rigid organization as displayed in the Clementsian model lacked reality. However, Clements had such a following that Gleason eventually turned to plant taxonomy where he made many important contributions (e. g. Gleason and Cronquist 1964). Later, even Tansley (1935) was critical of the Clementsian model. Despite these controversies, the general ideas of Clements held sway in the field of range management.

Dyksterhuis published his landmark paper on range condition in 1949, modeled closely after the Clementsian model of plant succession. He later amplified these concepts in a lengthy review paper (Dyksterhuis 1958). In this system, range condition was rated on the status of the vegetation in relation to climax. Seral stages were given the terms excellent, good, fair and poor depending on the percentage of climax vegetation represented by the present vegetation (Fig. 5.). Grazing was considered a disturbance factor, leading to regression along the same pathway as succession, except in the reverse direction. Relaxation of grazing pressure would again allow secondary succession to proceed toward climax (Fig. 6). This model dominated the thinking and writing in range management during the post- World War II period. From time to time, some workers questioned certain parts of the concept, but the idea of climax communities in equilibrium with the environment persisted.

Ecologists working with diverse groups of organisms also began to question equilibrial theories from different standpoints. This quotation by John Wiens, an ornithologist, writing about his paper "On competition and variable environments," (Wiens 1977) which had been selected as a "Citation Classic" by *Current Contents*, illustrates the changing concepts:

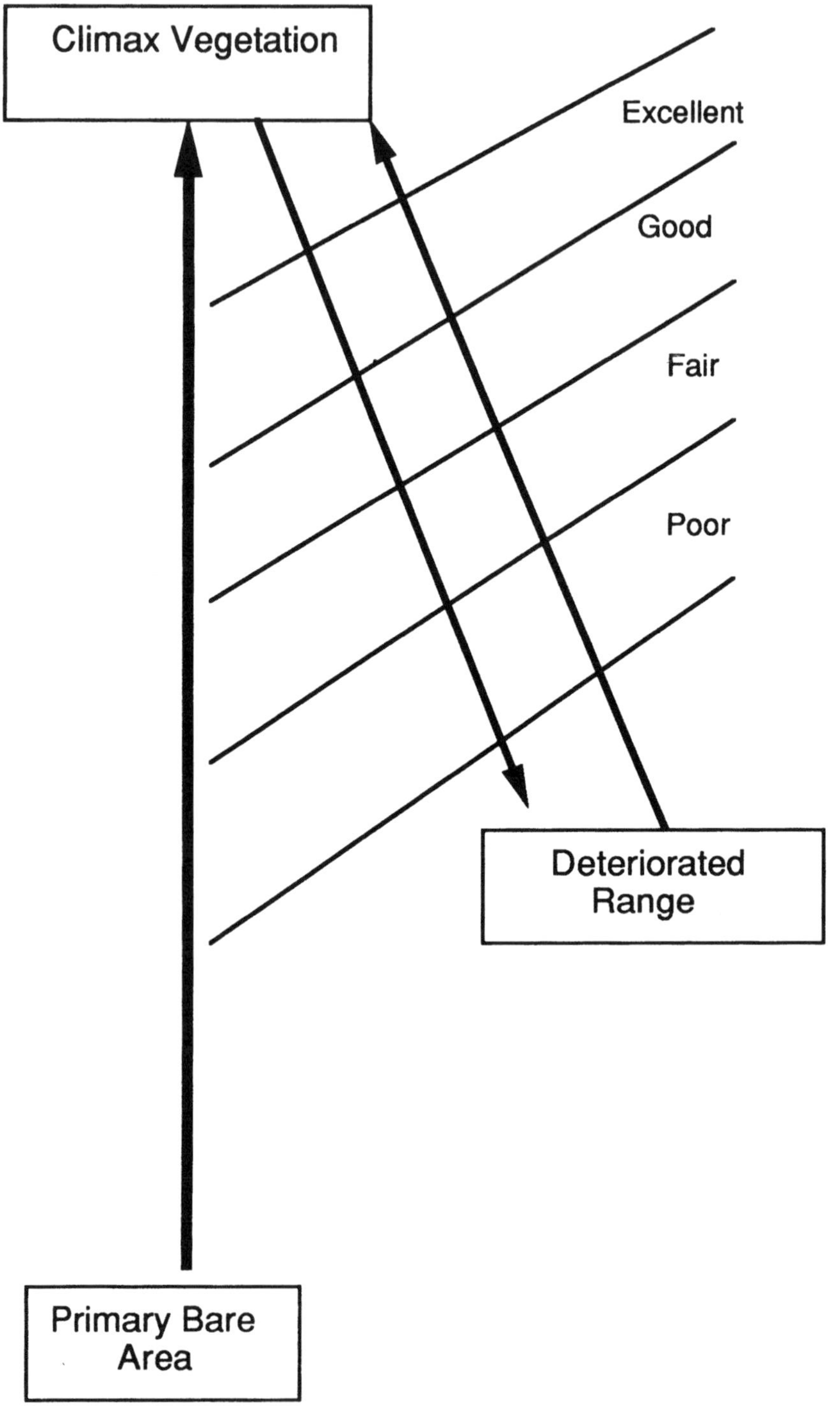

Figure 5. Relationship of secondary succession and range condition (from Thomas and Ronnigen 1964).

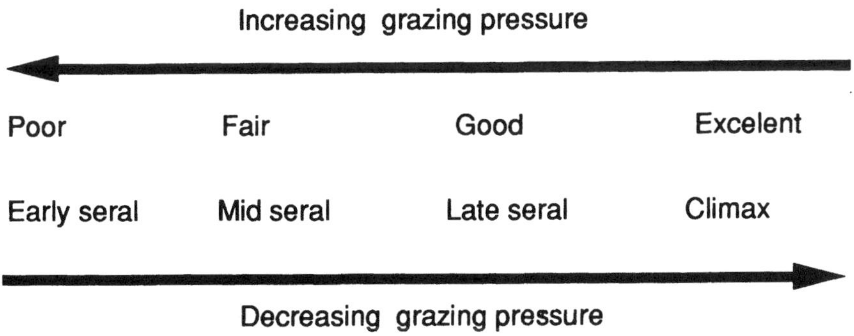

Figure 6. Successional changes and range condition classes induced by changes in grazing pressure according to equilibrium theory.

> During the 1960s and early 1970s, ecologists generally believed that natural communities were in equilibrium and their structure was determined by competitive interactions. I grew up academically, fervently embracing this view and conducted a PhD study of grassland birds in Wisconsin that showed (I thought) how subtle the niche partitioning produced by these interactions could be. After I moved west to a faculty position in Oregon, I began along-term study of breeding bird communities in the Great Basin shrubsteppe, partly to extend the grassland studies into more complex habitats and partly to escape the rain of the Willamette Valley. Despite my best efforts and the most probing analyses, the expected patterns failed to emerge. After several years of trying to fit the square peg of data into the round hole of theory, my student John Rotenberry and I realized (during a June snowstorm) that the conditions of resource limitation required by community theory probably occurred only intermittently in the shrubsteppe — much of the time the birds lived in a land of plenty (Wiens 1991).

Equilibrial theory suggests there are internal controls and feedbacks such that when one component of an ecosystem changes drastically (i. e. is out of balance with other components) internal negative feedbacks operate to return the system to its original condition. Non-equilibrial systems would change in response to outside forces, and may or may not return to the original condition.

Other ecologists were also working along similar lines, and held there were many instances of environmental variability and inconsistencies in community patterns (Andrewartha and Birch 1954 and DeAngelis and Waterhouse 1987). Botkin (1991) stated the case as follows: "Nature as we are coming to know it is a patchwork of complex systems with many things happening at once and with each system undergoing changes at many scales of time and space. Human beings, far from being alien interlopers who disturb the timeless rhythms of nature, are intrinsic elements of the natural order. Chance events seem to play an important role."

As early as 1974, Sutherland wrote about multiple stable points in natural communities, and two recent papers have made the leap of applying non-equilibrial dynamics to grazing ecosystems under a range livestock framework (Ellis and Swift 1988 and Westoby et al. 1989). According to the analyses of Westoby et al. (1989), each rangeland may have several successional pathways, depending on different environmental factors and how they interact. Domestic livestock are viewed as just one of these environmental components. They diagrammed 7 separate stages possible for a grassland in South Africa (Westoby et al. 1989), based primarily on different patterns of livestock grazing. They listed 5 mechanisms that may operate in their state and transition model:

1. Demographic inertia — Some event that happens infrequently may bring about long-lasting change in vegetation.
2. Grazing catastrophe — Influence of herbivory is discontinuous and variable and may not be reversible if stocking is too high.
3. Priority in competition — Alternative steady states may result when the outcome of competition depends on initial abundance of the competitors.
4. Fire positive feedback — Some species are fire types, i. e., they promote fire and are benefited by periodic burning.
5. Vegetation change which results in a persistent change in soil conditions.

Ellis and Swift (1988) had developed similar ideas earlier while working in northern Kenya. These rangelands were subject to varying climatic conditions and livestock grazing patterns of the nomadic herders. They also rejected an equilibrium concept for grazing ecosystems in Kenya: "The results of our work in Ngisonyoka Turkana, one of 15 sub-tribal territories in Turkana District, reveal anything but an equilibrial ecosystem." They contrasted some of the differences between the two types of systems. Heavy grazing tends to magnify the degree of instability of many ecosystems. Omar and Bartolome (1991) described 5 possible states for Arabian rangelands. There are least 3 situations where equilibrium conditions might not be possible: 1. situations where individual disturbances are too large or infrequent, 2. ephemeral events that have long-lasting disruptive effects, and 3. cases where climate changes interrupt any movement toward equilibrium that does occur. Recent work in South Africa tends to support the strong influence of the grazing factor (Danckwerts and Stuart -Hill 1991). It seems ironic that we appear now to be embracing many of the ideas, although somewhat modified, espoused by Gleason nearly 50 years ago.

The models described by Sutherland (1974), Westoby et al. (1989), Walker (1988) and others do not imply equilibrial situations never occur in rangeland ecosystems. Some of the alternate states may persist for some time. For example, Laycock (1991, and this volume) documented relative stable states for several situations in the western United States and Friedel (1991) described situations for Australian rangelands.

Recent analyses and approaches for range condition classification seem to call for abandonment of the traditional approaches; however, these approaches have been extremely valuable for the development of range management. Establishing successional patterns gave the rangeland ecologist some guidelines and bases for measuring departures. Clementsian concepts have been partially responsible for much of the progress made in application of ecological principles to management. However, it appears that some modifications need to be made as our understanding of natural processes has now changed. The traditional models fit some ranges well, others not so well, and still others hardly at all.

These discussions have deeper implications than simply academic debates. They are related directly to the process of recovery following excessive grazing. The top part of Figure 7 illustrates the traditional view of recovery: heavy grazing pressure tends to push vegetation toward poor range condition and early successional stages. Relaxation of grazing pressure tends to reverse the process toward excellent range condition and climax vegetation. Within limits, shifts along this gradient are largely linear and can be brought to bear merely with shifts in grazing pressure. As Anderson (1977) has stated, "Merely lessen the grazing intensity, shorten or change the season, and measurable improvement is apparent." It is this type of argument that is apparently behind such programs as Paul Ehrlich's Today program and such books as "Waste of the West: Public Lands Ranching" (Jacobs 1991). Moreover, the implication is that removal of livestock from rangelands will return them to pristine conditions. If livestock grazing is excessive (what constitutes excessive is sometimes open to debate), reduction in stocking will often provide some improvement such as greater plant density, cover and productivity and less soil erosion. Changing season of use or some other modification of grazing practice can result in dramatic improvements in riparian vegetation, as well as upland vegetation

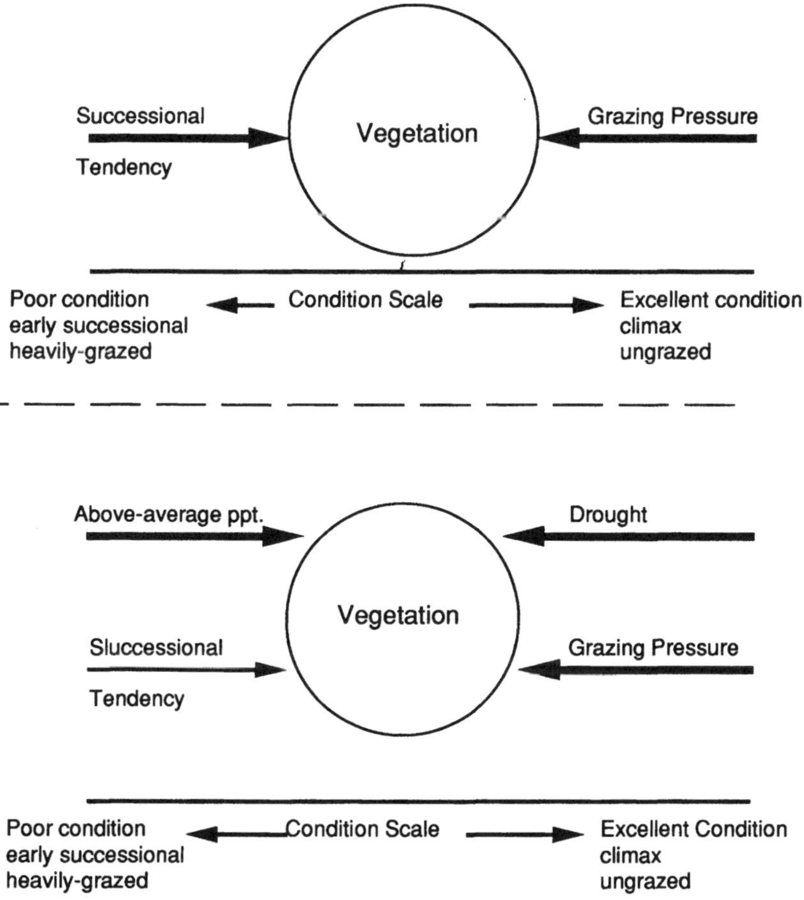

Figure 7. General scheme of range successional model (top) and model with incorporation of precipitation variability (bottom) (from Westoby et al. 1989).

(Elmore et al. this volume). However, the idea that recovery of pristine conditions can be restored simply by removing livestock is much too simplistic in light of other changes that have occurred such as introduction of alien species, changes in fire regimes, etc. Livestock constitute only one component of rangeland ecosystems, and many extrinsic factors, especially weather variations are instrumental in altering ecosystem components. The bottom part of Figure 7 shows the increase in complexity when climatic extremes are added. Heavy livestock grazing tends to lead to unstable conditions in most rangelands, especially arid and semi-arid areas.

Livestock grazing has often been implicated as a causal factor in the increase in shrubby vegetation at the expense of grassland (Gardner 1951, Martin 1975, Schlesinger et al. 1990, and York and Dick-Peddie 1969). Many of these authors believe livestock grazing played a significant role in the expansion of woody vegetation, but may not have been the only factor. Indeed, livestock likely played a major role in transporting mesquite seed throughout the Southwest (Buffington and Herbel 1965, Martin 1975). However, grassland vegetation in several long-term livestock grazing exclosures on the Jornada Experimental Range in southern New Mexico also have been replaced by shrubland. Gibbens and Beck (1987, 1988), working with long-term quadrat records from the Jornada, found black grama cover decreased on protected plots as well as on grazed plots, and shrubs also increased on protected quadrats.

Several authors have shown a correlation between shrub increase with changes in weather patterns. Hastings and Turner (1965) presented evidence that climatic changes favored dominance by shrubs and a decline in perennial grasses in the Southwest. Using sophisticated analytical procedures for climatic data, Neilson (1986) concluded that years favoring growth and reproduction of black grama have decreased during the recent past, and years favoring shrubs have increased. However, Conley et al. (1988), using essentially the same data set as Neilson, came to nearly the opposite conclusion: "We were unable to demonstrate any consistent statistical rationale for invoking climate as a causal mechanism in the observed vegetative change of the [*Mesilla*] valley. An examination of the history of grazing by introduced livestock leads to acceptance of the alternate hypothesis that this vegetation shift was caused by grazing effects from introduced livestock."

Livestock grazing probably played some role in the change from grassland vegetation to shrubland or savanna, but the exact role may never be completely understood in a historical perspective. In some cases, direct grazing impacts on palatable species may have promoted replacement by less palatable, often woody species. In other cases, livestock may have served mainly as a seed dispersal mechanism. Climatic shifts may have played a major role in vegetational shifts, including the recent precipitous increase in atmospheric CO_2 levels (Idso 1989). In addition, Wahl and Lawson (1970) estimated that summer precipitation in the Western United States has declined 30 to 50% in the last century. Bryson (1974) has translated this climatic change into a reduction of 50 to 75% in herbivore carrying capacity.

Even where grazing has been eliminated, vegetational changes can be slow. For example, in southern New Mexico, Beck and Tober (1985) reported that on areas supporting more than 9% canopy cover of creosotebush, black grama declined from a trace to 0% cover, bush muhly (*Muhlenbergia porteri*) increased from a trace to 1.9% basal cover, while shrub cover increased from 11.4 to 27.2% canopy cover, from 1956 to 1977 under protection from livestock grazing. Trying to unravel ecological events that began more than 75 years ago, and may still be occurring, is difficult.

One mechanism to reverse the shift from grassland to shrubland under protection from grazing in southern Arizona and New Mexico involves the perennial grass bush muhly. Bush muhly often occurs only under or within the canopy of a shrub (Haque et al. 1991 and Welsh and Beck 1976). Bush muhly apparently has the ability to grow at reduced light intensities in association with shrubs and may benefit from improved nutrient availability under shrub canopies (Tiedeman et al. 1971, Welsh and Beck 1976). In some conditions, bush muhly may actually cause the death of creosotebush plants by shading lower leaves (Welsh and Beck 1976). Beck and Tober (1985) found bush muhly responded positively to cattle removal at 2 out of 3 sites in southern New Mexico and to jackrabbit removal on one site.

Direct Grazing Impacts

Fire Effects

Many ecologists consider the reduced frequency and extent of wildfires, resulting from the emphasis on wildfire control, to be among the most pervasive influences of U. S. settlement by non-native peoples. Livestock grazing played a role in this modification. Livestock grazing reduces the amount of fine fuels necessary to start wildfires, thereby reducing the number of wildfires and reducing fire intensity when fires do occur.

West (1979) and West and Van Pelt (1987) have provided models showing the interactive role fire and grazing played in both sagebrush-grass and piñon-juniper vegetation. The timing and severity of fire in pristine sagebrush-grass communities is not known exactly, but Wright et al. (1979) suggested a fire frequency of about 50 years. Fuel was probably sparser on drier sites; thus fire frequency may have been less in some areas (West 1979). Because many sagebrush (*Artemisia*) species are not sprouters, it is likely that periodic fires in those plant communities favored perennial grass species, and kept sagebrush from increasing. With moderate livestock grazing, fuel was reduced and fire frequency may have declined. However, heavy livestock grazing soon contributed to the shift from perennial grasses to annuals, especially cheatgrass. Cheatgrass appears to be a fire type, producing large volumes of fuel that burns readily when dry. The Intermountain Region has experienced many severe fires during the summer following favorable spring growing seasons for cheatgrass. West (1979) and Miller et al. (This Volume) have outlined some of the concomitant changes involving the fire cycle of cheatgrass ranges.

Wilson et al. (1988) described the role of fire and the interaction of fire and climate in development of Australian plant communities. Chenopod shrubland communities near Deniliquin, New South Wales, can be either a grassland type or mixed chenopod shrubland-grassland, depending on the frequency of the fire and post-fire precipitation patterns. Their diagrams depict the various stages of vegetational development under possible combinations of fire and precipitation patterns.

Many ecologists (Wright et al. 1979, Young and Evans 1981, Pieper and Wittie 1990) suggest fire historically acted to maintain piñon-juniper woodlands relatively open with scattered large trees and has favored development of an herbaceous understory by killing developing tree seedlings. Livestock grazing in the 1880s reduced the fuel in these stands and, perhaps, reduced the fire frequency. As tree overstory increased, understory perennial grass cover decreased from heavy grazing pressure and lower light intensity reaching the herbaceous level (West and Van Pelt 1987). These fire cycles in piñon-juniper woodlands were also shown by Barney and Frischknecht (1974) and Erdman (1970). Such studies illustrate livestock grazing may have secondary as well as direct influences.

Nutrient Distribution

It seems obvious that grazing livestock would have an influence in changing nutrient cycling patterns. However, few studies have been conducted on rangelands to describe how livestock grazing alters nutrient distribution patterns. It is evident that nutrients can concentrate in areas where livestock concentrate, such as near water and areas of level terrain (Heady 1975, Pieper 1977). For example, Hilder (1966) reported sheep deposited nearly two-thirds of their feces on less than 10% of the pasture area. In some cases, cattle feces appear to be deposited more randomly than sheep feces (Heady 1975), but few studies have reported actual fecal distribution patterns on rangelands.

Nitrogen is mostly returned in the form of urine (Pieper 1977). Considerable work has been done on decomposition of feces, but little on distribution patterns. Tiedemann et al. (1986) constructed simulated nitrogen budgets under several livestock grazing regimes. They show a net loss of 3.2 kg nitrogen per acre when moderate livestock grazing is compared to no livestock grazing in Pacific Northwest grassland. Of this 0.9 kg were lost as volatilization of NH_3, 2 kg when carcasses were sold, 0.7 kg lost to increased erosion and 0.22 kg to increased leaching. No estimate of nutrient redistribution was made. The estimates were made on the assumption the standing crop of nitrogen in the soil was 2000 kg per ha. Other studies also showed a significant portion of the nitrogen in urine may be lost directly through volatilization (Simpson 1968, Williams 1970).

Estimates of other nutrient removal from grazed rangelands by sale of animals was small for most elements (Pieper 1977). Apparently ingestion by animals hastens the decomposition process (Barrow 1967).

Although livestock are involved in altering nutrient distribution, the results of this alteration are not understood very well. Native herbivores also alter nutrient distribution (Heady 1975, Klemmedson 1977). The impact could be relatively large in nutrient poor ecosystems, such as many desert systems, but it is not known at this time. Water quality of rangeland streams may be degraded by redistributing nutrients into already nutrient-rich riparian areas.

Grazing and Patch Dynamics

The study of ecological diversity at a fairly large, coarse-grained scale is a phenomenon of fairly recent times (Pickett and White 1986a). Patches of one type of vegetation may be embedded in a mosaic pattern in the more dominant type. Many authors have considered livestock grazing as one factor involved in creation of these patches, for example:

> Large herds of domestic livestock disrupt this tight connection of soil and plant processes and lead to a decline in the cover of black grama and other species in semiarid grasslands. Heavy grazing during the short summer wet season contributes to the loss of grass cover during moderate drought and to lowered competitive potential of grasses (Van Auken and Bush 1989). Trampling compacts the soil and reduces infiltration rates (Gamougoun, et al. 1984). Greater runoff results in erosion and increased transport of water, N, and other plant nutrients between geomorphological units in the basin. The net effect of these changes is to reduce the availability of soil moisture and nutrients in the landscape and to increase the heterogeneity of their horizontal distribution (Schlesinger et al. 1990).

This article, published in *Science*, placed the process of desertification in a global setting. It related changes in Chihuahuan Desert vegetation in southern New Mexico to desertification processes on a global scale. The historical timing of these changes induced by livestock grazing was not discussed, although the implication was that they are still occurring. In addition, no direct evidence was presented showing the role of domestic livestock grazing in the increase in shrubby vegetation.

Archer (1989, This Volume) and Brown and Archer (1989) have also discussed the role of livestock grazing in the process of mesquite encroachment into grassland vegetation in southern Texas. Grazing likely played a role in distribution of mesquite seed, and possibly, in opening stands of grass for mesquite establishment and expansion of individual clumps.

Wiens (1985) has discussed the role of domestic livestock in patch dynamics as follows:

> "In contrast to such native herbivores, domestic cattle are restricted in their movements both by their sedentary nature and by human enclosures. Much of their effect in generating patchiness thus occurs at a landscape scale, as a consequence of the interspersion of pastures or ranges that are subjected to different grazing intensities or seasonalities. Within pastures a finer scale or heterogeneity may be produced if the animals' grazing activities are spatially concentrated, as by topography, the distribution of palatable forage plants, or a localized water source. Where grazing is intense (on whatever scale), it has well-documented effects upon the stature of the vegetation, the relative coverages of plant species, water infiltration rates, and a variety of other system properties (Risser 1981). Not all consequences of grazing are detrimental, however. Domestic cattle, for example, produce seemingly endless quantities of dung, which is generally deposited in small heaps. These create local concentrations of nutrients returned to the system, and these enriched patches often exhibit clearly greater productivity or altered composition from the surrounding sward."

Other authors have discussed the development of mosaic patterns of certain species in a matrix of a different vegetation types (Anderson 1971, Atsatt and O'Dowd 1976, Hamilton et al. 1973 and Hill 1991). At the species level, several workers have developed the hypothesis that species richness is greatest under intermediate levels of disturbance (Connell 1978, Grime, 1973, 1979, Peet et al. 1983, and Pickett and White 1986b). Thus, moderate grazing may contribute to higher species diversity than either light or destructive grazing. In addition, Milchunas et al. (1988) have expanded on the intermediate disturbance hypothesis. They have described the role of past herbivory history as well as moisture gradients to predict diversity on climatically determined grasslands. Only on subhumid grasslands with long history of grazing does diversity reach a maximum at intermediate levels of grazing.

Impact on Cryptogamic Crusts

Cryptogamic crusts are composed of non-vascular plants of algae, lichens, mosses and diatoms. They may function in nitrogen fixation, water infiltration, and soil protection (Harper and Marble 1988, Johansen 1986). Apparently these crusts are susceptible to livestock concentrations. Anderson et al. (1982a) reported that mosses were the most susceptible and diatoms least susceptible to grazing. These authors stated that as much as 18 to 20 years might be required for recovery of cryptogamic crusts damaged by livestock grazing (Anderson et al. 982b). However, soil characteristics are also critical in influencing these crusts.

Examples Of Exclosure Studies

One approach for studying the impact of livestock grazing is to compare grazed areas with areas protected from grazing. While these studies have several limitations (Laycock, this volume), they do offer opportunities not available by other methods. The grazing intensity often is not stated in the study or is not known. The studies reviewed here are only examples to illustrate the variety of responses that have been reported in the literature.

Atwood (1987) sampled vegetation inside and outside 4 livestock exclosures of different ages in southern New Mexico. Figure 8 shows density of black grama for

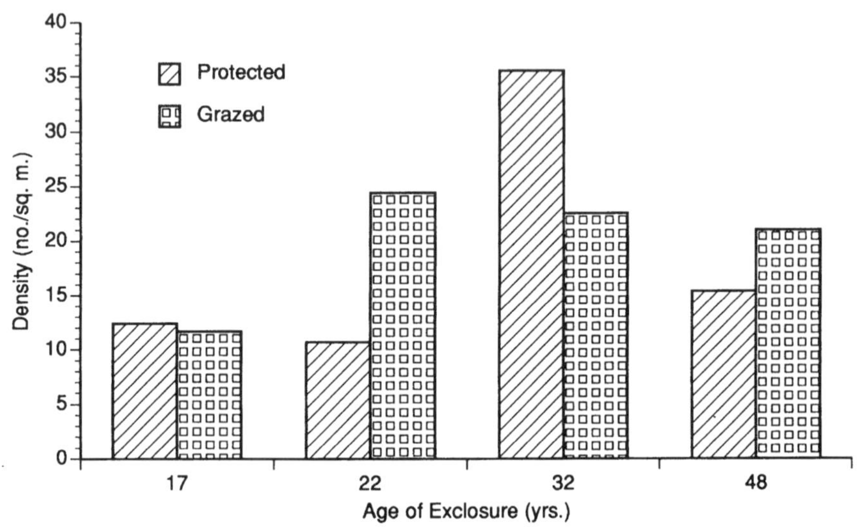

Figure 8. Density of black grama on grazed areas and livestock exclosures in southern New Mexico (from Atwood 1987).

these study locations. Density was similar inside and outside the 17-year -old exclosure, higher outside the exclosure for the 22 and 48 -year -old exclosures and much higher inside for the 32-year- old exclosure. Stocking rates during the period since the exclosures were established have varied but were probably not excessive at any location. Reasons for the different responses are not clear.

Canopy cover for short grass and mid-grass species apparently responded to protection from livestock grazing on the Audubon Ranch in south-central Arizona (Fig. 9) (Brady et al. 1989). The Ranch had not been grazed by livestock since 1968 and was described as being moderately grazed by livestock before that time. Mid grasses included sideoats grama (*Bouteloua curtipendula*), cane beardgrass (*Bothriochloa barbinodis*), plains lovegrass (*Eragrostis intermedia*) and short grasses included blue grama (*Bouteloua gracilis*), hairy grama and sprucetop grama (*Bouteloua chondrosiodies*). However, cover increased on both grazed and ungrazed areas since 1971, partly attributed to favorable precipitation patterns since then.

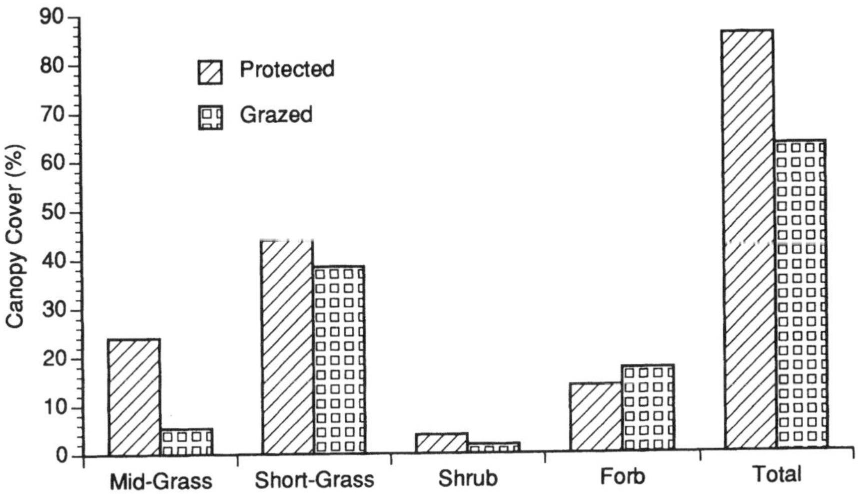

Figure 9. Density of plants under grazing and protection from grazing in southeastern Arizona (from Brady et al. 1989).

In south-central New Mexico, protection from grazing for 12 years generally increased end-of-season standing crop on different range sites (Pieper 1967). However, the response was more pronounced on some sites than others (Fig. 10).

The four-corners area of Colorado, Utah, Arizona, and New Mexico has been recognized as an area subjected to heavy livestock grazing for more than 100 years. However, 50 years of protection of the Chaco Canyon Culture National Historical Park in northwestern New Mexico resulted in little difference in biomass of Indian ricegrass (*Oryzopsis hymenoides*) or other grasses between the grazed area outside the park and the protected area within the park (Orodho et al. 1990) (Fig. 11). Indian ricegrass is rather sensitive to livestock grazing. On some sites, Indian ricegrass density was higher under protection from grazing than under livestock grazing, but not in all cases.

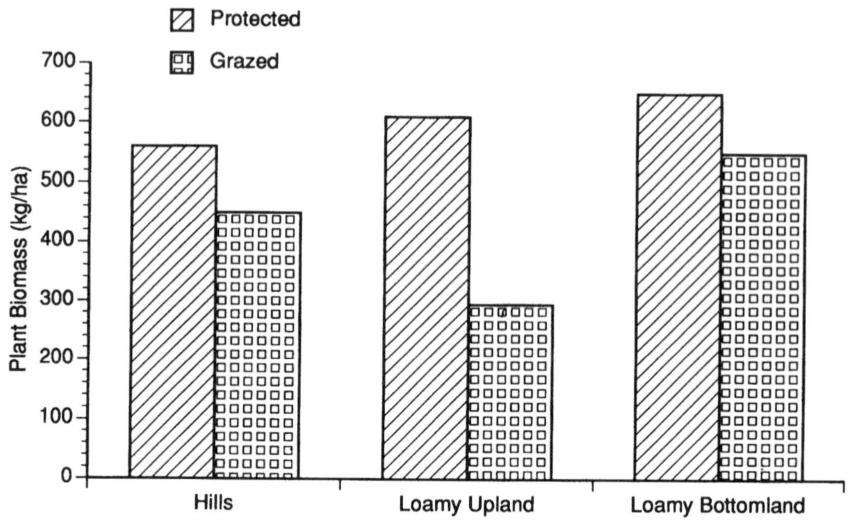

Figure 10. End-of-season-standing crop on livestock-grazed and protected sites in southcentral New Mexico (from Pieper 1967).

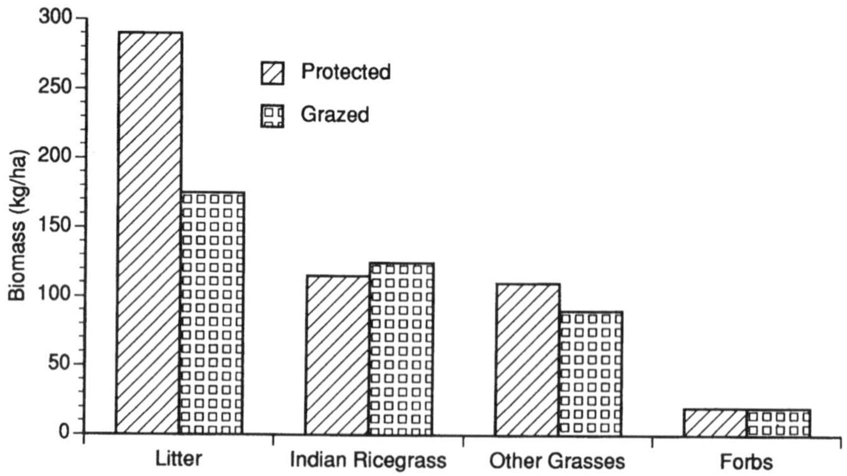

Figure 11. End-season-standing crop on livestock-grazed and protected sites in northwestern New Mexico (from Orodho et al. 1990).

These studies show it is difficult to generalize about the impact of livestock grazing on vegetation. Grazing does change some competitive relationships among plant species as outlined earlier, but extrinsic factors such as weather variations may play major roles in vegetation changes and may moderate grazing impacts. Grazing intensity and season of grazing, water distribution and a whole host of factors influence the grazing-vegetation interaction. Each case needs to be evaluated on its own characteristics to understand the mechanisms involved.

Sustainability Of Livestock Grazing

Is rangeland livestock production ecologically sustainable or will livestock grazing ultimately lead to rangeland deterioration, even at conservative stocking? Two examples illustrate the nature of vegetational changes over fairly long time spans. These two examples come from southern New Mexico where environmental conditions are severe and droughts common.

Figure 12 shows end-of-season perennial grass standing crop from 1939 through 1953 for one pasture on the Jornada Experimental Range grazed conservatively yearlong (Paulsen and Ares 1962). These data show that perennial grass standing crop varied from a high of 921 kg•ha^{-1} to a low of 156 kg•ha^{-1}. Standing crop declined in 1952 and 1953, the early part of the severe drought in the Southwest (Herbel et al. 1972). Drought is an extrinsic factor that interacts with the intrinsic processes and compartments in rangeland ecosystems. Drought can override internal controls in most situations. Examples include the drought of the 1930s in the Great Plains (Weaver and Albertson 1936), the drought of the 1950s in the Southwest (Herbel et al. 1972, Young 1956), and the Sahel in northern Africa in the early 1970s (Wallen and Gwynne 1978).

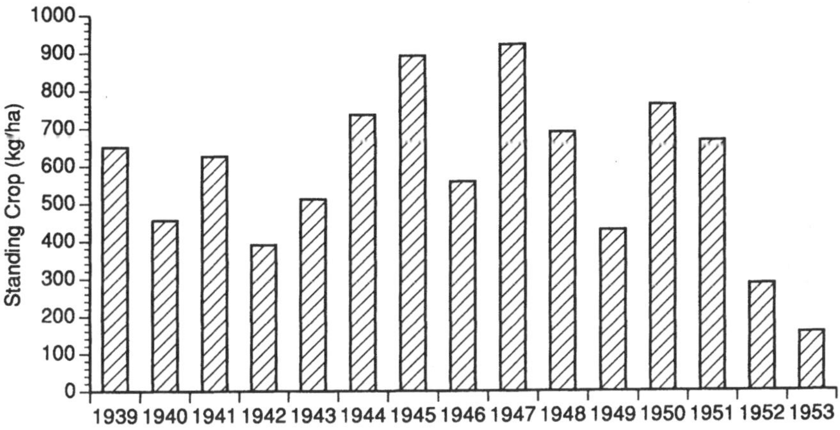

Figure 12. Perennial grass standing crop on Pasture 9 of the Jornada Experimental Range in southern New Mexico (from Paulsen and Ares 1962).

Figure 13 reflects end-of-season standing crop of grass species on a site similar to that reported by Paulsen and Ares (1962). This site also was grazed by cattle under conservative yearlong stocking (Beck and McNeeley 1991). Estimates of utilization of perennial grass species was less than 30% during early phases of the study. These data reflect favorable precipitation patterns during the 1980s (annual precipitation accounted for nearly 30% of the variation in plant biomass [Fig. 14]) under nearly constant grazing pressure. The pasture was stocked at the rate of 45 ha•AU yearlong (Beck and McNeeley 1991). These data also support the position that conservative livestock grazing can be sustained for as long as 25 years, even on relatively sensitive rangeland. Holechek (1991) has also discussed the sustainability of warm desert rangeland and reached a similar conclusion.

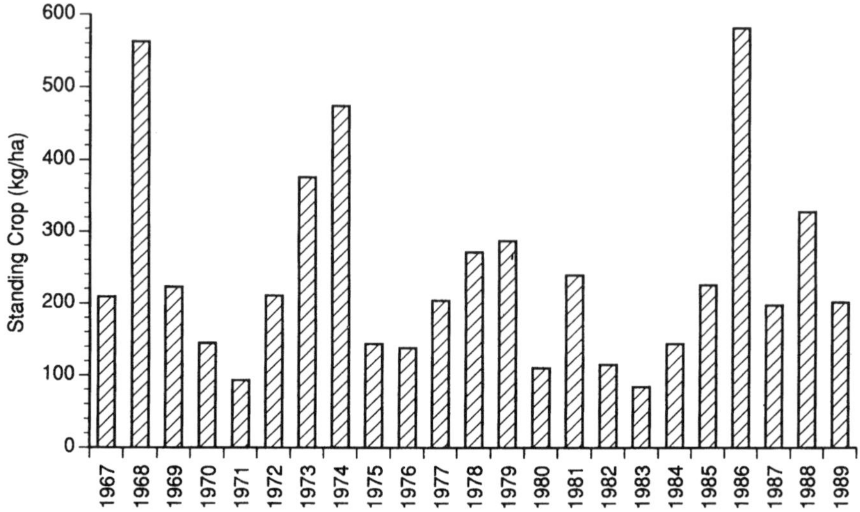

Figure 13. End-of-season standing crop of all species under light grazing in southern New Mexico (from Beck and McNeeley 1991).

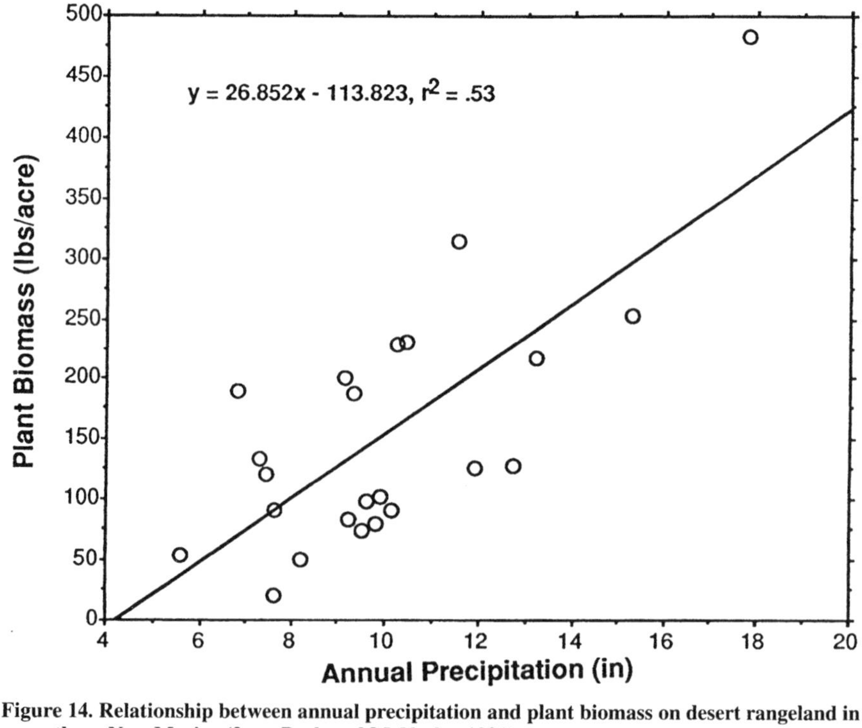

Figure 14. Relationship between annual precipitation and plant biomass on desert rangeland in southern New Mexico (from Beck and McNeeley 1991).

Evaluation of current rangeland conditions in the United States is difficult because of variations in the rangelands themselves as well as the variety of methods used to assess rangelands. The Society for Range Management has attempted to summarize the existing data on condition of rangelands in the United States but the data are difficult to interpret. Instead, I have attempted to show the size of rangelands of western states, which are considered in stable condition (Table 2). These data show more than 60% of western rangeland is in stable trend (i. e. there is no measurable change in range condition). Non-federal rangeland in stable condition appears to be about 10% greater than that on BLM rangeland. Data were not readily available for National Forest rangeland. However, only about 15 % of western rangeland is classified as having a declining trend. The Bureau of Land Management in Oregon estimated more than 60% of the rangeland was in satisfactory condition for the intended use (Walker 1991). These data need to be viewed with caution, but they do give us some idea of how ranges are changing.

Whether or not certain rangeland characteristics are desirable depends on the view of the beholder. One desiring a productive herd of pronghorn antelope might be willing to accept a range with abundant forbs and food resources for pronghorn (Howard et al. 1990), while others might settle for nothing less than tall waving grasses stretching from horizon to horizon. The present state of rangelands is a product of the interaction of many intrinsic (only one of which is livestock) and extrinsic factors.

Table 2. Percent of rangeland in stable trend on non-federal and public rangeland administered by BLM (SRM 1989)

State	BLM	Non-federal
West		
Colorado	59	68
Montana	82	71
Wyoming	57	73
Weighted Average	63	73
Great Basin		
Nevada	65	81
Utah	56	59
Weighted Average	62	72
Southwest		
Arizona	63	81
New Mexico	56	59
Weighted Average	60	71
Pacific		
Idaho	75	76
California	72	84
Oregon & Washington	66	68
Weighted Average	71	78

Summary And Conclusions

Much of the concern of those who feel current livestock grazing on public land is detrimental emanates from the following assumptions::

1. Much range deterioration (including shrub increase) in the West is a direct result of destructive livestock grazing.
2. Reduction or removal of livestock will solve the problem and return these rangeland to near-pristine conditions.
3. Livestock grazing is not sustainable on many fragile rangelands in the West.

Most range scientists believe that livestock grazing caused serious range deterioration during the period of open range grazing in the late 1800s and early 1900s before grazing control was applied. Even today destructive grazing is occurring on some western rangelands, but in general rangelands appear to be in the best condition that they have been in this century (Box and Sisson 1975). With the present state of knowledge, there appears to be no justification for destructive grazing on any rangeland today. On the other hand, removing all livestock grazing would not return rangelands to pristine conditions. In many cases, the changes would be subtle, and in the long run might even be negative in terms of biodiversity and other desirable characteristics.

Herbivory is a natural ecosystem process universal to rangelands worldwide. Although livestock are not native to western rangelands, these herbivores function similarly to native herbivores: livestock harvest plants, defecate, urinate and are involved in nutrient cycles; livestock compete with and complement other herbivores in rangeland ecosystems; livestock may stimulate primary production or depress it; livestock may serve as prey for several kinds of predators. Sometimes activities associated with livestock production operations have reduced predator populations. All of these processes operate within rangeland ecosystems, and often result in some type of equilibrium, however temporary, among the different components. Extrinsic factors such as drought, volcanic eruptions and increases in atmospheric CO_2 levels also influence rangeland ecosystems. These extrinsic factors often act in episodic fashion, causing widespread changes in organisms and landscape (e. g. the Mount St. Helens eruption, the drought of the 1930s in the Great Plains or that of the 1950s in the Southwest). Livestock grazing undoubtedly has played a role in vegetational change during the past 150 years, but the exact nature of this role is not clear and will probably never be settled to everyone's agreement.

Many rangelands still experiencing downward trend will respond favorably to lighter stocking, but some rangelands have probably reached a point where mere grazing management may not restore them to some previous condition. These ecological thresholds appear to be especially pertinent to sagebrush- grass, piñon-juniper woodland, and southwestern shrubby ecosystems (Laycock, this volume). In these cases, we need to decide what state we desire the vegetation to be in and if that state can be reached in a practical manner.

Conservative livestock grazing appears to be sustainable over the long- term. We have many examples where livestock have been grazed for many years and range conditions have improved (by most observer's standards). The recent series of examples illustrated at annual meetings of the Society for Range Management focusing on success stories provide evidence of the validity of the sustainability of livestock operations. Livestock grazing is only one of the many goods and services provided by rangelands (Lewis 1969), but controversies surrounding this one use often obscures the others.

Many rangeland soils are very susceptible to erosion and the tendency for some is to attribute many of these erosion problems to livestock grazing. We still have a difficult problem separating geological erosion from accelerated erosion. Where accelerated erosion is still occurring, remedial action should be considered, whether livestock grazing, cultivation, off road vehicle use or any other activity is involved.

In many cases, those on opposite sides of controversial issues overstate their cases. This seems to be the case in the present situation. We need to have the most reliable information possible, to consider all possible alternatives and compromises, and come to a rational decision. Extreme views will just lead us to louder confrontations. Livestock grazing can be destructive, but from a global ecological perspective, it provides much for many of the needs and pleasures of mankind.

Literature Cited

Aarssen, L. W. and D. L. Irwin. 1991. What selection: herbivory or competition. Oikos 60: 261-262.

Anderson, D. J. 1971. Pattern in desert perennials. J. Ecol. 59:555-560.

Anderson, D. C., K. T. Harper, and R. C. Holmgren. 1982a. Factors influencing development of cryptogamic soil crusts in Utah deserts. J. Range Manage. 35:180-185.

Anderson, D. E., K. T. Harper, and S. R. Rushforth. 1982b. Recovery of cryptogamic soil crusts from grazing on Utah winter ranges. J. Range Manage. 35:355-359.

Anderson, E. W. 1977. Effects of diet selectivity - a review of literature, p. 93-123. *In: The* impacts of herbivores on arid and semi-arid rangelands. Proc. 2nd U. S./Australia Rangeland Panel. Austral. Rangeland Soc. Perth, Western Australia.

Andrewartha, H. G. and L. C. Birch. 1954. The distribution and abundance of animals. Univ. Chicago Press. Chicago, IL.

Archer, S. 1989. Have southern Texas savannas been converted to woodlands in recent history? The Amer. Natur. 134:545-561.

Arthun, D. and J. L. Holechek. 1982. The North American bison. Rangelands 4:123-125.

Astatt, P. R. and D. J. O'Dowd. 1976. Plant defense guilds. Science 193:24-29.

Atwood, T. L. 1987. Influence of livestock grazing and protection from livestock grazing on vegetational characteristics of *Bouteloua eriopoda* rangeland. Ph. D. Dissertation. New Mex. State Univ., Las Cruces.

Barbour, M. G., J. H. Burk, and W. D. Pitts. 1987. Terrestrial plant ecology. 2nd Ed. Benjamin Cummings Pub. Co. Inc. Menlo Park, CA.

Barney, M. A. and N. C. Frischknecht. 1974. Vegetation change following fire in the pinyon-juniper type of west-central Utah. J. Range Mange. 27:91-96.

Barrow, N. J. 1967. Some aspects of the effects of grazing on the nutrition of pastures. J. Austr. Agr. Sci. 33:254-257.

Beck, R. F. and D. A. Tober. 1985. Vegetational changes on creosotebush sites after removal of shrubs, cattle and rabbits. New Mex. Agr. Exp. Sta. Bull. 717.

Beck, R. F. and R.P. McNeeley. 1991. Effects of continuous grazing on black grama range. Abstract. P 65. Annual Meeting, Soc. Range Manage. Washington, D.C.

Belsky, A. J. 1986. Does herbivory benefit plants? A review of the evidence. The Amer. Natur. 127:870-892.

Blaylock, D. M. 1982. Home, home on the range. Rangelands 4:220-221.

Botkin, D. B. 1991. A new balance of nature. Wilson Quarterly. Spring 1991.

Box, T. W. and W. B. Sisson. 1975. Public ranges are in better condition than they have been for 70 years - but more range managers could make then even better! Rangelands 2:44-46.

Brady, W. W., M. R. Stromberg, E. F. Aldon, C. D. Bonham, and S. H. Henry. 1989. Response of a semidesert grassland to 16 years of rest from grazing. J. Range Manage. 42:284-288.

Branscomb, B. L. 1958. Shrub invasion of a southern New Mexico desert grassland. J. Range Manage. 11:129-133.

Branson, F. A. 1953. Two factors affecting resistance of grasses to grazing. J. Range Manage. 6:165-171.

Branson, F. A. 1985. Vegetation changes on western rangelands. Range Monog. No. 2. Soc. Range Manage. Denver, CO.

Brown, J. R., and S. Archer. 1989. Woody plant invasion of grasslands: establishment of honey mesquite (*Prosopis glandulosa* var. *glandulosa*) on sites differing in herbaceous biomass and grazing history. Oecologia: 80:19-26.

Bryson, R.A. 1974. A perspective on climatic change. Science 184:753-760.

Buffington, L. C. and C. H. Herbel 1965. Vegetational changes on a semidesert grassland range from 1858 to 1963. Ecol. Monogr. 35:139-164.

Buwai, M. and M. J. Trlica. 1977. Multiple defoliation effects on herbage yield, vigor, and total non-structural carbohydrates of five range species. J. Range Manage. 30-164-171.

Caldwell, M. M. 1984. Plant requirements for prudent grazing, p. 117-152. *In*: Developing strategies for rangeland management. Westview Press, Boulder, CO.

Caldwell, M. M., J. H. Richards, D. A. Johnson, R. S. Nowak, and R. S. Dzurec. 1981. Coping with herbivory: photosynthetic capacity and resource allocation in two *Agropyron* bunchgrasses. Oecologia 50:14-24.

Clements, F. E. 1907. Plant physiology and ecology. Henry Holt and Co., New York.

Clements, F. E. 1916. Plant succession: an analysis of the development of vegetation. Carnegie Inst. Wash. Pub. 520. Washington, DC.

Clements, F. E. 1936. Nature and structure of the climax. J. Ecol. 24:252-284.

Conley, W., M. R. Conley and T. R. Karl. 1988. Historical increase of shrublands in the northern Chihuahuan desert: climate or grazing? Manuscript. Dept. Biology, New Mex. State Univ., Las Cruces, NM.

Connell, J. H. 1978. Diversity in tropical rain forests and coral reefs. Science 199:1302-1310.

Colville, F. J. 1898. Forest growth and sheep grazing in the Cascade Mountains of Oregon. U. S. Dept. Agr. For Div. Bull. 15.

Crawley, M. J. 1983. Herbivory - The dynamics of animal-plant interactions. Blackwell Sci. Publ., Oxford, England.

Crisp, D. J. (ed.). 1964. Grazing in terrestrial and marine environments. Blackwell Sci. Publ. Oxford, England.

Dahl, B. E. and D. N. Hyder. 1977. Developmental morphology and management implications, p. 257-290. *In*: R. E. Sosebee, (ed.). Rangeland plant physiology. Range Sci. Ser. No. 4. Soc. Range Manage. Denver, CO.

Danckwerts, J. E. and G. C. Stuart-Hill. 1991. Are rangeland dynamics under no more than weak control of the grazier? Abstract. Fourth Intern. Rangeland Congr. Montpellier, France.

DeAngelis, D. L. and J. C. Waterhouse. 1987. Equilibrium and nonequilibrium concepts in ecological models. Ecol. Monogr. 57:1-21.

Donart, G. B. 1984. The history and evolution of western rangelands in relation to woody plant communities, p. 1235-1258. *In*: Developing strategies for rangeland management. Westview Press. Boulder, CO,

Dyer, M. I., J. K. Detling, D. C. Coleman, and D. W. Hilbert. 1982. The role of herbivores in grasslands, p. 255-295. *In*: J. R. Estes, R. J. Tyrl, and J. N. Brunken (eds.). Grasses and grasslands -systematics and Ecology. Univ. Oklahoma Press, Norman.

Dyksterhuis, E. J. 1949. Condition and management of rangeland based on quantitative ecology. J. Range Manage. 2:104-115.

Dyksterhuis, E. J. 1958. Ecological principles in range evaluation. Bot. Rev. 24:253-272.

Egler, F. E. 1954. Vegetation science concepts. I. Initial floristic composition, a factor in old-field vegetation development. Vegetatio 4:412-417.

Ellis, J. E. and D. M. Swift. 1988. Stability of African pastoral societies: alternate paradigms and implications for development. J. Range Manage. 41:450-459.

Ellison, L. 1960. Influence of grazing on plant succession of rangelands. Bot. Rev. 26:1- 141.

England, R. E. and A. DeVos. 1969. Influence of animals on pristine conditions on the Canadian grasslands. J. Range Manage. 22:87-93.

Erdman, J. A. 1970. Piñon-juniper succession after natural wildfires in residual soils of Mesa Verde, Colorado. Sci. Bull. Biol. Ser. 11:1-24. BYU, Provo, UT.

Fleming, C. E. 1942. Over-night versus established bed-grounds on Nevada sheep ranges. Nev. Agr. Exp. Sta. Bull. 103.

Friedel, M. H. 1991. A viewpoint on range condition assessment and the concept of thresholds. J. Range Manage. 44:422-426.

Galbraith, W. A. and E. W. Anderson. 1971. Grazing history of the Northwest. J. Range Manage. 24:6-12.

Gamougoun, N. D. R. P. Smith, M. K. Wood, and R. D. Pieper. 1984. Soil, vegetation and hydrologic responses to grazing management at Fort Stanton, New Mexico. J. Range Manage. 37:538-541.

Gardner, J. L. 1951. Vegetation of the creosotebush area of the Rio Grande Valley in New Mexico. Ecol. Monogr. 21:379-403.

Gibbens, R. P. and R. F. Beck. 1987. Increase in the number of dominant plants and dominance-classes on a grassland in the northern Chihuahuan desert. J. Range Manage. 40:136-139.

Gibbens, R. P. and R. F. Beck. 1988. Changes in grass basal area and forb densities over a 64-year period on grassland types of the Jornada Experimental Range. J. Range Manage. 41:186-192.

Gleason, H. A. 1917. The structure and development of the plant association. Bull. Torrey Bot. Club. 44:463-481.

Gleason, H. A. 1926. The individualistic concept of the plant association. Bull. Torrey Bot. Club. 513:1-20.

Gleason, H. A. and A. Cronquist. 1964. The natural geography of plants. Columbia Univ. Press. New York.

Grime, J. P. 1973. Control of species density in herbaceous vegetation. J. Environ. Manage. 1:151-167.

Grime, J. P. 1979. Plant strategies and vegetation processes. John Wiley and Sons. New York.

Hamilton, B. A., K. J. Hutchinson, P. C., Annis, and J. B. Donnelly. 1973. Relationships between the diet selected by grazing sheep and the herbage on offer. Austr. J. Agr. Res. 24:272-277.

Haque, Z., A. Younga, K. C. McDaniel, and R. D. Pieper. 1991. Two-phase pattern in mesquite-herbland in southern New Mexico. Southwest Natur. 36:54-59.

Harper, K. T. and J. R. Marble. 1988. A role for nonvascular plants in management of arid and semiarid rangelands, p. 135-169. *In*: P. T. Tueller (ed.). Vegetation science applications for rangeland analysis and management. Kluwer Academic Publishers, Dordrecht, Netherlands.

Hart, R. H. and B. E. Norton. 1988. Grazing management and vegetation response, p. 493-525. *In*: P. T. Tueller, (ed.). Vegetation science applications for rangeland analysis and management. Kluwer Academic Publishers, Dordrecht, The Netherlands.

Hawley, A. W., D. G. Peden, and W. R. Stricklin. 1981. Bison and Hereford steer digestion of sedge hay. Can. J. Anim. Sci. 61:165-174.

Hastings, J. R. and R. M. Turner. 1965. The changing mile. Univ. Ariz. Press, Tucson, AZ.

Heady H. F. 1973. Structure and function of climax, p. 73-80. *In*: Arid shrublands. Proc. Third Workshop, U. S. Australia Rangelands Panel. Soc. Range Manage. Denver, CO.

Heady, H. F. 1975. Rangeland management. McGraw-Hill Book Co., New York.

Heady, H. F. 1984. Concepts and principles underlying grazing systems, p. 885-902. *In*: Developing strategies for rangeland management. Westview Press. Boulder, CO.

Herbel, C. H., F. N. Ares, and R. A. Wright. 1972. Drought effects of a semidesert grassland range. Ecology 53:1084-1093.

Hilbert, D. W., D. M. Swift, J. K Detling, and M. I. Dyer. 1981. Relative growth rates and the grazing hypothesis. Oecologia 51: 14-18.

Hilder, E. J. 1966. Distribution of excreta by sheep at pasture, p.977-981. *In*: Proc. Tenth Inter. Grassl. Congr. Helsinki.

Hill, A. 1991. Grazing as a management tool. Wildflower. Nat. Wildflower Res. Center. 4:13-18.

Holechek, J. L., R. D. Pieper, and C. H. Herbel. 1989. Range Management Principles and Practices. Prentice Hall. Englewood Cliffs, NJ.

Holechek, J. L. 1991. Chihuahuan desert rangeland, livestock grazing, and sustainability. Rangelands 13:115-120.

Holechek, J. L. 1981. Livestock grazing impacts on public lands: A viewpoint. J. Range Manage. 34:251-254.

Howard, V. W., Jr., J. L. Holechek, R. D. Pieper, K. Green-Hammond, M. Cardenas, and S. L. Beasom. 1990. Habitat requirements for pronghorn on rangeland impacted by livestock and net wire in eastcentral New Mexico. New Mex. State Univ. Agr. Exp. Sta. Bull. 750.

Humphrey, R. R. 1987. 90 years and 535 miles. Univ. New Mexico Press. Albuquerque.

Hyder, D. N. 1972. Defoliation in relation to vegetative growth, p. 304-317. *In*: V. B. Younger, and C. M. McKell. Biology and utilization of grasses. Academic Press, New York.

Hyder, D. N. 1974. Morphogenesis and management of perennial grasses the United States, p. 89-98. *In*: K. W. Kreitlow, (ed.). Plant morphogenesis as the basis for scientific management of range resources. U. S. Dept. Agr. Misc. Pub. 1271

Idso, S. B. 1989. A problem for paleoclimatology? Quartenary Res. 31:423-434.

Jacobs, L. 1991. Waste of the West: public lands ranching. Lynn Jacobs, P. O. Box 5784, Tucson, AZ.

Janzen, D. H. 1986. Chihuahuan Desert nopolleras: defaunated big mammal vegetation. Ann. Rev. Ecol. Syst. 17:595-636.

Jardine, J. T. and M. Anderson. 1919. Range management on the national forests. U. S. Dept. Agr. Bull. 790.

Jameson, D. A. 1963. Responses of individual plants to harvesting. Bot. Rev. 29:532-594.

Johansen, J. R. 1986. Importance of cryptogamic soil crusts to arid rangelands: Implications for short duration grazing, p. 127-136. *In*: J. A. Tideman (ed.). Short duration grazing. Washington State Univ., Pullman.

Kautz, J. E. and G. M. Van Dyne. 1978. Comparative analyses of diets of bison, cattle, sheep, and pronghorn antelope on shortgrass prairie in northeastern Colorado, U. S. A, p. 438-443. *In*: D. N. Hyder (ed.). Proc. First Intern. Rangeland Congr. Soc. Range Manage. Denver, CO.

Kirsch, L. M. and A. D. Kruse. 1972. Prairie fires and wildlife. Proc. Tall Timbers Fire Ecol. Conf. No. 12. Tall Timbers Res Sta., Tallahassee, FL.

Klemmedson, J. O. 1977. Physical effects of herbivores on arid and semiarid rangeland ecosystems, p. 187-210. *In: The* impact of herbivores on arid and semi-arid rangeland. Proc. 2nd U. S./Australia Rangeland Panel. Austr. Rangeland Soc., Perth, West. Austr.

Klemmedson, J. O. and J. G. Smith. 1964. Cheatgrass (*Bromus tectorum* L). Bot. Rev. 30:226-262.

Krueger, W. C. 1988. Rangelands- nature, history and ownership, p. 1-7. *In*: B. A. Buchanan (ed.). Rangelands. Univ. New Mex. Press, Albuquerque, NM.

Kennedy, P. B. and S. B. Doten. 1901. A preliminary report on the summer ranges of western Nevada sheep. Nev. State Univ. Agr. Exp. Sta. Bull. 51.

Larson, F. 1940. The role of the bison in maintaining the short grass plains. Ecology 21:113-121.

Laycock, W. A. 1991. Stable states and thresholds of range condition on North American rangelands-a viewpoint. J. Range Manage. 44:427-433.

Lewis, J. K. 1969. Range management viewed in the ecosystem framework., p. 97-187. *In*: G. M. Van Dyne (ed.). The ecosystem concept in natural resource management. Academic Press, New York.

Mack, R. N. and J. N. Thompson. 1982. Evolution in steppe with few large, hooved mammals. The Amer. Natur. 119:757-773.

Martin, C. S. 1975. Ecology and management of southwestern semi-desert grass-shrub ranges: The status of our knowledge. U. S. Dept. Agr. For. Ser. Res. Paper RM-156.

Martin, P. S. and H. D. Wright. 1967. Pleistocene extinctions, the search for a cause. Yale Univ. Press. New Haven.

Martin, P. S. 1968. The meaning of empty niches. AIBS Meeting, Pleistocene Man- environment Relationships Symposium. Columbus, OH.

Maschinski, J. and T. G. Whitham. 1989. The continuum of plant responses to herbivory: the influence of plant association, nutrient availability, and timing. Amer. Natur. 134: 1-19.

Mabbutt, J. A. and C. Floret (eds.). 1980. Case studies on desertification. UNESCO. Paris.

McCarty, E. C. and R. Price. 1942. Growth and carbohydrate content of important mountain forage plants in central Utah as affected by clipping and grazing. U. S. Dept. Agr. Tech. Bull. 818.

McCormick, J. 1968. Succession. Via 1:22-35, 131-132.

McNaughton, S. J. 1979. Grazing as a optimization process: grass-ungulate relationships in the Serengeti. Amer. Natur. 113:691-703.

McNaughton, S. J. 1983. Compensatory plant growth as a response to herbivory. Oikos 40:329-336.

McNaughton, S. J. 1984. Grazing lawns: animals in herds, plant form and coevolution. The Amer. Natur. 124:863-886.

McNaughton, S. J. 1985. Ecology of a grazing ecosystem: the Serengeti. Ecol. Monogr. 55:259-294.

McNaughton, S. J. 1991. Evolutionary ecology of large tropical herbivores, p. 92-116. *In*: Price, P. W, T. M. Lesinson, G. W. Fernandes, and W. W. Benson. Plant animal interactions: evolutionary ecology in tropical and temperate regions. John Wiley & Sons Inc, NY.

Menke, J. W. and M. J. Trlica. 1981. Carbohydrate reserve, phenology, and growth cycles on nine Colorado range species. J. Range Manage. 34:269-277.

Milchunas, D. G., O. E. Sala, and W. K. Lauenroth. 1988. A generalized model of the effects of grazing by large herbivores on grassland community structure. Amer. Nat. 132:87-106.

Mueggler, W. F. 1972. Influence of competition on the response of bluebunch wheatgrass to clipping. J. Range Manage. 25:88-92.

Neilson, R. P. 1986. High-resolution climatic analysis and Southwest biography. Science 232:27-34.

Nelson, A. 1898. The red desert of Wyoming and its forage resources. U. S. Dept. Agr. Div. Agros. Bull. 13.

Nelson, E. W. 1934. The influence of precipitation and grazing upon black grama grass range. U. S. Dept. Agr. Tech Bull. 409.

Omar, S. A. and J. Bartolome. 1991. Range and state/transition model for the *Rhanterium* and *Haloxylon* steppe Arabian rangelands. Abstract. Fourth Intern. Rangel. Congr. Montpellier, France.

Orodho, A. O., M. J. Trlica, and C. D. Bonham. 1990. Long-term heavy-grazing effects on soil and vegetation in the four corners region. The Southwestern Natur. 35:90-14.

Owen, D. F. 1980. How plants may benefit from the animals that eat them. Oikos 35:230- 235.

Owen, D. F. 1990. The language of attack and defense. Oikos 37:133-135.

Owen, D. F. and R. G. Wiegert. 1976. Do consumers maximize plant fitness? Oikos 27:488- 492.

Paige, K. N. and T. G. Whitham. 1987. Overcompensation in response to mammalian herbivory: the advantage of being eaten. The Amer. Natur. 129:407-416.

Paulsen, H. A. Jr. and F. N. Ares. 1962. Grazing values and management of black grama and tobosa grasslands and associated shrub ranges of the Southwest. U. S. Dept. Agr., For. Ser. Tech. Bull. 1270.

Peet, R. K. D. C. Glenn-Lewin, and J. W. Wold. 1983. Predictions of man's impact on vegetation, p. 324-338. *In*: W. Holzner, M. J. A. Werger, and I. Ikusiam (eds.). Junk Pub., The Hague.

Peterson, C. S. 1985. Grazing in Utah: A historical perspective, p. 5-14. *In*: Symposium of cultural, physical, and biological characteristics of the range livestock industry in the Great Basin. Soc. Range Manage. Denver, CO.

Peden, D. G., G. M. Van Dyne, R. W. Rice, and R. M. Hansen. 1974. The trophic ecology of *Bison bison* L. on shortgrass plains. J. Applied Ecol. 11:489-498.

Picket, S. T. A. and P. S. White (eds.). 1986a. The ecology of patch disturbances and patch dynamics. Academic Press. San Diego, CA.

Pickett, S. T. A. and P. S. White. 1986b. Patch dynamics: a synthesis, p. 371-384. *In*: Pickett, S. T. A. and P. S. White (eds.). The ecology of patch disturbances and patch dynamics. Academic Press. San Diego, CA.

Pieper, R. D. 1967. Comparisons of vegetation on grazed and ungrazed pinyon-juniper grassland sites in southcentral New Mexico. J. Range Manage. 21:51-53.

Pieper, R. D., C. W. Cook, and L. E. Harris 1959. Effects of intensity of grazing upon nutritive content of the diet. J. Anim. Sci. 18:1031-1037.

Pieper, R. D. 1977. Effects of herbivores on nutrient cycling and distribution, p. 249-275. *In*: The impact of herbivores on arid and semi-arid rangelands. Proc. 2nd U. S./Australia Rangeland Panel. Austr. Rangeland Soc., Perth, West. Austr.

Pieper, R. D. 1984. Plant community response as a basis for grazing systems. Abstract. 37th Annual meeting, Soc. Range Manage., Denver, CO.

Pieper, R. D. and R. K. Heitschmidt, 1988. Is short-duration grazing the answer? J. Soil and Water Cons. 43:133-137.

Pieper, R. D. and R. D. Wittie. 1990. Fire effects in southwestern chaparral and piñon-juniper woodlands, p. 87-93. *In*: J. S. Krammes (tech coord.). Effects of fire management on southwestern natural resources. U. S. Dept. Agr. For. Ser. Gen Tech. Rep. RM-191.

Platou, K. A. and P. T. Tueller. 1985. Evolutionary implications for grazing management systems. Rangelands 7:57-61.

Pulyarkin, V. A. and A. N. Rakitnikov. 1986. Agricultural development and desertification control, p. 33-38. *In*: Arid land development and the combat against desertification: an integrated approach. Centre for International Projects GKNT. Moscow, USSR.

Risser, P. G., E. C. Birney, H. D. Blocker, S. W. May, W. J. Parton, J. A. Wiens. 1981. The true prairie ecosystem. Hutchinson Ross Pub. Co., Stroudsburg, PA.

Sampson, A. W. 1917. Succession as a factor in range management. J. For. 15:593-596.

Sampson, A. W. 1919. Plant succession in relation to range management. U. S. Dept. Agr. Bull. No. 791.

Sampson, A. W. and E. C. McCarty. 1930. The carbohydrate metabolism of *Stipa pulchra*. Hilgardia 5:61-100.

Schickedanz, J. G. 1980. History of grazing in the Southwest, p. 1-9. *In*: K. C. McDaniel and C. D. Allison (eds). Proc. Grazing Manage. Systems for Southwestern Rangelands Sym. New Mex. State Univ., Las Cruces, NM.

Schlesinger, W. H., J. F. Reynolds, G. L. Cunningham, L. F. Huenneke, W. M. Jarrell, Ross A. Virginia, W. G. Whitford. 1990. Biological feedbacks in global desertification. Science: 247:1043-1048.

Seton, E. T. 1927. Lives of game animals. Vol. 3. Ungulates. Doran and Co. New York.

Silvertown, J. W. 1982. Introduction to plant population ecology. Longman House, London.

Simpson, J. R. 1968. Losses of urea nitrogen from the surface of pasture soils, p. 89-93. Trans. 9th Inter. Congr. Soil. Sci. Adelaide, Australia.

Smeins, F. E. 1983. Origin of the brush problem - a geological and ecological perspective of contemporary distributions, p. 5-16. *In*: K. C. McDaniel (ed.). Proceedings brush management symposium. Texas Tech Univ. Press. Lubbock, TX

Smith, J. G. 1895. Forage conditions of the prairie region. U. S. Dept. Agr. Yearbook, Washington, D. C.

Smith, J. G. 1899. Grazing problems in the southwest and how to meet them. U. S. Dept. Agr. Div. Agros. Bull. 16.

Society for Range Management. 1989. Assessment of rangeland condition and trend of the United States. Soc. Range Manage. Denver, CO.

Sollod, A. E. 1990. Rainfall, biomass and the pastoral economy of Niger: assessing the impact of drought. J. Arid. Environ. 18:97-107.

Sosebee, R. E. and C. Wan. 1988. Ecophysiology of range plants, p. 29-67. *In*: P. T. Tueller (ed.). Vegetation science applications for rangeland management. Kluwer Academic Publishers, Dordrecht, The Netherlands.

Strange, L. R. N. 1980. Africa pastureland ecology. Food and Agr. Organization of the United Nations. Rome.

Stoddart, L. A. and D. A. Smith. 1943. Range management. McGraw-Hill Book Co., New York.

Sutherland, J. P. 1974. Multiple stable points in natural communities. The Amer. Natur. 108:859-873.

Tansley, A. G. 1935. The use and abuse of ecological concepts and terms. Ecology 16:284-307.

Thomas, G. W. and T. S. Ronningen. 1965. Rangelands: our billion acre resource. Agr. Sci. Rev. 3:10-17.

Thornber, J. J. 1910. The grazing ranges of Arizona. Univ. Ariz., Agr. Exp. Sta. Bull. 65. Tucson, AZ.

Tiedemann, A. R., J. O. Klemmedson, and P. R. Ogden. 1971. Response of four perennial southwestern grasses to shade. J. Range Manage. 24: 442-447.

Tiedemann, A. R., H. R. Sanderson, and N. J. Cimon. 1986. Future site productivity considerations of short-duration, high-intensity grazing, p. 137-144. *In*: J. A. Tiedeman (ed.). Short Duration Grazing. Washington State Univ., Pullman.

Tobey, R. C. 1981. Saving the prairie. The life cycle of the founding school of American plant ecology, 1895-1955. Univ. Calif. Press. Berkeley,. CA.

Tucker, C. J., H. E. Dregne, and W. W. Newcomb. 1991. Expansion and contraction of the Sahara Desert from 1980 to 1990. Science 253:299-301.

Turner, R. B., C. E. Poulton, and W. L. Gould. 1963. Medusahead-a threat to Oregon rangeland. Ore. State. Univ. Agr. Exp. Sta. Spec. Rep. 149.

Vale, T. H. 1982. Plants and People. Vegetation change in North America. Assoc. Amer. Geogr. Washington, D. C.

Valentine, K. A. 1970. Influence of grazing intensity on improvement of deteriorated black grama range. New Mex. State Univ. Agr. Exp. Sta. Bull. 553.

Van Auken, O. W. and J. K. Bush. 1989. *Prosopis glandulosa* growth: influence of nutrients and simulated grazing of *Bouteloua curtipendula.* Ecology 70:512-516.

Van Dyne, G. M., N. R. Brockington, Z. Szocs, J. Duek, and C. A. Ribic. 1980. Large herbivore subsystem., p. 269-537. *In*: A. J. Breymeyer and G. M. Van Dyne (eds). Grasslands, systems analysis and man. Cambridge University Press, Great Britain.

Wagner, F.H. 1978. Livestock grazing and the livestock industry, p. 49-78. *In*: H. P. Brokaw (ed.). Wildlife and America. Council on Environmental Quality. Washington, DC.

Wahl, E. W. and T.L. Lawson. 1970. The climate of the mid-ninteenth century United States compared to the current normals. Monthly Weather Rev. 98:259-265.

Walker, B. H. 1988. Autecology, synecology, climate, and livestock as agents of rangeland dynamics. Austr. Rangel. J. 10:69-75.

Walker, L. 1991. Majority of BLM range sites meet land use objectives. *In*: BLM news. June 1991. Portland, OR.

Wallen, C. C. and M. D. Gwynne. 1978. Drought - a challenge to rangeland management, p. 21-31. *In*: D. N. Hyder (ed.). Proc. First Intern. Rangeland Congr. Soc. Range Manage. Denver, CO.

Weaver, J. E. and F. W. Albertson. 1936. Effects of the great drought on the prairies of Iowa, Nebraska, and Kansas. Ecology 17:567-639.

Welsh, R. G. and R. F. Beck. 1976. Some ecological relationships between creosotebush and bush muhly. J. Range Manage. 29:472-475.

West, N. E. 1979. Basic synecological relationships of sagebrush-dominated lands in the Great Basin and the Colorado Plateau, p. 33-41. *In:* The sagebrush ecosystem: a symposium. Utah State Univ. Coll. Natur. Resources, Logan, UT.

West, N. N. 1988. Plant synecology in the service of rangeland management, p. 11–27. *In*: P. T. Tueller, (Ed.). Vegetation science applications for rangeland analysis and management. Kluwer Academic Publishers, Dordrecht, Netherlands.

West, N. W. and N. S. Van Pelt. 1987. Successional patterns in piñon-juniper woodlands, p. 43–52. *In*: R. L. Everett (compiler). Proceedings - Piñon-juniper Conference. U. S. Dept. Agr. For. Ser. Gen. Tech. Rep. INT - 215.

Westoby, M., B. Walker, and I. Noy-Meir. 1989. Opportunistic management for rangelands not at equilibrium. J. Range. Manage. 42:266-274.

Wiens, J. A. 1977. On competition and variable environments. Amer. Sci. 65:590–597.

Wiens, J. A. 1985. Vertebrate responses to environmental patchiness in arid and semiarid ecosystems, p. 193. *In*: S. T. A. Pickett and P. S. White. The ecology of natural disturbance and patch dynamics. Academic Press. San Diego, CA.

Wiens, J. A. 1991. Ecological crunches and competition. Current Contents. April 8, 1991.

Williams, C. R. 1970. Pasture nitrogen in Australia. J. Austr. Inst. Agr. Sci. 36:199-205.

Williamson, S. C., J. K. Detling, J. L Dodd, and M. I. Dyer. 1989. Experimental evaluation of the grazing optimization hypothesis. J. Range Manage. 42:149-152.

Wilson, A. D., K. C. Hodgkinson and J. C. Noble. 1988. Vegetation attributes and their application to the management of Australian rangelands, p. 253-294. *In*: P. T. Tueller (ed.). Vegetation science applications for rangeland analysis and management. Kluwer Academic Publishers, Dordrecht, The Netherlands.

Wooton, E. O. 1908. The range problem in New Mexico. New Mex. Coll. Agr. and Mechanic Arts. Agr Exp. Sta. Bull 66., Las Cruces.

Wright, H. A, L. F. Neuenschwander, and C. M. Britton. 1979. The role and use of fire in sagebrush-grass and piñon-juniper plant communities-a state-of-the-art review. U. S. Dept. Agr. For. Ser. Gen. Tech. Rep. INT-58.

York, J. D. and W. A. Dick-Peddie. 1969. Vegetation changes in southern New Mexico during the past hundred years, p. 156-166. *In*: W. G. McGinnies and B. J. Goldman (eds.). Arid lands in perspective. Univ. Ariz. Press, Tucson.

Young, J. A. and R. A. Evans 1981. Demography and fire history of a western juniper stand. J. Range Manage. 34:501-506.

Young, J. A. and F. Tipton. 1990. Invasion of cheatgrass into arid environments of the Lahontan Basin, p. 37-40. *In*: Proceedings - Symposium on cheatgrass invasion, shrub die-off, and other aspects of shrub biology and management. U. S. Dept. Agr. For. Ser. Ben. Tech. Rep. INT -176.

Young. J. A. and R. A. Evans. 1979. Historical perspective regarding the sagebrush ecosystem, p. 1-13. *In*: The sagebrush ecosystem: a symposium. College of Natur. Res. Utah State Univ. Logan, UT.

Young, J. A., R. A. Evans, and P. T. Tueller. 1976. Great Basin plant communities — Pristine and grazed, p. 187-215. *In*: Elation, R. (ed.). Holocene environmental change in the Great Basin. Nevada Archeological Survey Res. Paper No. 5.

Young, J. A. and R. A. Evans. 1985. The grand experiment -ranching in the Great Basin., p, 1–4. *In*: Symposium of cultural, physical, and biological characteristics of the range livestock industry in the Great Basin. Soc. Range Manage. Denver, CO.

Young, J. A., R. A. Evans, and J. Major. 1972. Alien plants in the Great Basin. J. Range Manage. 25:194-201.

Young, V. A. 1956. The effects of the 1949-1954 drought on the ranges of Texas. J. Range Manage. 9:139-142.

Riparian and Watershed Systems: Degradation and Restoration

Wayne Elmore and Boone Kauffman

Abstract

Diaries of travelers and emigrants that passed through western rangelands between the years 1812-1880 reported dense and continuous stands of willows interspersed with wide, wet meadows along stream systems. By the turn of the century these riparian corridors were fragmented into thinner and less continuous bunches in many places. Damage to riparian systems was completed after World War II when extensive channelization and phreatophyte control was standard practice. At the same time the uplands, in many areas of the west, were becoming afforested with juniper and extensive stands of sage. This was probably a result of fire suppression and overgrazing. This afforestation has resulted in seeps and small streams drying up, or experiencing much lower summer base flows. The reduction of water, coupled with a lowering of the adjacent streamside water table from channel downcutting, has created conditions favoring the establishment of xeric plant communities. Steams and their floodplains are dynamic, and riparian vegetation density and contiguousness depends on the geomorphic setting. Unconstrained steams (wide valleys or floodplains less than twice the active channel width) have the greatest potential for elevating water tables and restoring native riparian plant communities. In-channel habitat restoration may aid in raising the water table, however, care must be exercised in its use. Riparian exclosures (and livestock reductions) are still the preferred methods for recovery of the shrub component of riparian vegetation by many groups. However, livestock grazing can be present in some areas while streams are improving. Grazing strategies that link the uplands, the riparian areas, and the stream channels must be used if we are going to have an effective and self-sustaining riparian restoration program.

Historical

Historical accounts by the first Euro-americans to venture into the western rangelands of North America often included descriptions of dense stands of willows (*Salix* spp.) that were interspersed with wetland meadows along meandering stream systems (Ogden 1824-26). Since that time a myriad of land uses has dramatically changed the character of western riparian ecosystems. The first major anthropogenic (i.e. human caused) effects that influenced the composition, structure, and function of many western stream systems came through the extirpation of beaver (*Castor canadensis*) by trappers (Kauffman 1987, Kovalchik and Elmore 1991, Naiman et al. 1988). The elimination of beavers from streams altered site hydrology with subsequent changes in stream nutrient processes and riparian vegetation dynamics (Kovalchik and Elmore 1991, Naiman et al. 1986, Naiman et al. 1988). The beaver plays a functional role as a keystone species in many riparian ecosystems. Their activities influence stream hydrology including residence times of water in riparian zones, stream temperatures, and nutrient cycles; provide habitats for aquatic organisms; and vegetation dispersal, dynamics, and structure of riparian zones. Their elimination can result in decreased surface water (i.e. loss of ponds), channel diversity, and conversion of wetland plant communities to xeric plant communities.

Following the severe overtrapping of North American beaver populations, the second, and perhaps most influential, human perturbation (i.e. human caused activities which cause a deviation from natural processes) of western riparian ecosystems appeared — the introduction of livestock. It was estimated that by the end of the century there were 26 million cattle and 20 million sheep in the Western United States (Wilkenson 1992). Grazing during this time period was likely year-long resulting in extensive degradation of riparian/stream systems throughout the West. The grazing by livestock continues to influence riparian zones. While the total animal unit months of livestock use has declined on public lands (U.S. Forest Service and BLM), the numbers of cattle have remained relatively stable on BLM lands since the 1940's and on U.S. Forest Service lands since the late 1950's (Wagner 1978). However, over all land ownerships in the 11 western range states the number of cattle and the total forage demand for livestock has steadily risen throughout the 20th Century (Wagner 1978).

Following the recognition of the influences of overgrazing on rangelands, the first legal attempt to reduce livestock numbers on Federal lands came in 1934 with the passage of the Taylor Grazing Act (Wilkenson 1992). The Depression resulted in lower beef and lamb prices and the Dust Bowl was a testament to a half century of land abuse. Congress was also being lobbied by a few far sighted members of the livestock industry who recognized that changes were needed to perpetuate the forage resource (Wilkenson 1992). The results were the establishment of the Grazing Service and the first regulations for grazing on approximately 160 million acres of federal land.

Degradative influences came not only through extractive uses such as beaver trapping, grazing, or mining, but also through ill-conceived engineering activities that permanently modified riparian/stream systems. With the advent of mechanized equipment in the 1930's came the first attempts to modify stream channels through the use of instream structures, channelization, bank modification, and rip-rap. Objectives have included flood control, irrigation development, and wetland conversion.

Riparian management activities following World War II were largely directed at manipulating or controlling the flow and utilization of water. The important interactions between the riparian and aquatic systems, bound by water, was not recognized as an integral factor for maintaining the productivity of these systems. As such, once complex riparian stream systems were dramatically altered through dredging, diking, and channelization (Figure 1). Eradication of phreatophytic vegetation was implemented with the ill-conceived hypothesis that this would reduce transpiration loss and increase the amount of usable water. Channels were cleared

Figure 1. Channelization was a popular method used to increase the drainage of floodplains and wetland areas.

of woody plants and coarse woody debris in order to increase the velocity of streams and reduce flood damage. Floodplains became inactive and without vegetation to buffer energy, erosion of streams increased (Figures 2 and 3). The result was a dramatic simplification of riparian stream ecosystems with decreased interactions, decreased biological diversity, and declines in productivity (Green and Kauffman 1991).

Paradoxically, similar activities were implemented with the objective of fish habitat "enhancement". Fish populations and stream habitat conditions were declining dramatically. Engineering approaches were implemented with few considerations or little understanding of fish habitat needs or riparian ecosystem dynamics. Thousands of structures (e.g., check dams) were installed in streams throughout the West and midwest (Heady 1977, White 1990). Fifty years later similar engineering activities or stream manipulations continued to be applied in attempts to restore degraded riparian zones and stream channels for salmonid recovery. At the same time natural woody debris was being removed from many streams to improve fish passage. Again, little consideration for the ecological dynamics of riparian/stream ecosystems were included in the implementation process. In terms of restoring fish populations, and accelerating riparian recovery these approaches have largely been ineffectual, and in some cases counter productive (Beschta, Platts, and Kauffman 1991; Sedell and Beschta 1991, Li et al. 1992, and Ahlborn et al. 1992). However, some success have been achieved in increasing spawning area, number of fish redds (i.e. egg deposition areas), and juvenile survival (House et al 1989, House et al 1990, Crispin et al 1993).

For stream recovery projects to be successful, managers must begin by halting those human activities that are causing the decline or preventing recovery (Kauff-

Figures 2 & 3. A wide, entrenched channel system that has incised 5 to 20 feet into silty-clay deposits. Figure 2 shows the area in 1968. Figure 3 shows the same area 16 years later, after exclusion from grazing. Note the expanded riparian area, as the water table influences vegetation composition across the entire bottom.

man, Beschta and Platts 1993). Riparian zones are characterized as areas with inherently high levels of natural disturbance. As such, they often have a remarkable resilience once degradative land use activities cease. Rather than engineering approaches that many times disregard ecological processes, a more feasible approach is to facilitate the natural recovery of riparian biotic and hydrological features by restoring natural ecosystem processes (Elmore and Beschta 1988).

On western forests and rangelands where grazing by livestock resulted in declines in riparian condition, exclusion was implemented as a technique of stream restoration. This includes removal of livestock from degraded rangelands or the construction of corridor fencing. Most early exclosures were corridor fences that were built close to the stream channels resulting in very narrow areas of protection and high susceptibility to flood damage.

Livestock exclusion has consistently resulted in the most dramatic and rapid rates of ecosystem recovery (Beschta, Platts and Kauffman 1991, Figures 4 & 5 Busse 1989). In Figure 4, the pre-exclusion population of all Salicaceae-dominated stands from the Crooked River National Grasslands, Oregon, clearly indicate the young size classes were greatly underrepresented; successful establishment of willow and cottonwood communities was not occurring. Persistence of these relatively short-lived species was not likely under the prevailing livestock management scenarios given the underrepresentation of young size classes in an environment of high fluvial disturbances. Following livestock exclusion in the same riparian zones, the population is represented by numerous young stands establishing on coarse-

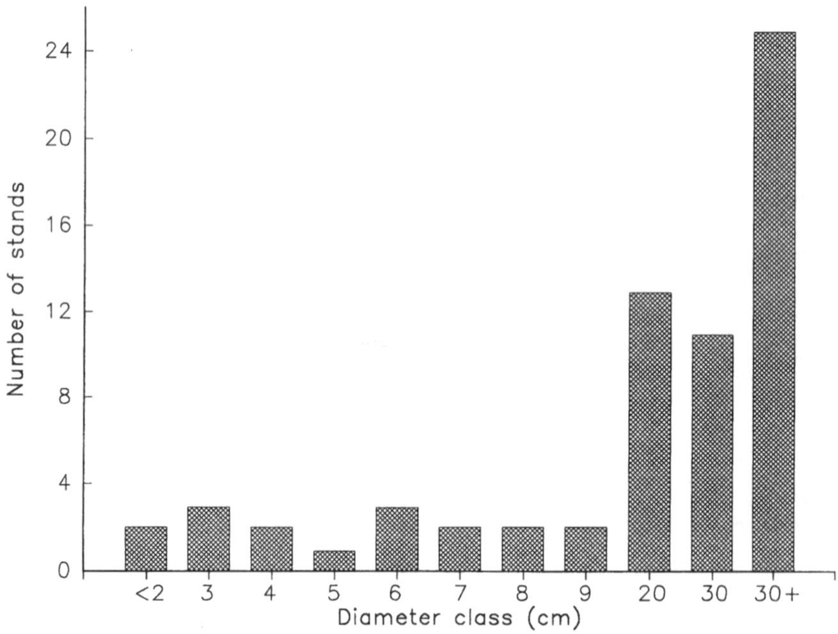

Figure 4. Diameter class distribution of the Salicacea stands present prior to riparian exclosure construction (n=65) on the Crooked River National Grasslands, Oregon. Data are from Busse (1988).

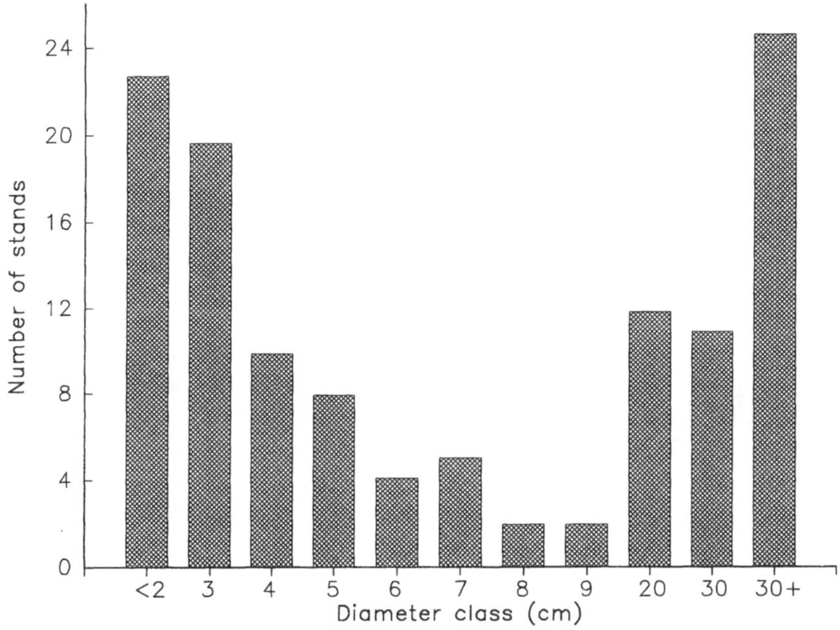

Figure 5. Diameter class distribution of the Salicaceae in livestock exclosures (n=125) at the Crooked River National Grasslands, Oregon, following cessation of grazing for 1-29 years. Data are from Busse (1988).

textured gravel bars (Figure 5). It is apparent that following cessation of improper livestock grazing, barriers to the re-establishment of these riparian-obligate communities were eliminated.

There are many problems with using exclosures as a solution to halt overuse by livestock. They include wildlife access, cost of construction, and maintenance. In addition, the management outside of the exclosures frequently is not changed, subsequently, unfenced riparian zones and uplands continue to decline or fail to recover.

Present

Restoration

The closer an ecosystem is managed to allow for natural ecological processes to function, the more successful that management strategy will be. While a number of factors are negatively influencing riparian ecosystems today (e.g., grazing, roads, dams, farming, urbanization, irrigation withdraw, mining etc.) in this chapter we focus on those areas where livestock influences are the ecological perturbation needing attention.

Attempts at the development of livestock grazing strategies that allowed for riparian restoration were begun in the late 1970's and early 1980's. To be successful, management must consider the complexity of riparian ecosystem function and the integral role that vegetation plays in these functions (Swanson et al. 1982,

Elmore and Beschta 1987). Stream/riparian functions are those processes influenced by the interaction of soil, water (hydrology), and vegetation (Elmore 1992). They include the physical filtering and retention of sediment and nutrients by vegetation. The processes associated with high water, sediment retention, and vegetation structure are intimately linked to bank building, biotic composition, recharge of subsurface aquifers, and hydroperiod.

Every stream has a certain suite of natural disturbances, or stresses that are related to soil type, stream gradient, climate, geology and, other physical features. It must be recognized that every management activity (road construction, logging, grazing, etc.) will either alter these natural disturbance processes or exacerbate their effects. The response of a given stream to a given suite of management stresses depends on the inherent level of ecological resilience of the ecosystem and how far the management system departs from a natural ecosystem equilibrium. Resilience is defined as the degree and rapidity of ecosystem recovery following a natural disturbance or human perturbation. Some streams with high levels of natural stress (such as those with bentonitic soils and high erosion potential) are rapidly degraded by almost any additional disturbances from human activities. In contrast, streams of low natural stress (such as those with low gradients and sandy loam soils) are resilient to a much higher level of management perturbations. The inherent natural disturbance factors, as well as the capacity of ecological resilience of riparian zones, and the suite of ecological processes necessary for intact ecosystems to function must be properly integrated in management strategies for recovery or maintenance to occur (Figure 6) (Elmore and Beschta 1987, Elmore 1992).

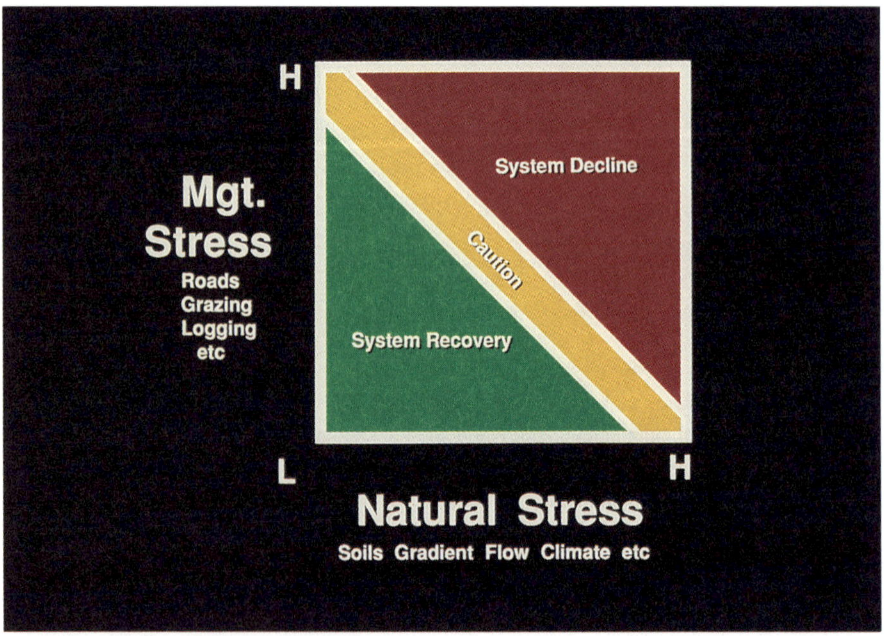

Figure 6. Management Stress verses Natural Stress.

Grazing Management

There are many grazing strategies that have been devised to achieve specific ecological or management goals. Depending upon the unique features of the ecosystem and goals of management, successful strategies range from loosely supervised season-long use, to intensively managed multi-pasture rotational systems (Elmore 1992). The effectiveness of a given system with respect to sustainability and restoration of ecosystem structure, and diversity will depend upon the ecological characteristics of the stream system. This is important because failures in management often occur when a grazing strategy developed for a certain stream system is applied in another stream/riparian reach with different ecological and/or management characteristics. With the exception of exclusion, there is no single grazing management strategy that has been proven to consistently improve degraded western riparian areas (Kinch 1989, Clary and Webster 1989, Marlow 1985).

The difficulties in devising proper grazing strategies is further complicated by the inherent complexity of riparian zones. Green (1991) quantified a highly variable response by plant communities to livestock grazing within the same stream reach. Under moderate, late season use, productivity and diversity of riparian meadows was maintained. However, woody plant succession and growth was negatively influenced on gravel bars. Meyers (1989) evaluated 34 grazing strategies in Montana and found 74% were unsuccessful in accommodating a positive riparian vegetation response within a 10-to-20 year period. Upland areas, however, did show a positive response on most of the areas. Marlow (1985) addressed the failure of grazing strategies to restore riparian ecosystems and recommended the length of the grazing period should be based on the areas cattle are actually using, not the entire area. Often this may be the community type of interest within the riparian zone. Green (1991) described phenomena in northeastern Oregon where livestock preference and utilization was greatest (40-70%) for dry and moist meadows in late season grazing schemes. Willow-dominated gravel bars were intermediate in preferences with a utilization of 20-45% and the understory of mature cottonwood and alder communities were utilized at levels of ≤20%. Therefore, management strategies should focus on the specific components of the ecosystem in need of restoration. In this study the levels of herbaceous utilization in meadows were not an adequate indicator of willow recovery. The productivity of meadows was maintained, but these levels of utilization did influence willow and cottonwood establishment.

Throughout river basins of the Pacific Northwest, salmonid populations have precipitously declined or have been completely extirpated (Armour 1977; Benhke 1977; Williams et al 1989). Many key riparian plant communities (e.g., gallery forests of black cottonwood, quaking aspen, alders, or willows) have been eliminated from stream reaches (Busse 1989; General Accounting Office 1988; Chaney et al 1990; Kauffman, Beschta and Platts 1993). There is a general acceptance by managers today that most riparian areas are in an unacceptable condition and that approaches to restoration in the past have had limited success. Hansen (1993), Kauffman, Beschta, and Platts (1993), Kinch (1989), and Skovlin (1984) have outlined ecological approaches for developing successful grazing management or ecological restoration strategies for riparian\wetlands restoration. It is recommended that these strategies be developed with interdisciplinary teams representing several disciplines. These would include, but not be limited to, range specialists, wildlife and or fisheries biologists, soil scientists, botanists, and foresters depending on the resources present in each area. Generalized recommendations include:

1. Each grazing strategy must be tailored to a particular stream or stream reach. Management objectives and components of the ecosystem that are of critical value must be identified (i.e. woody species recovery, streambank restoration, increased stream channel diversity, etc.). Other information that should be identified includes present vegetation, potential of the site for recovery, the desired future condition and the current factors that are causing habitat degradation or limiting its recovery.
2. Describe the relationships between ecological processes that must function for riparian recovery. Assess factors affecting present condition (i.e., management stress versus natural stress) and what conditions will be needed to allow the stream to resume natural functions. Anthropogenic factors that are causing stream degradation must be identified and changed.
3. Design attainable goals, objectives, and management activities that will achieve the desired future condition of the riparian/stream ecosystem.
4. Implement the plan.
5. Design and implement a monitoring plan that will evaluate management, allowing for corrections or modifications in management, if necessary.
6. Implement a strong compliance and use supervision program.

Generalized responses of riparian ecosystems to livestock grazing strategies have been discussed by Elmore (1992), Platts and Nelson (1989 Fig. 7), Kovalchik and Elmore (1991 Fig. 8), Buckhouse and Elmore (1991 Fig. 9), and Meyers (1989 Fig. 10). The ratings by these authors are based on observations in different riparian/stream systems, and therefore should be viewed with caution. However they do contain similarities for assessing management potential for success in the northern Rocky Mountains or Pacific Northwest. All of these reviews recognized that the rates of recovery and resultant vegetation composition for each riparian/stream system are dependent on many factors including site potential, current ecological condition, stream geomorphology, and climate.

Some generalized characteristics that are commonly observed are reviewed here.

Three Pasture Rest Rotation

This grazing system typically has provided total annual rest for each pasture on a regular basis. This strategy was designed for upland vegetation to promote plant vigor, seed production and seedling establishment, root production, and litter accumulation for herbaceous plants in upland ecosystems. In general, woody riparian species have been observed to decline under this strategy. Like most grazing systems, this strategy was designed for the physiological needs of herbaceous plants and can be very successful on low gradient sedge (*Carex*), rush (*Scirpus*), and grass-dominated sites. It is usually inappropriate for early seral shrub dominated sites because heavy utilization of riparian areas will retard establishment and growth of willow dominated communities (Figures 11 and 12). However, it has been successful in maintaining late seral condition on woody dominated sites in

Figure 7. Platts 1989. Evaluation and rating of grazing strategies based on the author's personal observations, as related to stream-riparian habitats.

Strategy	Level to which riparian vegetation is commonly used	Control of animal distribution (allotment)	Steam-bank stability	Brushy species condition	Seasonal plant regrowth	Stream-riparian rehabilitative potential	Rating
Continuous season-long (cattle)	Heavy	Poor	Poor	Poor	Poor	Poor	1[a]
Holding (sheep or cattle)	Heavy	Excellent	Poor	Poor	Fair	Poor	1
Short duration-high intensity (cattle)	Heavy	Excellent	Poor	Poor	Poor	Poor	1
Three herd-four pasture (cattle)	Heavy to moderate	Good	Poor	Poor	Poor	Poor	2
Holistic (cattle or sheep)	Heavy to light	Good	Poor to good	Poor	Good	Poor to excellent	2-9
Deferred (cattle)	Moderate to heavy	Fair	Poor	Poor	Fair	Fair	3
Seasonal suitability (cattle)	Heavy	Good	Poor	Poor	Fair	Fair	3
Deferred-rotation (cattle)	Heavy to moderate	Good	Fair	Fair	Fair	Fair	4
Stuttered deferred-rotation (cattle)	Heavy to moderate	Good	Fair	Fair	Fair	Fair	4
Winter (sheep or cattle)	Moderate to heavy	Fair	Good	Fair	Fair to good	Good	5
Rest-rotation (cattle)	Heavy to moderate	Good	Fair to good	Fair	Fair to good	Fair	5
Double rest-rotation (cattle)	Moderate	Good	Good	Fair	Good	Good	6
Seasonal riparian preference (cattle or sheep)	Moderate to light	Good	Good	Good	Fair	Fair	6
Riparian pasture (cattle or sheep)	As prescribed	Good	Good	Good	Good	Good	8
Corridor fencing (cattle or sheep)	None	Excellent	Good to excellent	Excellent	Good to excellent	Excellent	9
Rest rotation with seasonal preference (sheep)	Light	Good	Good to excellent	Good to excellent	Good	Excellent	9
Rest or closure (cattle or sheep)	None	Excellent	Excellent	Excellent	Excellent	Excellent	10

[a]Rating scale based on 1 (poorly compatible) to 10 (highly compatible with fishery needs).

some areas where the shrubs have grown out of the reach of grazing animals (Laura Gutzwiller pers. comm.). This, however, may ultimately result in a population of only mature decadent stands and woody vegetation with no ongoing replacement of younger stands (Figure 4).

Winter Grazing

This system prevents utilization during the growing period annually because grazing occurs only during plant dormancy. Dormant woody riparian species can be negatively affected by browsing or trampling in areas where winter temperatures are moderate or livestock movements are restricted. However, dramatic recovery rates have been observed where light use occurs because of cold drainage patterns and livestock avoidance of the riparian zone or availability of alternate livestock water systems away from streams (Figure 13 & 14). A full understanding of expected livestock use patterns is necessary using this strategy or land use objectives may not be achieved. For instance, in wide valley bottoms in Montana, winter use resulted in an increase in sedge (Carex) communities but continued overuse of woody riparian vegetation. Livestock did not move to adjacent hillsides but continued to concentrate in the bottoms. Soils were frozen resulting in little bank damage (Lew Meyers pers. comm.).

Early Growing Season

Regrowth of riparian vegetation is facilitated by the absence of summer utilization. Utilization on woody plants, (e.g., willow and alder), may be low because upland grasses are green and more palatable. If periods of use allow for adequate

Figure 8. Meyers 1989. Characteristics of successful and unsuccessful grazing systems with means and 95% confidence intervals.

Characteristics	Grazing systems	
	Successful	**Unsuccessful**
Number of grazing systems	9	25
Stocking rates (Hectares/AUM)	4.9 (2.3)	3.8 (1.1)
Days of post-grazing regrowth (up to 9/15)	34.9 (17.8)	20.8 (7.3)
Percentage of grazing treatments providing residual cover through rest or regrowth	74.9 (12.8)	37.8 (13.2)
Duration (days) of hot season (7/1-9/15) treatments	12.5 (10.5)	33.4 (10.4)
Duration (days) of all grazing treatments	28.2 (3.7)	59.3 (8.1)
Duration (days) of fall grazing treatments (8/15-1/10)	21.0 (9.1)	36.5 (8.1)
Percentage of grazing treatments with fall use	31.1 (20.7)	51.1 (9.8)

Figure 9. Generalized relationships between grazing system, steam system characteristics, and riparian vegetation response (adapted from Buckhouse and Elmore 1991).

Grazing System	Steep Low Sediment Load		Steep High Sediment Load		Moderate Low Sediment Load		Moderate High Sediment Load		Flat Low Sediment Load		Flat High Sediment Load	
No Grazing	Shrubs	+	Shrubs	+	Shrubs	+	Shrubs	+	Shrubs	+	Shrubs	+
	Herbs	+	Herbs	+	Herbs	+	Herbs	+	Herbs	+	Herbs	+
	Banks	0	Banks	0 to +	Banks	0	Banks	+	Banks	+	Banks	+
Winter or Dormant Season	Shrubs	+	Shrubs	+	Shrubs	+	Shrubs	+	Shrubs	+	Shrubs	+
	Herbs	+	Herbs	+	Herbs	+	Herbs	+	Herbs	+	Herbs	+
	Banks	0	Banks	0 to +	Banks	+	Banks	+	Banks	+	Banks	+
Early Growing Season	Shrubs	+	Shrubs	+	Shrubs	+	Shrubs	+	Shrubs	+	Shrubs	+
	Herbs	+	Herbs	+	Herbs	+	Herbs	+	Herbs	+	Herbs	+
	Banks	0	Banks	0 to +	Banks	+	Banks	+	Banks	+	Banks	+
Deferred or Late Season	Shrubs	–	Shrubs	–	Shrubs	–	Shrubs	–	Shrubs	–	Shrubs	–
	Herbs	+	Herbs	+	Herbs	+	Herbs	+	Herbs	+	Herbs	+
	Banks	0 to –	Banks	0 to –	Banks	0 to +	Banks	+	Banks	+	Banks	+
Three-Pasture Rest Rotation	Shrubs	–	Shrubs	–	Shrubs	–	Shrubs	–	Shrubs	–	Shrubs	–
	Herbs	+	Herbs	+	Herbs	+	Herbs	+	Herbs	+	Herbs	+
	Banks	0 to –	Banks	0 to –	Banks	0 to +	Banks	+	Banks	+	Banks	+
Deferred Rotation	Shrubs	–	Shrubs	–	Shrubs	–	Shrubs	–	Shrubs	–	Shrubs	+
	Herbs	+	Herbs	+	Herbs	+	Herbs	+	Herbs	+	Herbs	+
	Banks	0 to –	Banks	0 to –	Banks	+ to 0	Banks	+	Banks	+	Banks	+
Early Rotation	Shrubs	+	Shrubs	+	Shrubs	+	Shrubs	+	Shrubs	+	Shrubs	+
	Herbs	+	Herbs	+	Herbs	+	Herbs	+	Herbs	+	Herbs	+
	Banks	0 to –	Banks	0 to +	Banks	+ to 0	Banks	+	Banks	+	Banks	+
Rotation	Shrubs	–	Shrubs	–	Shrubs	–	Shrubs	–	Shrubs	–	Shrubs	–
	Herbs	+	Herbs	+	Herbs	+	Herbs	+	Herbs	+	Herbs	+
	Banks	0 to –	Banks	0 to –	Banks	0 to +	Banks	+	Banks	+	Banks	+
Season-Long	Shrubs	–	Shrubs	–	Shrubs	–	Shrubs	–	Shrubs	–	Shrubs	–
	Herbs	–	Herbs	–	Herbs	–	Herbs	–	Herbs	–	Herbs	–
	Banks	0 to –	Banks	0 to –	Banks	–	Banks	–	Banks	–	Banks	–
Spring and Fall	Shrubs	–	Shrubs	–	Shrubs	–	Shrubs	–	Shrubs	–	Shrubs	–
	Herbs	–	Herbs	–	Herbs	–	Herbs	–	Herbs	–	Herbs	–
	Banks	0 to –	Banks	0 to –	Banks	–	Banks	–	Banks	– to 0	Banks	0 to +
Spring and Summer	Shrubs	–	Shrubs	–	Shrubs	–	Shrubs	–	Shrubs	–	Shrubs	–
	Herbs	–	Herbs	–	Herbs	–	Herbs	–	Herbs	–	Herbs	–
	Banks	0 to –	Banks	0 to –	Banks	–	Banks	– to 0	Banks	– to 0	Banks	0 to +

Note. – = decrease: + = increase: 0 = no change. Stream gradient: 0 to 2% = flat; 2 to 4% = moderate; >4% = steep.

regrowth and do not correspond to the seasons of willow reproduction, this grazing strategy can be very beneficial to riparian areas, especially in establishing woody plants. However, the influence of livestock grazing on willows during flowering and early seedling establishment has also not been quantified. Conversely, this system can be detrimental to upland grasses if grazing strategy results in utilization during the critical part (shoot elongation) of their growing season. In addition, this system may not be applicable on finer textured soils or those riparian zones associated with steep gradients. Under these scenarios, soil compaction, accelerated streambank losses or increased erosion rates may result.

Figure 10. Kovalchik and Elmore 1991. Generalized relationships between grazing system and willow and sedge response on willow-dominated plant associations.

	Systems highly compatible with willow management		
Corridor fencing	Willows ⇑ Sedges ⇑	Riparian pasture	Willows ⇑ Sedges ⇑
Spring grazing	Willows ⇑ Sedges ⇑	Winter grazing	Willows ⇔ to ⇑ Sedges ⇑
	Systems moderately compatible with willow management		
Two-pasture rotation	Willows ⇔ to ⇑ Sedges ⇑	Three-pasture rotation	Willows ⇔ to ⇓ Sedges ⇑
Three-pasture deferred rotation	Willows ⇔ to ⇓ Sedges ⇔ to ⇑		
	Systems incompatible with willow management		
Spring-fall grazing	Willows ⇓ Sedges ⇔ to ⇓	Deferred grazing	Willows ⇔ to ⇓ Sedges ⇔ to ⇓
Late-season grazing	Willows ⇓ Sedges ⇓	Season-long grazing	Willows ⇓ Sedges ⇓

⇑ = highly compatible
⇓ = incompatible
⇔ = no change

Deferred Grazing (after seed ripe) and Deferred Rotation (three pasture)

These strategies can be beneficial to sedge/grass communities especially if sufficient regrowth is allowed to provide for adequate riparian function (i.e., enough regrowth to retain a capacity to trap sediment) during the next high flow event. Late season grazing strategies have been found to be very beneficial in maintaining species diversity and productivity of meadow systems (Kauffman et al. 1983, Green 1991). Levels of utilization of woody riparian species must be carefully monitored because use always occurs during the summer months when livestock tend to concentrate in riparian areas. Levels of utilization that maintain the diversity and productivity of the meadow communities have been found to retard woody plant succession on gravel bars (Green 1991).

Rotation Grazing (two pasture)

This strategy consists of an early use treatment one year followed by a deferred use period the next. It provides total growing season rest every other year for herbaceous plants. However this system can be detrimental to woody riparian vegetation if the summer use that occurs every other year exceeds the regrowth of one year of nonuse. It can be acceptable on low gradient wide valley bottom sedge-rush-grass sites especially if adequate regrowth time is allowed at the end of the deferred use period.

Spring-Summer Grazing or Season Long Grazing

This strategy typically provides no rest during the growing period for plant vigor, reproduction, or litter accumulation. It generally has resulted in heavy utilization

Figure 11 & 12. Ten years of continued channel degradation in a high gradient, high energy stream system under three-pasture rest rotation grazing. Willows, which show a downward trend from 1976 (fig. 11) to 1986 (fig 12), are needed to help stabilize this laterally unstable channel.

Figure 13 & 14. Willow and herbaceous vegetation response after eleven years. Grazing was changed in 1976 (figure 13) from season long grazing to winter grazing (1987 fig. 14).

of woody riparian vegetation. Trampling damage, soil compaction, and accelerated streambank erosion are likely. This strategy is most commonly associated with the wide spread decline of riparian and watershed conditions in the West. This strategy should be eliminated in most areas where it is still being used.

Fall Grazing

This strategy provides growing period rest every year however can be detrimental to riparian vegetation depending on site specific criteria (elevation, magnitude of peak flows, etc.). Interactions between high flows over floodplains can be negatively influenced if there is a lack of residual vegetation to facilitate sediment retention. Woody riparian species are commonly used heavily in the fall if green herbaceous materials are limiting. Riparian areas can improve with this strategy if fall use occurs when temperatures are cool, fall green up has occurred, and utilization is closely monitored. Upland areas also benefit depending on when livestock are removed and if rest and regrowth are allowed.

Riparian Exclusion

The complete elimination of livestock provides total protection from domestic grazing animals. The rate of recovery of the stream features and riparian vegetation, both woody and herbaceous, has been the most rapid under this management scenario. Beschta, Platts and Kauffman (1991) reported willow and cottonwood growth in ungrazed exclosures to be as high as 60-100 cm per year. However, simply excluding the riparian area does not address the needs of the upland vegetation or the overall condition of the watershed. Unless a landscape-level approach is taken, important ecological linkages between the uplands and aquatic systems cannot be restored and riparian recovery will likely be limited. Exclusion should only be used in conjunction with an upland management plan designed to restore the entire landscape or when there are situations where the most rapid recovery possible is necessary (i.e., for habitat restoration of federally listed threatened or endangered salmonids).

Riparian Pasture

This strategy is applicable in areas where riparian zones encompass a large enough area to manage them separately from uplands. The riparian area may be managed separately or in combination with other allotments or pastures. The use of the forage in a riparian pasture is typically designed to specifically meet certain stream/riparian restoration goals. Because the pasture is separate from the rest of the allotment, it can be grazed or rested depending on current conditions and stream riparian needs. It can provide flexibility in the design of strategies for both woody and herbaceous species.

Conclusion

Whether riparian zones are currently improving in ecological condition is a debatable question. In areas where progressive management steps have been taken, trends towards improvement have occurred. However, given the declines in anadromous and resident native salmonid fish populations, lack of reproduction of riparian woody species and the preponderance of streams greatly lacking in channel diversity, it is apparent that many riparian zones are not receiving adequate management. Riparian restoration has been, and will continue to be among the most important natural resource issues affecting land use decisions on public as well as private lands. In addition, livestock grazing in riparian zones will continue to be one of the most controversial of land management issues.

The major short-comings of grazing strategies that fail to result in the restoration of degraded riparian zones are: (1) They are applied with a cookbook approach with little analysis of riparian effects or recognition of the complex and heterogenous nature of riparian zones; (2) the grazing strategies utilized were developed for uplands and/or for herbaceous recovery with little consideration of woody vegetation, streambank integrity, or riparian function; or (3) they were developed for increased livestock production with similar shortcomings in consideration of riparian values. Few grazing strategies or monitoring programs have been developed for the maintenance of riparian woody species such as willows, cottonwood or alder nor have they been developed with considerations of streambank structure. Most riparian management strategies are oversimplified by suggesting a single level of desired utilization. This may result in maintenance or improvement of one component of the ecosystem but continued declines in others. New approaches to grazing management should be implemented to include a watershed-level perspective. This would often result in cooperative efforts of all ownerships in a watershed. Strategies that are developed with interdisciplinary groups and are designed to allow flexibility addressing the entire stream and riparian ecosystem are likely necessary for long-term recovery and sustainability to occur. To manage riparian areas from an ecosystem perspective we must determine the functions and needs of an ecosystem before we can devise a sustainable mix of uses among the wide array of competing natural resource interests. Management decisions should be based on the ecological needs or functions of the system that will result in a sustained high level of biological diversity and site productivity. Management should not be driven by the needs or objectives of special interests.

If we are going to successfully manage our riparian areas from an ecosystem perspective then we must consider the following when we develop our objectives and design our plans.

1. We must consider the linkages and processes that are associated with full floodplain function — preferably the 100 year floodplain. Further, objectives should focus on the reconnection of the ecological linkages between uplands, floodplains, and the aquatic zone. Land management activities that degrade or sever these linkages should be modified or discontinued.
2. Riparian zones are extremely complex. The complexity is far greater than scientists and land managers are capable of understanding. We must manage riparian ecosystems within the context of the environment in which they are located, recognizing their unique value, and remembering that what works for one may not work for another.

3. Headwater streams have not received levels of management attention necessary to achieve landscape-level goals. We must recognize the importance of all stream systems, regardless of size, particularly in consideration of cumulative effects of land management in watersheds.
4. Restoration activities within the stream channel and the riparian management zone should reestablish natural ecological processes and communities. Revegetation utilizing offsite or exotic species are similarly counter to goals of the restoration of the inherent biological diversity of the riparian ecosystems. Approaches that sever linkages, retard recovery or degrade riparian stream function should be discontinued. These are usually engineering approaches that give little attention to the ecological function of a riparian/stream ecosystem.
5. We must manage to maintain connectivity across landscapes and minimize ecosystem fragmentation. No other landscape feature is as effective as riparian zones in linking ecosystems.

Society has a responsibility to ensure that future generations have the opportunity to benefit from intact riparian/stream ecosystems. The restoration or maintenance of long-term ecosystem structure, function, and productivity should be a primary consideration among land managers. We have learned much from the degradation that has occurred to watersheds and riparian ecosystems over the last century. This legacy has lead us to a better understanding of ecosystem function and processes and has identified the needs for restoration. Now is the time to initiate management strategies that will allow our riparian and watershed systems to approach their productive potential.

Literature Cited

Ahlborn, G., W.C. Platts, T. Hanes, and S. Jensen. 1992. Evaluation of watershed and riparian treatments, US Forest Service, S.W. Region. USDA Forest Service, Rocky Mtn. Forest and Range Experiment Station, Fort Collins, CO.

Armour, C.L. 1977. Effects of deteriorated range streams on trout. USDI, Bureau of Land Management, Boise, ID.

Behnke, R.J. 1977. Fish faunal changes associated with land-use and water development. Great Plains-Rocky Mountain Geographical Journal 6(2):133-136.

Beschta, R.L., W.S. Platts, and J.B. Kauffman. 1991. Field review of fish habitat improvement projects in the Grande Rhonde and John Day River basins of Eastern Oregon. U. S. Dept. of Energy, Bonneville Power Administration, Project 91-069. Portland, OR. 53 p.

Buckhouse, J.C., and W. Elmore. 1991. Grazing practice relationships: predicting riparian vegetation response from stream systems. T.E. Bedell, editor in: Watershed management guide for the interior Northwest. Oregon State University Publication EM 8436.

Busse, K.G. 1989. Ecology of *Salix* and *Populus* species of the Crooked River National Grassland. M.S. Thesis. Oregon State Univ., Corvallis, OR. 116 p.

Chaney, E., W. Elmore, and W.S. Platts. 1990. Livestock grazing on western riparian areas. Publication of US Environmental Protection Agency, Region 8, Denver, CO.

Claire, E.W. and R.L. Storch. 1983. Streamside management and livestock grazing: an objective look at the situation. Pages 111-128 in J. Menke, editor. Workshop on livestock wildlife-fisheries relationships in the Great Basin. U.S. Forest Service, Berkeley, CA.

Clary, W.P., and B.F. Webster. 1989. Managing grazing of riparian areas in the Intermountain Region. Gen. Tech. Rep. INT-263. Ogden, UT. U.S. Department of Agriculture, Forest Service, Intermountain Research Station. 11 p.

Crispin, Val, R. House, and D. Roberts. 1993. Changes in instream habitat large woody debris and coho salmon production after restructuring a costal Oregon stream. In Press. North American Journal of Fisheries Management. American Fisheries Society.

Elmore, W., and R.L. Beschta. 1987. Riparian areas: perceptions in management. Rangelands 9:260-265.

Elmore, W. and R.L. Beschta. 1988. Fallacy of structures and fortitude of vegetation. D. Abell editor in, Proceedings of Calif. Riparian Systems Conference, Sept. 1988. U.S. Forest Service, Pacific S.W. Forest and Range Experiment Station. PSW-110. pp 116-119.

Elmore, W. 1992. Riparian responses to grazing practices. Pages 442-457 in Robert J. Naiman, editor. Watershed management. Springer-Verlag, New York, USA.

Green, D.M. 1991. Soil conditions along a hydrologic gradient and successional dynamics in a grazed and ungrazed montane riparian ecosystem. PhD Diss. Oregon State Univ., Corvallis, OR. 236 p.

Green, D. and J.B. Kauffman. 1989. Nutrient cycling at the land-water interface: The importance of the riparian zone. pp. 61-67. *IN*: Gresswell, R. I., B. A. Barton and J. L. Kershner (eds.) Practical approaches to riparian resource management. USDI-BLM, Billings, MT.

Gutzwiller, L. pers. comm. 1993. Fisheries Biologist, Bureau of Land Management, Elko, NV.

Hansen, P.L. 1993. Developing a successful riparian-wetland grazing management plan. In press: Proceedings of Riparian Management: common threads and shared interests. Western regional conference on river management. Albuquerque, NM. Feb. 4-6, 1993. USDA Forest Service, Intermountian Forest and Range Experiment Station, Ogden, UT.

Heady, B.H. 1977. Case study of a watershed rehabilitation project: Alkali Creek, Colorado. USDA Forest Service, Rocky Mountain Forest Experiment Station, Fort Collins, CO. RM-189.

House, Robert, V. Crispin, and R. Monthey. 1989. Evaluation of stream rehabilitation projects — Salem District 1981-1988. USDI Bureau of Land Management, Oregon State Office, Tech. Note OR-6, Portland, OR.

House, Robert, and V. Crispin. 1990. Economic analysis of the value of large woody debris as salmonid habitat in costal Oregon streams. USDI, Bureau of Land Management, Oregon State Office, Tech. Note OR-7, Portland, OR.

Kauffman, J.B. and W.C. Krueger. 1984. Livestock impacts on riparian ecosystems and streamside management implications . . . A review. J. Range Manage. 37:430-437.

Kauffman, J.B., R.L. Beschta, and W.S. Platts. 1993. Fish habitat improvement projects in the Fifteen Mile Creek and Trout Creek Basins of central Oregon. Field review of management recommendations. U.S. Dept. of Energy, Bonneville Power Administration Report. Portland, OR. 51 p.

Kinch, G. 1989. Grazing management in riparian areas. USDI, Bureau of Land Management, S.C., Denver, CO. Tech. Ref. 1737-4. 41 p.

Kovalchik, B.L. and W. Elmore. 1991. Effects of cattle grazing systems on willow-dominated plant associations in central Oregon. Symposium on ecology and management of riparian shrubs communities, Sun Valley, ID, May 29-31, 1991. p. 111-119.

Li, H., T.N. Pearsons, C.K. Tart, J.L. Li, and R. Gaither. 1992. Approaches to evaluate habitat improvement programs in streams of the John Day Basin — Completion Report. Oregon Coop Fisheries Research Unit. Corvallis, OR. Unpubl. Report. 126 p.

Marlow, C. 1985. Controlling riparian zone damage with little forage loss. Montana Agriculture Research, Fall 1985. Montana Agricultural Experiment Station, Bozeman, MT.

Meyers, L. 1989. Grazing and riparian management in southwestern Montana. R.E. Gresswell editor in: Practical approaches to riparian resource management. 1989 May 8-11, Billings, MT. USDI, Bureau of Land Management. p. 117-120.

Meyers, L. pers. comm. Riparian Program Lead, USFS, Tonto National Forest, Phoenix, AZ.

Naiman, R.J., C.A. Johnston, and J.C. Kelley. 1988. Alteration of North American streams by beaver. Bioscience 38:753-762.

Naiman, R.J., J.M. Melillo, and J.E. Hobbie. 1986. Ecosystem alteration of boreal forest streams by beaver (*Castor canadensis*). Ecology 67:1254-1269.

Ogden, P.S. 1950. Ogden's snake country journals, 1824-26. Hudson Bay Records Society (1950).

Platts, W. S. and R. L. Nelson. 1989. Characteristics of riparian plant communities with respect to livestock grazing. R.E. Gresswell in: Practical approaches to riparian resource management; 1989 May 8-11; Billings, MT. USDI, Bureau of Land Management. p. 73-81.

Sedell, J.R. and R.L. Beschta. 1991. Bringing back the Bio in Bioengineering. American Fisheries Symposium 10:160-175.

Skovlin, J. M. 1984. Impacts on grazing of wetlands and riparian habitat: a review of our knowledge. *IN*: Developing strategies for rangeland management. National Research Council. National Academy of Sciences Boulder, CO: Westview Press: 1001-1103.

Storch, R. L. 1979. Camp Creek, a catalyst in positive management direction for streamsides. John Day, OR: U.S. Department of Agriculture, Forest Service, Malheur National Forest. 13 p. Unpublished paper.

Swanson, W. S., S.V. Gregory, J.R. Sedell, and A.G. Campbell. 1982. Land-water interactions: the riparian zone. Pages 267-291 in R.L. Edmonds, editor. Analysis of coniferous forest ecosystems in the western United States. Hutchinson Ross, Stroudsburg, Pennsylvania, USA.

U.S. Government Accounting Office. 1988. Public rangelands: some riparian areas restored but widespread improvement will be slow. GAO\RCED-88-105, Washington, D.C.

White, R.J. 1989. We're going wild: A thirty year transition from hatcheries to habitat. Trout, Summer 1989. Trout Unlimited, Vienna Virginia.

Wilkenson, C.F. 1992. Crossing the next meridian. Land water and the future of the west. Island Press, Wash. D.C.

Williams, J.E., J.E. Johnson, D.A. Hendrickson, (and others). 1989. Fishes of North America endangered, threatened, or of special concern: 1989. Fisheries. 14:2-20

Livestock Grazing: A Tool To Improve Wildlife Habitat

Kieth E. Severson and Philip J. Urness

Abstract

Livestock can influence vegetation by altering its composition, structure, and productivity. Most examples of how these changes affect wildlife habitats, however, have been negative. Livestock effects are usually characterized by the often misunderstood term "overgrazing," and generally originate from situations involving severely abused rangelands. This paper examines livestock impacts from a different perspective. Given that livestock influence vegetation, are there situations in which these changes could be manipulated or directed to create habitats that are favored by certain wild species? There are four ways livestock can be used in this respect: 1) alter composition of the vegetation, 2) increase productivity of selected species, 3) increase the nutritive quality of forage, and 4) increase diversity of habitat by altering its structure. The need for careful planning relative to how changing habitats for a featured species will likely adversely affect others, implications of controlled overgrazing relative to ecosystem responses, evolutionary history of grazing, and effects on livestock are discussed. We conclude that existing knowledge regarding the use of prescribed grazing by livestock as a tool to enhance wildlife habitats is meager, but enough information exists to warrant applications in some areas and investigate its applicability in others.

Introduction

The dynamic processes involved in biotic community changes are a concern of wildlife biologists when managing wildlife habitats. Changes in plant composition, productivity, and community structure will influence wildlife populations at the individual species and faunal assemblage levels. In managed situations, control of succession or retrogression is desirable because we do not want random (i.e., natural or unplanned) events to overwhelm or dictate our goals. Connolly and Wallmo (1981) suggested, for example, that livestock grazing has been credited for improving wildlife habitats under some conditions, but in most cases, these benefits have been fortuitous rather than planned. Reliance upon fortuitous events is not a practical way to manage wildlife habitats. The purpose of this paper is to review existing information and to examine the feasibility of using planned (prescribed) livestock grazing as a tool to direct community processes to improve wildlife habitats.

The Problem

Problems associated with livestock-wildlife interactions have commanded the attention of managers and researchers for nearly 100 years. Impacts have been controversial and issues divisive. Over 35 years ago, John Hall, then Director of the Arizona Game and Fish Department, stated, "The problem of Livestock versus Wildlife has plagued us since the turn of the century . . . Instead of decreasing in intensity, this battle has become broader and more bitter as the years go by. It would seem that as we learn more about the proper management of both livestock and wildlife on our ranges, some progress would be made towards solving the problem of conflicting use" (Hall 1955). We cannot fault Mr. Hall's logic; however, after 35 years of additional work, schisms are still growing and controversies are becoming more bitter. A conflict that was once centered on livestock and wild ungulates and confined to the West has expanded to include all forms of wildlife and has become a national issue (Severson and Medina 1983).

Our knowledge base has increased; an extensive literature has detailed adverse impacts on a broad spectrum of wildlife resulting from livestock production activities and from large-scale alterations of rangeland communities for the purpose of increasing livestock forage. Most references to the damaging effects of livestock on wildlife have been derived from situations involving heavily or severely overgrazed ranges, particularly on public lands. Much research has, therefore, been directed at mitigating damaging effects or assessing competitive interactions (Severson 1990). We do not intend to rework this material but rather to acknowledge it as a given and to focus attention on the more positive, but neglected, goal-oriented manipulation of livestock to benefit one or more wildlife species.

Can livestock be used as a tool to improve wildlife habitats? The idea of using livestock in such a manner is not new. Aldo Leopold stated, "Cover is controlled by controlling the plant succession in the right direction at the right time and place. Cow, plow, axe, and fire reverse succession. Fencing, fire-protection, and planting advance it." (Leopold 1933, p. 322). Leopold considered these as "tools" (p. 305). Well into the first half of this century, these "tools" were considered to adversely affect populations of native animals and have been, with justification, credited by contemporary scientists and managers for today's problems. Unrestricted logging, grazing, wildfires, and plowing of nonarable lands have created a multitude of resource management problems. Interestingly, managed timber harvest, prescribed fire, and plowing (including its rangeland counterparts; brush crushing, bulldozing, chaining, etc.) have evolved into recognized and accepted tools. Grazing, however, has not been granted this status. Grazing by livestock is still largely regarded as detrimental when habitats of wild animals are considered (Severson 1990).

The Rationale

Skeptics are bound to ask why, with all the evidence that livestock production activities have adversely affected wild animal populations, would anyone even consider using livestock as a tool to manage habitat for those populations? We can suggest two possible reasons. First, given that habitat management is a desired and necessary function, the tools available to managers are limited. Managing wildlife habitat is an expensive exercise and budgets have chronically been limited. Increasing labor and fuel costs have negated the use of mechanical methods (e.g., root plowing, chaining, etc.) except in special circumstances. Chemical methods, such as fertilizers and herbicides, also are expensive and have the further disadvantage of generating adverse side-effects on the environment. Fire is a useful tool, but applications are limited by potential air pollution and the requirements for exact conditions. Livestock grazing would not be intended to replace any existing methods, and has its own limitations, but could be considered as another possible tool for altering plant communities.

Second, if we accept the need for tools, how can livestock be used? We know that livestock can, via grazing and trampling activities, influence the vegetation by altering its composition, productivity, and structure. We have already alluded to the fact that most examples of these influences have emphasized the negative aspects of livestock presence. These impacts are often described by the onerous and often misunderstood term—overgrazing—a phenomenon that will be discussed later in the paper. But, examining grazing from a different perspective, if livestock do indeed influence vegetation, are there situations or ways in which these influences could be directed to create habitat conditions that result in favorable or positive responses by some wild species? We must recognize that managed change will seldom, if ever, result in uniformly positive changes for all species involved. While some may be favored by the change in habitat, others will be adversely affected. Implications of such tradeoffs will be discussed in a separate section.

The Status Of Our Existing Knowledge

There are four primary ways that livestock can be used to manipulate vegetation and thereby influence wildlife habitats. One is by altering the composition of the vegetation in an area, while another is by increasing the productivity of selected species in the forage base. A third is by increasing the nutritive quality of forage and the fourth is by increasing the diversity of habitat by altering its structure. We will discuss each of these separately but it should be realized that they are not mutually exclusive. If, for example, the goal is to increase the productivity of a certain species by increasing growth of individual plants, the composition in the community may be changed if that species also becomes more abundant and replaces other species. Similarly, attempts to increase habitat diversity by altering structure may result in the increase of some plant species, particularly trees and shrubs, and the decline of others.

Altering the composition of vegetation

One of the first studies that highlighted how effects of grazing by one class of animal might benefit another was done in Utah. Smith (1949) noted significant differences in the composition of vegetation in pastures that were grazed either by livestock or wintering mule deer (*Odocoileus hemionus*). He determined that a pasture used by deer, but protected from livestock grazing for 11 years, contained an abundant and vigorous cover of perennial grasses and forbs, but that shrubs were dead or weakened by excessive browsing by deer. In an adjacent pasture heavily grazed by cattle and horses in summer, shrubs were abundant and vigorous but perennial herbaceous plants were scarce. Urness (1990) remeasured the pastures in 1982 and found that the downward trend in shrub density on the deer pasture continued until the shrubs disappeared. The livestock had been removed in 1957 and that pasture had been receiving heavy use by deer during most winters. It had the same appearance the deer pasture had in 1948. These results provide an indication of how effects of grazing by one kind of animal might benefit another by imposing a balancing of use on all forage components, thus preventing shifts in competitive advantage.

Neal (1982), working in southern Oregon, found that heavy, spring grazing with cattle reduced all bunchgrasses except Sandberg bluegrass (*Poa sandbergii*). Heavy grazing was described as removal of 75% to 100% of the annual production of bunchgrasses and spring grazing was from April 15 to that time when one-third of feeding time was spent browsing bitterbrush (*Purshia tridentata*). This grazing regime also produced conditions favorable for the establishment of new bitterbrush plants, and resulted in increases in big sagebrush (*Artemisia tridentata*), low sagebrush (*A. arbuscula*), and Sandberg bluegrass; all species Neal (1982) considered very important to deer as cover and/or food in winter and early spring.

The literature relative to upland birds and livestock grazing has tended to focus on the consequences of management practices associated with effort to maximize forage and livestock production. For example, sage grouse (*Centrocercus urophasianus*) are affected by broad-scale attempts to reduce sagebrush to favor grass production via herbicides, prescribed fire, mechanical treatments, etc. Guidelines to maintain habitats for these birds focus on retention of sagebrush (Braun et al. 1977, Dunn and Braun 1986). The same concerns and principles regarding sagebrush hold for managing habitats of pronghorn (*Antilocapra americana*) (Sundstrom et al. 1973). What is often overlooked is that many extensive stands of sagebrush, particularly big sagebrush, present today are the result of early excessive grazing by livestock (Hull and Hull 1974, Young et al. 1976), particularly during spring (Laycock 1967). The implication here is that if past sagebrush control operations have negatively affected sage grouse or pronghorn populations, sagebrush could be reestablished by judicious use of spring grazing by livestock.

Management of bobwhite (*Colinus virginianus*) in the southern Great Plains is somewhat complicated by the fact that the optimum seral stage for bobwhite appears to vary inversely with site productivity. In more mesic environments, better bobwhite habitat is provided by earlier seral stages while in more xeric areas, best habitat is composed of later seral stages (Spears 1991). Even in mesic habitats, however, bobwhite require a variety of vegetation types for year-long habitat; too much vegetation can be detrimental at times (e.g., for brood movements) and too little can be detrimental at others (e.g., for nesting cover). Guthery (1986) noted that rangelands in high seral stages produced excellent nesting cover but little food, whereas rangelands in low seral stages produced excellent food supplies but poor nesting cover. Recommendations included the establishment of small overgrazed areas within pastures. These could be created by the use of proper placement of salt or supplemental feeding areas that would induce heavy, but localized, "spot" grazing (Guthery 1986) or by the use of a grazing system such as short-duration grazing which rapidly rotates a single livestock herd through a series of relatively small pastures. Under these circumstances, some pastures may be rested or grazed more than others to create a mosaic of seral stages in close proximity (Guthery et al. 1990).

Increasing productivity of selected species

Based on the data presented by Smith (1949) and general observations on conditions in the Intermountain West through the 1950's, a decision was made to study the question of whether carefully managed livestock grazing, in terms of intensity and timing, could reverse the downward trends in shrub vigor (Urness 1990). A two-part study first demonstrated that early summer clipping of all understory vegetation around individual bitterbrush plants increased the current annual growth of twigs by 58%. The second part was a grazing trial which determined that mountain foothill shrub ranges in northern Utah could be moderately grazed by cattle from late May to early July without reducing bitterbrush yields (Smith and Doell 1968). Cattle using these ranges past mid-July started consuming significant quantities of bitterbrush.

Jensen et al. (1972) conducted a similar study with sheep and obtained results similar to those reported for cattle; livestock use on mule deer winter range is tolerable only during the early part of the plant growing season (before mid-July) if maximum forage is to be retained for over-wintering mule deer. Smith et al. (1979) determined that wintering mule deer, feeding on pastures grazed by sheep, consumed more herbaceous material than those feeding in a pasture without sheep. Despite these differences in diet, there were no significant differences in overall nutritive quality of forage ingested by deer (Fulgham et al. 1982).

Reiner and Urness (1982) examined the use of grazing by horses and determined that horses never utilized bitterbrush, but consumed grass species throughout the summer. Thus, horses offer a clear advantage over cattle or sheep because the time of removal from pastures is not critical. They further determined that production of bitterbrush on heavily grazed pastures was 46% greater than in the ungrazed pasture, but only on the early season pastures that had sufficient soil moisture for growth.

Spanish goats also have been used to increase production of forage on mule deer winter ranges dominated by Gambel oak (*Quercus gambelii*) brush (Riggs and Urness 1989, Riggs et al. 1990). Normal succession in Gambel oak communities is a progressive thickening of oak clones and reduction of other shrubs which are more important to wintering mule deer (Urness 1990). Gambel oak is a strong root sprouter and cannot be managed effectively by prescribed fire, herbicides, or mechanical means (Kufeld 1983). Riggs and Urness (1989) found that cumulative effects of goat browsing included declines in productivity of serviceberry (*Amelanchier alnifolia*) (which was not a major component of the vegetation) and Gambel oak, and an increase in that of big sagebrush. Riggs et al. (1990) concluded that deer responded to these changes by increasing the proportion of big sagebrush in their diets, which increased dry matter digestibility and resulted in less fiber and tannins. Urness (1990) concluded that Spanish goats can effectively manipulate oakbrush communities to: 1) increase production and use of a nutritionally superior winter browse (big sagebrush) by deer, and 2) interrupt the successional trajectory that leads to increased oak at the expense of other shrub associates.

A point to emphasize is that opportunities to enhance production of a forage species are better if there is limited overlap in forage selection, as occurred with Spanish goats and mule deer in Gambel oak habitats due to seasonal separation. In contrast, Severson and DeBano (1991) noted that while Spanish goats could be used to reduce shrub cover in Arizona chaparral, the shrubs most affected were those preferred by mule and white-tailed deer (*Odocoileus virginianus*). These chaparral habitats normally are nutritionally marginal for deer (McCulloch and Urness 1973) and reducing forage diversity by eliminating these favored species from the forage base would likely have a negative impact on the native cervids.

Increasing the nutritive quality of forage

Where dietary overlap is greater, such as among elk (*Cervus elaphus*), bighorn sheep (*Ovis canadensis*), and cattle, opportunities to use livestock to manage habitat are more restricted. However, there are several indications that increases in forage quality and accessibility to preferred forage can be achieved with livestock management.

Anderson and Scherzinger (1975) developed a method intended to increase nutritive contents of winter elk forage by grazing with livestock in late spring or early summer. The hypothesis was that short-term, high-intensity use of grasses before soil moisture is depleted would allow regrowth of new tillers. Growth of these tillers would be arrested by summer heat and drought at a stage below flowering. This would fix nutrients at a level significantly higher than in grasses that reached maturity (Anderson et al. 1990). Anderson and Scherzinger (1975) applied this theory on wildlife management units in eastern Oregon and documented dramatic increases in both cattle and elk use on the units concomitant with improved range conditions; however, they did not include information on nutritional analyses. While final results were impressive, the authors did suggest that the success could not be attributed solely to the cattle grazing system. Other management techniques were initiated at the same time (e.g., development of more and better water, redesigning the entire pasture system, some areas were plowed and reseeded, some roads were permanently closed, and the area was closed to all motorized vehicles during winter). Since nutritive values of forages were not measured, a logical question persists: What proportion of the increases in carrying capacity can be attributed to the method of grazing and what proportion to improved overall management?

Skovlin et al. (1983) attempted a limited field test of the concept in eastern Oregon that, unfortunately, also did not include a nutritional analysis. The design was somewhat confounded by introducing fertilization and burning, with and without spring cattle grazing; hence, only cattle-grazed control plots were analogous. Even here, results may not be directly comparable to those of Anderson and Scherzinger (1975) because cattle grazing was started earlier (mid-April as compared to early May). They measured elk use by counting pellet groups on relatively small subplots and found that elk use of fertilized plots increased the first year but no significant carryover effect was noted. Fall burning and spring cattle grazing did not increase elk use; in fact, cattle grazing decreased elk use in one of three years.

A study by Dragt and Havstad (1987) in west-central Montana did include chemical analysis of three grass species. They measured nutrient contents of plants that were grazed at six phenological stages throughout the summer, then collected for analyses during the following winter. They concluded that there was no deleterious effect of summer grazing by cattle upon the winter forage quality. Neither was there any evidence that it was improved. While the studies appear related, there was no indication they were testing the hypothesis developed by Anderson and Scherzinger (1975).

More promising results were obtained by Pitt (1986) who found that bluebunch wheatgrass (*Agropyon spicatum*) plants clipped at boot, emergence, flowering, and seed formation produced lower levels of acid detergent fiber and higher values of crude protein and phosphorus than control plants. While he stated his study verified the hypothesis of Anderson and Scherzinger (1975), Pitt cautioned that these nutritional benefits were realized by defoliation during stages when the plant was most susceptible to grazing damage. Pitt suggested that Anderson and Scherzinger's recommendation be implemented judiciously and applied within a rotation that permits rest from grazing every second or third year.

Rhodes and Sharrow (1990), working in Oregon, determined that grazing by sheep from May to September reduced forb and shrub (but not grass) biomass available to Roosevelt elk (*C. e. roosevelti*) and black-tailed deer (*O. h. columbianus*) during fall, but the grasses and forbs that remained had increased crude protein and digestible dry matter contents. They suggested that grazing by sheep during summer can improve habitats by increasing the nutritive quality of forage in the fall.

The variability of results reported from the preceding five studies indicates much remains to be learned relative to using livestock grazing to increase the nutritive content of forage species. Further study is obviously necessary to clarify the mechanisms affecting results. Since native ungulate performance on winter ranges is often limited by forage quality rather than quantity (Wallmo et al. 1977), we suggest that such efforts would be warranted.

Another potential use of livestock grazing is to remove the old growth that accumulates around ungrazed plants, particularly grasses. These excessive amounts of leached residues are of poor nutritive quality and, during spring, can impede initiation of new growth by acting as an insulation layer. They also can make the new, nutritious growth less available to grazers when it does start growing. Jourdonnais and Bedunah (1990) examined the use of prescribed fire and fall cattle grazing to reduce litter accumulations and increase elk use on winter range in Montana. They determined that more litter remained with cattle grazing than with burning, but both treatments reduced litter significantly relative to controls. A noted advantage of cattle grazing was that it created a mosaic of heavy to lightly grazed areas while maintaining some litter cover on the soil surface. Elk use of treatments was greater on the burned and cattle grazed areas compared with controls. Willms et al. (1980) had conducted a similar study using fall burning and clipping as treatments and found that deer and cattle preferred the burned treatment to the clipped treatment, but both were preferred over controls.

Increasing diversity of habitat by altering its structure

Altering structure of habitats can be a relatively straightforward concept. Hammerquist-Wilson and Crawford (1981) and Schulz and Guthery (1988), for example, noted that bobwhite use increased in pastures managed by rapid rotation type grazing systems (e.g., high intensity-low frequency and short duration grazing systems) as compared to continuous grazing systems. They attributed results to increased bare ground and decreased amounts of perennial grasses. Bare ground is an important habitat component for bobwhites because it provides a desirable surface for movements and feeding. Dense, tall grass impedes quail movement.

Cattle also have been found to have a beneficial effect on wildlife in high elevation shrub-willow (*Salix* spp.) riparian habitats by creating tunnels (Krueger and Anderson 1985). The formation of tunnels occurs as cattle move through dense willow stands breaking out lower branches. Tunnel width increased as height increased and floors were covered with a mat of grasses or sedges. These tunnels benefit birds and mammals by increasing structural diversity and the amount of grassland habitat. Direct effects were increased diversity of insects, a primary food of birds, and overhead protection from avian predators while flying to and from nests. Krueger and Anderson (1985) cautioned, however, that so much riparian habitat has been excessively damaged by overgrazing that low density shrub-willow habitats are most often readily available. Further, once structural changes have been made by grazing practices, recovery will take a long time, longer than 10-12 years in one reported case (Knopf and Cannon 1982).

Grazing by cattle is often considered a negative effect on waterfowl production, but most studies arriving at this conclusion emphasized grazing effects on upland waterfowl nesting (Braun et al. 1978). A start has been made to investigate various grazing systems as they affect nest density and success (Sedivec et al. 1990). Although there is a need for greater replication over space, it appears some grazing systems promise to improve duck production over ungrazed areas. A critical point made by Sedivec et al. (1990), however, is that no grazing was recommended prior to late May, by which time at least 50% of duck nests are initiated.

The effects of grazing on the quality of wetland habitats used by breeding waterfowl also have received much attention during general investigations, but little by experimental design (Kantrud 1990).

Kantrud (1990), noted that structure of vegetation in prairie wetlands can be changed by fluctuating water levels, fire, and grazing and suggested the effects of burning and grazing are the two community-oriented management systems most in need of study. Control of water levels was acknowledged as the most important, but is possible on only a small portion of wetlands. His thesis is based on the documented fact that wetlands containing an interspersion of cover and open water are preferred by waterfowl. Supportive data include production of better food resources, increased isolation of conspecifics, greater nest densities and hatching success, openings in shoreline vegetation make nest sites on adjacent uplands more accessible to females, some openings or bare areas along shores are important for preening and resting, and waterfowl are less susceptible to predation and/or predator pressure may be buffered by presence of other prey. Therefore, as habitat homogeneity increases, waterfowl use decreases. In the absence of disturbing factors such as fire and grazing by large ungulates, tall robust plants such as bullrushes (*Scirpus* spp.) and cattails (*Typha* spp.) tend to dominate wetlands in extensive, unbroken, monotypic stands. Kantrud (1990) urges long-term research programs on effects of burning and grazing as a high priority need to increase waterfowl productivity on private and public wetlands.

Important Considerations And Points To Ponder

Overgrazing

The term "overgrazing" has been used several times in this paper. This phenomenon does not have a particularly good reputation; it has been used to describe the totally negative results of too many livestock and/or native ungulates on too small an area for too long a time. Overgrazing is an herbivore-induced retrogressive change in the species composition of vegetation on a site. It is generally a slow, often insidious, process caused by continued overuse of the forage over a period of years. It is, however, a relative term. There are varying degrees of overgrazing. A 'moderately' overgrazed range site may exhibit only a slightly altered flora, while a 'heavily' overgrazed range may support a completely different plant complex (Severson and Medina 1983).

Is this process always undesirable? We can only conclude that it is not. Vegetation change caused by grazing may enhance the value or productivity of vegetation for some uses (Smith 1989). We have previously discussed cases where overgrazing of perennial grasses is necessary to realize a viable shrub component in some grassland systems to benefit mule deer, pronghorn, and sage grouse. Situations such as these led Urness (1990) to conclude that overgrazing is bad only if it leads succession away from management objectives or if it degrades site integrity. Smith (1989) suggested that the term overgrazing should be employed only when site potential deteriorates, a generally undesirable situation in the long run. We agree that the concepts and philosophies regarding this phenomenon should be revised. Grazing influences represent a continuum, and the point at which overgrazing occurs is a subjective assessment, a feature which has contributed greatly to current confusion and controversy. Ambiguities associated with its use likely have contributed significantly to problems involving livestock-wildlife interactions (Bryant 1982). Further, the misconceptions and dogmas associated with its current usage indicate that the term overgrazing is outmoded. A revision of concepts should include a more appropriate and meaningful terminology (Pieper and Beck 1990).

Identification of the point where site potential deteriorates may be difficult to determine. Considerable differences exist among sites relative to successional and/or retrogressional processes. There is little specific evidence, for example, that indicates that succession will be a reversible, linear response from retrogression once the grazing stress has been removed (Smith 1989). Some sites are far more fragile than others; a small retrogressive change may occur rapidly, but successional responses, even with full protection, may occur very slowly. Further, on some rangelands, especially in arid or semiarid areas, some seral stages may become static for very long periods (Laycock 1989). Once retrogressive processes cross these thresholds, succession may not occur within a realistic time. Therefore, while we consider management of retrogression via livestock overgrazing a logical and proper procedure, it cannot be considered or initiated capriciously. Its proper implementation will require an intimate knowledge of the ecological processes controlling rangeland biota.

There may be situations where maintenance of very low seral stages would be needed. Such cases would likely be rare but could occur with a featured species such as one declared threatened, endangered, or sensitive. Black-tailed prairie dogs (*Cynomys ludovicianus*) tend to be more abundant on areas grazed heavily by livestock (Koford 1958) and can contribute to further range deterioration by their own feeding and digging activities. Distribution of the endangered black-footed ferret (*Mustela nigripes*) generally is limited to prairie dog towns and recommendations for ferret habitat focus on maintenance of a sufficient number and size of prairie dog towns to support ferrets (Hillman et al. 1979). Since heavy livestock grazing or removal thereof can regulate prairie dog populations (Uresk et al. 1981), its application to increase or maintain areas of prairie dog town via severe overgrazing may be warranted.

The determination of which rangeland sites can be successfully manipulated by grazing livestock to enhance wildlife habitat and which ones warrant extreme caution can be difficult. On some North American rangelands the evolution of herbivores and plants was simultaneous (Vallentine 1990). While grazing history alone may not be a good predictor of plant-animal interactions, mechanisms between plants and grazing animals are well developed in grasslands with a long evolutionary history of grazing (Milchunas et al. 1988). Therefore, it is conceptually easier to develop methods of using livestock herbivory to alter vegetation in a predetermined manner on rangelands with an evolutionary history of grazing because there is a greater margin for error. The hypothesis developed by Kantrud (1990) regarding the use of fire and grazing to manage waterfowl habitat in the prairie pothole region is logical and proper for that area. The vegetation of the prairie pothole country (Great Plains) evolved under consistent influence from fire (Higgins 1986) and grazing by bison (*Bison bison*) (Platou and Tueller 1985). Application of similar principles to the high-elevation wetlands of the Southwest (cienegas) should be cautiously applied because these wetlands evolved under less frequent fire and little grazing by large herbivores (Severson 1990).

An explicit evolutionary history of grazing by large ungulates should not be considered a prerequisite for success, however. The vegetation of the Great Basin did not evolve under significant pressure by large herbivores (Platou and Tueller 1985, Urness 1989, Young et al. 1976), yet one of the more successful tests of using livestock grazing to improve a specified condition for a wildlife species was developed in Great Basin/Rocky Mountain transitional vegetation (Urness 1990).

Tradeoffs

It would not be difficult, while perusing the literature on livestock-wildlife interactions, to find the statement that "livestock grazing destroys wildlife habitat." Livestock grazing does not destroy wildlife habitat any more than does timber harvest or fire, but it certainly has the capability to alter it; improving it for some species and degrading it for others. In our quest to improve habitat conditions for a featured species, be it mule deer or bobwhite quail, we will undoubtedly create adverse and beneficial habitat conditions for other species in the process.

Kie and Loft (1990) used a wildlife habitat relationships database to evaluate effects of livestock-induced changes in annual grassland habitat in California from a tall-herb to short-herb structural class. They concluded that this structural height reduction would have a positive effect on 52 wildlife species, no effect on 171, and a negative effect on 29. In general, birds and reptiles would be favored by short-herb conditions, while mammals and amphibians would be adversely affected. In a similar analysis of wet meadow habitat in California, they suggested that reduction in herbage height would benefit 59 wildlife species, have no effect on 169, and negatively affect 37 species. More birds and reptiles were favorably affected by height reductions; and most mammals, which depended on herbaceous vegetation for cover, and amphibians were either not affected or were adversely affected. Kie and Loft (1990) cautioned, however, that this information should be used only as a first approximation for such comparisons because of potential errors from faulty assumptions or actual errors in the database.

Hanley and Page (1982) reported a situation where livestock grazing decreased plant structural diversity on xeric sites and increased it on more mesic sites. The xeric habitats were dominated by shrubs and reduction of herbaceous vegetation by livestock resulted in a reduction in plant life diversity. Conversely, mesic habitats were dominated by herbaceous vegetation with few shrubs. The herbs were reduced by livestock grazing, but the shrub component increased, which resulted in an overall increase in plant life diversity up to a point. Further excessive use by livestock could then result in a decline in diversity. As a result, some rodent species (e.g., least chipmunk (*Eutamias minimus*) and Great Basin pocket mouse (*Perognathus parvus*) decreased in xeric communities and increased in mesic communities. Microtine rodents, which were dependent primarily on the herbaceous component, declined with any livestock grazing.

An assessment of bird and small mammal populations in grazed and ungrazed semidesert grasslands in southeastern Arizona revealed that more birds were present in the grazed area in summer (Bock et al. 1984). Four species preferred the more open grazed area whereas only two were significantly more abundant in ungrazed habitats. While there were no differences in total number of birds in winter, species composition again differed considerably. Conversely, small mammals were more abundant in ungrazed habitats. One species, Merriam's kangaroo rat (*Dipodomys merriami*), was significantly more common in grazed areas compared to six species in ungrazed habitats (Bock et al. 1984).

Generally similar results were found in a comparison of grazed and ungrazed (grazing excluded for 14 years) riparian habitats in Idaho (Medin and Clary 1990). Three species of shorebirds normally associated with the low vegetational profiles characteristic of grazed areas bred only in those habitats. Conversely, other species were more common in ungrazed habitat. Species richness and diversity of small mammals was higher in ungrazed habitat. Deer mice (*Peromyscus maniculatus*) were more frequently trapped in grazed habitat, whereas montane voles (*Microtus montanus*) were more common on ungrazed areas.

Such tradeoffs are not limited to terrestrial systems. Rinne (1988) determined that total numbers of fishes were consistently higher in streams traversing grazed riparian areas than in ungrazed reaches. Trout were more abundant in stream sections in the ungrazed areas and chubs and suckers more abundant where streambanks were grazed. Since human interest is generally oriented towards trout, a management decision based on this information would not be difficult to make. However, if one of the nonsalmonid species were rare and potentially could be classified as threatened or endangered, livestock could become a viable tool to improve its habitat.

There is an abundance of literature reporting effects of livestock grazing on wildlife populations from which information on tradeoffs can be gleaned. We have mentioned only a few. It is important to know what tradeoffs in wildlife species composition will occur when a planned habitat change directed at a featured species if only to ensure that another key or featured species is not adversely affected.

Planning

The formation of understandable and realistic goals and the development of plans to attain those goals are keys to using livestock to improve wildlife habitats. It does, however, present some unique planning problems and requires expertise from several disciplines including rangeland ecology, wildlife biology, and livestock production.

Two very important considerations pertaining to the planning process have already been mentioned: 1) A thorough knowledge of the specific rangeland ecosystem and seral dynamics of each site within that system will be paramount and 2) an intimate understanding of how wildlife species other than the featured or target species will be affected by habitat changes is equally important.

Since livestock grazing is the tool being used, it is important to examine some aspects of management that should be considered. In a recent symposium on using livestock to enhance wildlife habitat all six papers emphasized the proper timing of grazing relative to season and/or plant phenology and attaining precise utilization levels on some forage species while avoiding use on others (Severson 1990). Timing and degree of utilization are important in several respects. Kie and Loft (1990) stated that defoliation of plants by grazing animals will have different effects on plant vigor and reproduction depending on the life stage of the individual plants. Urness (1990) and Anderson et al. (1990) offered specific examples concerning timing and degree of grazing and described how these factors can influence production of another species and nutritive content of forages, respectively. Guthery et al. (1990) suggested that individual pastures in a short-duration grazing system could be rested or grazed longer to create a variety of seral stages in close proximity. Kantrud (1990) recommended that research on effects of fire and grazing on waterfowl habitat should stress seasonality, frequency, intensity, and interactions.

Kie and Loft (1990) stressed that critical periods in the life cycle of wildlife species also should be considered. Grazing at the wrong time can cause destruction of bird nests and excessive grazing can result in loss of hiding cover for newborn animals. Sedivec et al. (1990) stated, as an example, that cattle grazing should not begin (in North Dakota) until after the third or fourth week in May to avoid disturbing nesting waterfowl.

A grazing system is the manner in which grazing and nongrazing periods are arranged within the grazing season, either within or between years (Vallentine 1990). An assessment of how particular grazing systems affect wildlife or how they could be used to enhance wildlife habitat would be beyond the scope of this paper. Bryant (1982) reviewed 214 studies to evaluate effects of grazing systems on wildlife and noted that systems tend to benefit wildlife when compared to year-long or season-long grazing. There are exceptions, but generally any grazing system is better for wildlife than none at all. Trade-offs will still occur, however. Not all wildlife species will respond the same to any one system. Some will benefit, others will not (Bryant 1982). Other information is available describing how various systems could be used to improve wildlife habitat (Holechek et al. 1982, Guthery et al. 1990, Sedivec et al. 1990). Based on the knowledge available, we suggest that managers keep an open mind relative to the kind of grazing system to employ. Standardized methods may suffice in some situations but since optimization of wildlife rather than red meat production is a goal, innovative adjustments may be necessary. Since objectives differ with species and locality, managers may want to develop site-specific systems which, while likely based on standard systems, may have significant differences. Managers must remember that each situation is unique and will require a separate analysis and likely a different prescription (Holechek et al. 1982). Systems such as the short-duration grazing system discussed by Guthery et al. (1990) offer promise because they provide the advantage of maximizing flexibility.

Management of livestock grazing for maximum wildlife benefits or to increase diversity of vegetation may be different than management for maximum livestock production (Holechek et al. 1982) and may involve nontraditional approaches in season of use and intensity of stocking (Urness and Austin 1989). The livestock being used to manage habitats will likely be owned by a private individual intent on making a living from livestock production. Therefore, it would help if effects on livestock production were compatible with that intent (Longhurst et al. 1976). If livestock production levels were adversely affected it would still be possible to offset losses with a lease fee (i.e., a wildlife management agency or interest group leases a herd). Either way, it will be necessary to know how livestock production will be affected.

Unfortunately, there is little specific information available. Malechek et al. (1978) studied production of the sheep used to improve mule deer winter range. They found no differences between short-term and season-long grazing, but noted that diet quality, forage intake, and lamb weights were lower under heavy as compared to moderate stocking rates. Sedivec et al. (1990), in a study with non-replicated treatments, found that livestock production (measured as average gain of total calf body weight) was higher on the same specialized grazing systems that yielded better waterfowl and sharp-tailed grouse (*Pedioecetes phasianellus*) production. These results may be unique to this site, however. Generally, current or short-term livestock production is similar or higher under continuous grazing systems when compared to any rotational or resting systems (Wilson 1986).

A final point to ponder is one that pertains primarily to native ungulates, particularly elk and deer. We must recognize that regulation of numbers, including the native ungulates, is an essential part of a successful management program. If numbers cannot be controlled, any effort towards habitat improvement will have been wasted (Longhurst et al. 1983). Controlling cervid numbers via hunting and even thinking in terms of periodic heavy removal to simulate a nongrazing phase of a grazing system may be in order. Neither prescription grazing nor cervid population control will suffice alone; they must be considered as a package (Urness 1990).

Conclusions

While the idea of using livestock as a tool to improve wildlife habitat is not new; its actual implementation into management practice is—so new, in fact, that it is not yet well understood, much less accepted. While the general community interested in natural resources appears willing to consider the concept, there are two separate factions involved. Those that believe the concept is going to solve all problems relative to livestock-wildlife interactions and those that see it as nothing more than a feeble attempt to camouflage past, current, and future rangeland abuse by livestock. We suggest that neither interpretation has merit. We hope that the idea of using prescribed livestock grazing as a tool to improve habitat presents a new, positive approach to looking at the problems of livestock-wildlife interactions, but it certainly will not solve all of them. Its application may indeed be very limited. To those who view the concept as an attempt to exonerate mismanagement of livestock, we can only urge that you view this in the context of the planning considerations described above. The emphasis embodied by the idea of using livestock as a tool to manage wildlife habitat is one of establishing definite goals and developing precise plans to achieve those goals; such themes do not imply bad or lackadaisical management. Serious acceptance of the concept will require increased management effort, not less.

It is clear, however, that in order for the concept to succeed, several misconceptions must be overcome. One is a bias against livestock by many professional wildlife biologists and wildlife interest groups. Examples of obvious livestock mismanagement should not be used to circumvent planned efforts to improve conditions for featured wildlife species. Second, livestock interests should be more willing to put wildlife habitat management goals on a par with livestock production goals. As Anderson and Scherzinger (1975) stated, "... the livestock operator must understand the objectives and principles involved [of an elk management system] and strive to achieve them by his livestock management." While interaction problems such as competition and predation do occur, they are generally the result of improper management decisions. A healthy and diverse wildlife component indicates a healthy and diverse rangeland ecosystem—a credible goal regardless of specific interests. Finally, a primary problem is the almost insurmountable idea that overgrazing is totally bad. While site degradation should be avoided, managers must be given the latitude to maintain or create a variety of seral stages (e.g., sites in something less than a high seral stage) to meet specific objectives.

Presently, knowledge regarding the use of livestock as a tool to improve wildlife habitats is quite meager. As mentioned previously, it cannot be universally applied in all areas; wildlife habitat and range managers, as well as members of the scientific community, have a clear responsibility to define where and how it can be used as well as to determine where it cannot or should not be used (Severson 1990). We do believe, however, that enough information exists to apply techniques in some areas (e.g., mule deer winter range in many areas of the West) and to investigate its applicability in others (e.g., waterfowl habitat in the Great Plains).

Acknowledgments

We would like to thank Drs. Fred S. Guthery, Texas A&I University; John G. Kie, USDA Forest Service, Pacific Southwest Forest and Range Experiment Station; James M. Peek, University of Idaho, and two anonymous colleagues for reviewing this paper.

Literature Cited

Anderson, E.W., D.L. Franzen, and J.E. Melland. 1990. Forage Quality as influenced by prescribed grazing, p. 56-70. *In*: K.E. Severson, Tech. Coord., Can Livestock Be Used as a Tool to Enhance Wildlife Habitat? USDA Forest Service Gen. Tech. Rep. RM-194.

Anderson, E.W. and R.J. Scherzinger. 1975. Improving quality of winter range forage for elk by cattle grazing. J. Range Manage. 28:120-125.

Bock, C.E., J.H. Bock, W.R. Kenney, and V.M. Hawthorne. 1984. Responses of birds, rodents, and vegetation to livestock exclosure in a semidesert grassland site. J. Range Manage. 37:239-242.

Braun, C.E., T. Britt, and R.O. Wallestad. 1977. Guidelines for maintenance of sage grouse habitats. Wildl. Soc. Bull. 5:99-106.

Braun, C.E., K.W. Harmon, J.A. Jackson, and C.D. Littlefield. 1978. Management of national wildlife refuges in the United States: Its impacts on birds. Wilson Bull. 90:309-321.

Bryant, F.C. 1982. Grazing, grazing systems, and wildlife. Tech. Article T-9-297, Texas Tech Univ., Lubbock (Paper presented at the Great Plains Agric. Council Annual Meeting, North Platte, NE, June 1982).

Connolly, G.E. and O.C. Wallmo. 1981. Management challenges and opportunities, p. 537-545. *In*: O.C. Wallmo, ed. Mule and black-tailed deer of North America. The Wildlife Management Institute, Washington, D.C.

Dragt, W.J. and K.M. Havstad. 1987. Effects of cattle grazing upon chemical constituents within important forages for elk. Northwest Sci. 61:70-73.

Dunn, P.O. and C.E. Braun. 1986. Summer habitat use by adult female and juvenile sage grouse. J. Wildl. Manage. 50:228-235.

Fulgham, K.O., M.A. Smith, and J.C. Malechek. 1982. A compatible grazing relationship can exist between domestic sheep and mule deer, p. 458-478. *In*: J.M. Peek and P.D. Dalke, eds. Wildlife-livestock relationships symposium proceedings. Univ. of Idaho, Moscow.

Guthery, F.S. 1986. Beef, brush, and bobwhites, quail management in cattle country. Caesar Kleberg Wildl. Res. Instit., Texas A&I Univ., Kingsville, TX.

Guthery, F.S., C.A. DeYoung, F.C. Bryant, and D.L. Drawe. 1990. Using short duration grazing to accomplish wildlife habitat objectives, p. 41-55. *In*: K.E. Severson, Tech. Coord., Can Livestock Be Used as a Tool to Enhance Wildlife Habitat? USDA Forest Service Gen. Tech. Rep. RM-194.

Hall, R.M. 1955. Livestock and big game relationships. J. Range Manage. 8:4-6.

Hammerquist-Wilson, M.M. and J.A. Crawford. 1981. Response of bobwhites to cover changes within three grazing systems. J. Range Manage. 34:213-215.

Hanley, T.A. and J.L. Page. 1982. Differential effects of livestock use on habitat structure and rodent populations in Great Basin communities. Calif. Fish and Game 68:160-174.

Higgins, K.F. 1986. Interpretation and compendium of historical fire accounts in the Northern Great Plains. U.S. Dept. Interior, Fish and Wildl. Ser. Resource Pub. 161. Washington, D.C.

Hillman, C.N., R.L. Linder, and R.B. Dahlgren. 1979. Prairie dog distribution in areas inhabited by black-footed ferrets. Amer. Mid. Nat. 102:185-187.

Holechek, J.L., R. Valdez, S.D. Schemnitz, R.D. Pieper, and C.A. Davis. 1982. Manipulation of grazing to improve or maintain wildlife habitat. Wildl. Soc. Bull. 10:204-210.

Hull, A.C., Jr. and M.K. Hull. 1974. Presettlement vegetation of Cache Valley, Utah and Idaho. J. Range Manage. 27:27-29.

Jensen, C.H., A.D. Smith, and G.W. Scotter. 1972. Guidelines for grazing sheep on rangelands used by big game in winter. J. Range Manage. 25:346-352.

Jourdonnais, C.S. and D.J. Bedunah. 1990. Prescribed fire and cattle grazing on an elk winter range in Montana. Wildl. Soc. Bull. 18:232-240.

Kantrud, H.A. 1990. Effects of vegetation manipulation on breeding waterfowl in prairie wetlands - a literature review, p. 93-123. *In:* K.E. Severson, Tech. Coord., Can Livestock Be Used as a Tool to Enhance Wildlife Habitat? USDA Forest Service Gen. Tech. Rep. RM-194.

Kie, J.G. and E.R. Loft. 1990. Using livestock to manage wildlife habitat: Some examples from California annual grassland and wet meadow communities, p. 7-24. *In*: K.E. Severson, Tech. Coord., Can Livestock Be Used as a Tool to Enhance Wildlife Habitat? USDA Forest Service Gen. Tech. Rep. RM-194.

Knopf, F.L. and R.W. Cannon. 1982. Structural resilience of a willow riparian community to changes in grazing practices, p. 198-207. *In*: J.M. Peek and P.D. Dalke, eds. Wildlife-livestock relationships symposium proceedings. Univ. of Idaho, Moscow.

Koford, C.B. 1958. Prairie dogs, whitefaces, and blue grama. Wildl. Monogr. No. 3.

Krueger, H.O. and S.H. Anderson. 1985. The use of cattle as a management tool for wildlife in shrub-willow riparian systems, p. 300-304. *In*: R. Johnson, C. Ziebell, D. Patton, P. Ffolliott, and R. Hamre, Tech. Coord. Riparian ecosystems and their management: Reconciling conflicting uses. First North American Riparian Conference. USDA Forest Service Gen. Tech. Rep. RM-120.

Kufeld, R.C. 1983. Responses of elk, mule deer, cattle, and vegetation to burning, spraying, and chaining of Gambel oak rangeland. Colo. Div. of Wildl. Tech. Publ. 34.

Laycock, W. A. 1967. How heavy grazing and protection affect sagebrush-grass ranges. J. Range Manage. 20:206-213.

Laycock, W.A. 1989. Secondary succession and range condition criteria: Introduction to the problem, p. 1-15. *In*: W.K. Lauenroth and W.A. Laycock, eds, Secondary succession and the evaluation of rangeland condition. Westview Press, Boulder, CO.

Leopold, A. 1933. Game management. Charles Scribner's Sons, New York.

Longhurst, W.M., E.O. Garton, H.F. Heady, and G.E Connolly. 1976. The California deer decline and possibilities for restoration. [Paper presented at the annual meeting of the West. Sec., The Wildl. Soc., Fresno, CA, Jan. 30-31, 1976. Photocopy].

Longhurst, W.M., A.L. Lesperance, M. Morse, R.J. Mackie, D.L. Neal, H. Salwasser, D. Swickard, P.T. Tueller, P.J. Urness, and J.D. Yoakum. 1983. Livestock and wild ungulates, p. 42-64. *In*: Proceedings: Workshop on Livestock and Wildlife-fisheries Relationships in the Great Basin. Agr. Sci. Spec. Pub. 3301, Univ. Calif., Berkeley.

Malechek, J.C., K.J. Kotter, and C.H. Jensen. 1978. Nutrition and production of domestic sheep managed as manipulators of big game. J. Range Manage. 31:91-96.

McCulloch, C.Y. and P.J. Urness. 1973. Deer nutrition in Arizona chaparral and desert habitats. Arizona Game and Fish Dept. Spec. Rept. 3.

Medin, D.E. and W.P. Clary. 1990. Bird and small mammal populations in a grazed and ungrazed riparian habitat in Idaho. USDA Forest Service Res. Pap. INT-425.

Milchunas, D.G., O.E. Sala, and W.K. Lauenroth. 1988. A generalized model of the effects of grazing by large herbivores on grassland community structure. Amer. Nat. 132:87-106.

Neal, D.L. 1982. Improvement of Great Basin winter range with livestock grazing, p.61-73. *In:* J.M. Peek and P.D. Dalke, eds. Wildlife-livestock relationships symposium proceedings. Univ. of Idaho, Moscow.

Pieper, R.D. and R.F. Beck. 1990. Range condition from an ecological perspective: Modifications to recognize multiple use objectives. J. Range Manage. 43:550-552.

Pitt, M.D. 1986. Assessment of spring defoliation to improve fall forage quality of bluebunch wheatgrass (*Agropyron spicatum*). J. Range Manage. 39:175-181.

Platou, K.A. and P.T. Tueller. 1985. Evolutionary implications for grazing management systems. Rangelands 7:57-61.

Reiner, R.J. and P.J. Urness. 1982. Effects of grazing horses managed as manipulators of big game winter range. J. Range Manage. 35:567-571.

Rhodes, B.D. and S.H. Sharrow. 1990. Effect of grazing by sheep on the quantity and quality of forage available to big game in Oregon's Coast Range. J. Range Manage. 43:235-237.

Riggs, R.A. and P.J. Urness. 1989. Effects of goat browsing on Gambel oak communities in northern Utah. J. Range Manage. 42:354-360.

Riggs, R.A., P.J. Urness, and K.A. Gonzalez. 1990. Effects of domestic goats on deer wintering in Utah oakbrush. J. Range Manage. 43:229-234.

Rinne, J.N. 1988. Grazing effects on stream habitat and fishes: Research design considerations. N. Amer. J. Fish. Manage. 8:240-247.

Schulz, P.A. and F.S. Guthery. 1988. Effects of short duration grazing on northern bobwhites: A pilot study. Wildl. Soc. Bull. 16:18-24.

Sedivec, K.K., T.A. Messmer, W.T. Barker, K.F. Higgens, and D.R. Hertel. 1990. Nesting success of upland nesting waterfowl and sharp-tailed grouse in specialized grazing systems in southcentral North Dakota, p. 71-92. *In*: K.E. Severson, Tech. Coord., Can Livestock Be Used as a Tool to Enhance Wildlife Habitat? USDA Forest Service Gen. Tech. Rep. RM-194.

Severson, K.E. (Tech. Coord.). 1990. Can Livestock Be Used as a Tool to Enhance Wildlife Habitat? USDA Forest Service Gen. Tech. Rep. RM-194.

Severson, K.E. and L.F. DeBano. 1991. Influence of Spanish goats on vegetation and soils in Arizona chaparral. J. Range Manage. 44:111-117.

Severson, K.E. and A.L. Medina. 1983. Deer and elk habitat management in the Southwest. J. Range Manage. Monogr. No. 2.

Skovlin, J.M., P.J. Edgerton, and B.R. McConnell. 1983. Elk use of winter range as affected by cattle grazing, fertilization, and burning in southeastern Washington. J. Range Manage. 36:184-189.

Smith, A.D. 1949. Effects of mule deer and livestock upon a foothill range in northern Utah. J. Wildl. Manage. 13:421-423.

Smith, A.D. and D.D. Doell. 1968. Guides to allocating forage between cattle and big game on big game winter range. Publication 68-11. Utah State Div. of Fish and Game, Salt Lake City.

Smith, E.L. 1989. Range condition and secondary succession: A critique, p. 103-141. *In:* W.K. Lauenroth and W.A. Laycock, eds, Secondary succession and the evaluation of rangeland condition. Westview Press, Boulder, CO.

Smith, M.A., J.C. Malechek, and K.O. Fulgham. 1979. Forage selection by mule deer on winter range grazed by sheep in spring. J. Range Manage. 32:40-45.

Spears, G.S. 1991. Optimal seral stage for northern bobwhite as influenced by site productivity. MS Thesis. Texas A&I University, Kingsville.

Sundstrom, C., W.G Hepworth, and K.L. Diem. 1973. Abundance, distribution and food habits of the pronghorn. Wyo. Game and Fish Commission Bull. No. 12.

Uresk, D.W., J.G. McCracken, and A.J. Bjugstad. 1981. Prairie dog density and cattle grazing relationships, p. 199-201. *In*: Proceedings Fifth Great Plains Wildlife Damage Control Workshop, Univ. Nebraska, Lincoln.

Urness, P.J. 1989. Why did bison fail west of the Rockies? Utah Science 50:175-179.

Urness, P.J. 1990. Livestock as manipulators of mule deer winter habitats in northern Utah, p. 25-40. *In*: K.E. Severson, Tech. Coord., Can Livestock Be Used as a Tool to Enhance Wildlife Habitat? USDA Forest Service Gen. Tech. Rep. RM-194.

Urness, P.J. and D.D. Austin. 1989. The effects of grazing and browsing animals on wildlife habitats. Utah Sci. 50:104-107.

Vallentine, J.F. 1990. Grazing management. Academic Press, Inc., New York.

Wallmo, O.C., L.H. Carpenter, W.L. Reglin, R.B. Gill, and D.C. Baker. 1977. Evaluation of deer habitat on a nutritional basis. J. Range Manage. 30:122-127.

Willms, W., A.W. Bailey, and A. McLean. 1980. Effect of burning or clipping (*Agropyron spicatum*) in the autumn on the spring foraging behaviour of mule deer and cattle. J. Appl. Ecol. 17:69-84.

Wilson, A.D. 1986. Principles of grazing management systems. Proc. Int. Rangelands Cong. 2:221-225.

Young, J.A., R.A. Evans, and P.T. Tueller. 1976. Great Basin plant communities— pristine and grazed, p. 186-215. *In*: R. Elson, ed. Holocene environmental change in the Great Basin. Nev. Arch. Surv. Res. Pap. No. 6. Reno, NV.

Implications of Grazing vs. No Grazing on Today's Rangelands

W.A. Laycock

Abstract

Grazing by livestock or other herbivores affects different rangeland communities in different ways. In general, more mesic communities that evolved under grazing, such as most grassland communities, may be rather resistant to deterioration caused by grazing. However, if they do become degraded they usually improve rather quickly when grazing is reduced or stopped. This predictable response fits the present conceptual range condition model reasonably well. On many arid and semi-arid rangeland communities, grazing too heavily can cause rapid degradation. However, improvement in range condition when grazing is reduced or removed often is extremely slow or absent. The current conceptual model used to determine range condition does not take into account this lack of response to a release from grazing. Examples are given of apparent "stable states" of range condition for many arid and semi-arid vegetation types, including a number of shrub-dominated communities. In these situations, even complete removal of livestock grazing results in little or no improvement in range condition for many years or even decades. This lack of conformity to the present range condition model for many vegetation types is not very well understood by managers and often leads to inappropriate conclusions about proper management strategies to improve range condition. Examples of possible alternative conceptual models, including the "stable state" and "threshold" concepts are presented.

Introduction

Much of range management and range science focuses on the effects of herbivore grazing on individual plants and plant communities. Thus, it is impossible to cover the subject in one paper. The purpose of this book is to examine all aspects of herbivory. The present chapter will compare the effects of herbivore grazing to that of protection from grazing in a number of different circumstances and rangeland communities.

Currently, there is much interest and controversy about the effects of grazing by domestic livestock, especially on public lands in the western United States (Gillis 1991). Ehrlich (1990) stated that "much of the millions of acres of land controlled by the U.S. Forest Service and the Bureau of Land Management is severely overgrazed." He stated that overgrazing causes unacceptable erosion and reduces the diversity of plants and animals in both riparian and non-riparian habitats. Such unsubstantiated charges of widespread overgrazing, erosion, threats to endangered species and reduction of wildlife habitat and species diversity have prompted some environmental organizations to advocate the elimination of livestock grazing from public lands (Wuerthner 1990, Jacobs 1991).

Methods Used To Study Effects Of Grazing

The most common method used to study grazing effects is to compare the vegetation, soil, etc., of protected areas with those on adjacent grazed areas. These comparisons most often involve relict areas protected by natural features, relatively large research natural areas (Federal Committee on Ecological Reserves 1977) or fenced exclosures (Daubenmire 1940, Laycock 1975, Michalk and Norton 1980).

Vegetation conditions at the time of exclosure construction and the microclimate of adjacent protected and grazed areas are assumed to be identical although this assumption may not always be valid (Costello and Turner 1941). Small exclosures (less than 0.5-1.0 ha) may have large differences in microclimate compared to adjacent grazed ranges. Fence materials, acting as a barrier to wind movement, can substantially affect moisture or soil deposition, humidity, temperature, snow deposition, and other factors (Daubenmire 1940). The smaller the exclosure or cage used to protect an area from grazing, the greater is the potential influence from the fence. Owensby (1969) reported that total forage production was 42% (720 kg/ha) higher in small (1 sq. m) cages on tall grass prairie compared to uncaged and ungrazed adjacent areas. Other researchers have found smaller (Williams 1951) or no (Heady 1957) differences between caged and uncaged areas. Tueller and Tower (1979) stated that temporary cages, moved each year, are required to determine annual forage production of shrubs accurately.

In addition to microclimate differences, the following aspects must be considered in comparing data from inside and outside exclosures: similarity of sites, small mammal or insect populations, nutrient tie-up in litter accumulations, invasion by non-native species, influences of fire protection and others (Laycock 1975). Also, an accurate record of grazing seasons and intensities outside an exclosure especially areas close to the exclosure often have not been kept, making the comparisons difficult to interpret. In some areas, like high elevation rangelands, small exclosures seem to attract rodents, especially pocket gophers (McClean and Tisdale 1972, Laycock 1975).

Michalk and Norton (1980) stated that reference areas (including exclosures), in addition to providing information to determine condition and trend, provide a way for range managers to separate the impacts of grazing from those caused by climatic variations, insects and other factors. They concluded that "comparisons with reference areas are needed to keep grazing from being blamed for the detrimental effects caused by something else." For example Brown and Heske (1990) found that after 12 years of excluding three species of kangaroo rats from plots of Chihuahuan desert shrub habitat in southeastern Arizona, tall perennial and annual grasses increased threefold and smaller rodent species colonized the exclosures. In this case, seed predation and soil disturbance by kangaroo rats were the major factors suppressing biological diversity and biogeochemical processes rather than livestock grazing.

Many early studies of the effects of protection from grazing, such as that by Costello and Turner (1941), showed substantial differences between grazed and protected areas. Release from the rather heavy grazing during the previous 50 or more years often resulted in rapid vegetation changes in the exclosures. Newly established exclosures often show small or no differences between grazed and ungrazed areas, especially in arid and semi-arid rangeland communities. In the last 50 years, reduced grazing intensities and better management have improved or stabilized vegetation conditions outside as well as inside the exclosures, resulting in little differences between grazed and ungrazed areas.

Exclosures have also been used to study the effects of grazing or browsing by big game and rodents (Young 1958). Tall fences are required to exclude large, jumping game animals while special fence designs keep out burrowing rodents, like pocket gophers (Keith 1961). Interpreting the condition of an area protected from grazing or browsing by native ungulates may be somewhat controversial. Several large elk-proof exclosures were built in aspen stands in Grand Teton National Park in Wyoming in the 1950's. Heavy browsing outside the exclosures effectively suppressed aspen reproduction while aspen thrived inside the exclosures. Several exclosures were removed by the Park Service in the 1970's because, according to Cole (1971) and a letter from Park superintendent quoted by Beetle (1974), they represented a departure from natural conditions, were a significant aesthetic intrusion and cause of public confusion and misinformation.

Trottier (1976) expressed a more logical perspective on exclosures in national parks. "Exclosures as they now exist in the national parks of western Canada are permanent baseline monitoring stations where vegetation change over time in concert with environmental factors can be recorded. It is essential that park managers keep abreast with changes in the ecosystem they are striving to protect, otherwise it becomes impossible to assess the effectiveness of protective measures. Only by knowing what is going on in park ecosystems which are essentially protected from unnatural disturbance can we intelligently rate the impact of man-induced changes on the unprotected world. . . . The role of existing baseline range monitoring stations such as range exclosures . . . is obviously fundamental to research. . . . It seems unwise therefore to neglect or remove these important research tools."

Costello and Turner (1941), summarizing data from 139 exclosures and adjacent grazed areas, generally found better range conditions inside exclosures with some exceptions. They concluded: "Expected results from enclosures and protected areas are not always obtained. This is not the fault of the enclosure method. Lack of contrast between grazed and ungrazed areas deserves as much attention as outstanding differences between these areas."

Objectives

Differences or lack of differences between grazed and ungrazed rangelands will be the major focus of this paper. Effects of both heavy grazing and the lack of grazing will be documented for the major vegetation types in North America. When grazing is removed, improvements in range condition predicted by our current model, sometimes occur. Where lower successional stable states are formed and the "expected" improvements do not take place when grazing is removed, the implications of this lack of change on the management of these rangelands will be discussed. Discussion will also focus on existing and new range condition models, concepts that may help explain situations where the current model is unsatisfactory, and available information on the effects of grazing on species diversity.

Determination Of Range Condition

Current Range Condition Concepts

In the United States, range condition generally is assessed on what has been called a "climax" (Friedel 1991) or "successional" (Westoby et al. 1989) model of vegetation changes. The concepts underlying this model relate directly to the climax and plant succession concepts of Clements (1916) and the application of these concepts to rangelands by Sampson (1919). As indicated by Pieper (1993), the range condition model currently used was proposed by Dyksterhuis (1949). This model assumes that: 1) a vegetation type or range site has only one stable state ("climax"), 2) any retrogressive change caused by improper grazing results in an unstable state which can be reversed by reduction, manipulation, or elimination of grazing, 3) the pathway of vegetation change as rangelands improve (secondary succession) is identical and the reverse of that followed in retrogression and 4) for any given community, all changes in condition can occur continuously and reversibly along a continuum from poor to excellent condition (Westoby et al. 1989). As I will indicate below, the first three assumptions appear to be valid for some rangeland vegetation types. However, even for these types, the changes may not be linear, as implied by assumption number four. For other vegetation types all four assumptions appear to be invalid.

The "climax" based model, commonly used in the U.S., has been strongly defended as the only logical choice for management (Meeker and Merkel 1984, Pendleton 1989). Malin (1955) questioned the use of the concepts of succession and climax and raised "an honest doubt as to whether the idea of climax vegetation is even legitimate or practical value." In recent years, a number of U.S., Australian, and other range scientists (Archer and Smeins 1992, Brown 1992, Ellis and Swift 1988, Friedel 1988, 1991; Hart and Norton 1988; Laycock 1989a, 1991; Smith 1978, 1989; Svejcar and Brown 1991; Westoby et al. 1989; Wilson 1989 and Wilson and Tupper 1982) have discussed the inadequacies of the current range condition model, emphasizing the need to develop better techniques to describe and interpret range condition. Attempts to modify the existing model (Foran et al. 1978, Hart and Norton 1988, Schlatterer 1989) have been unsuccessful due to problems caused by inappropriate assumptions discussed earlier.

Heady (1973) presented the idea of multiple climaxes resulting from various human activities such as grazing, introduction of new herbivores or plants, fertilization, etc. and stated "Permanent alterations in rangeland ecosystems should be recognized as such when their removal becomes impossible in practical terms. This has been recognized by the Potential Natural Vegetation concept proposed by the Range Inventory Standardization Committee (1983) and now used by both the Forest Service and Bureau of Land Management instead of the term "climax."

New Range Condition Concepts

In spite of Heady's (1973) concepts of multiple climaxes, a major problem with the present range condition model and range condition standards is the assumption that vegetation composition will move toward "the" climax as grazing pressure is reduced or eliminated. Examples will illustrate that this assumption is incorrect for many arid and semi-arid rangeland vegetation communities. Instead, lower successional stable communities may persist for decades following the elimination of grazing, with little or no change (Laycock 1991). Discussion of some alternative range condition conceptual models will be presented after the sections on individual vegetation types.

Comparisons Of Grazing Vs. No Grazing

Tall Grass Prairie

On tall grass prairie, the more palatable tall grasses decrease and the mid-grasses and/or short grasses become dominant with prolonged heavy grazing. When grazing pressure is reduced, but not removed completely, the stand again becomes dominated by the tall grass species (Penfound 1964). This sequence of degradation caused by heavy grazing and improvement with removal of grazing is explained quite well by the current range condition model.

However, when grazing is completely removed from tall grass prairie in relatively good condition, changes often occur which are not explained by the "climax" model. After elimination of grazing, Cosby (undated) described invasion of different areas of tall grass prairie in North Dakota by Kentucky bluegrass (*Poa pratensis*) and smooth brome (*Bromus inermis*) causing native dominant grasses to decline. Even where alien species do not invade ungrazed tall grass areas, the rapid accumulation of large volumes of litter retards growth in the spring and may eliminate many seedlings (Weaver and Fitzpatrick 1934). Periodic manipulation by grazing, mowing or burning usually is necessary to maintain desired "climax" tall grass plant communities.

Northern Mixed Grass and Palouse Prairie

On Idaho fescue-bluebunch wheatgrass area in southwestern Montana, Evanko and Peterson (1955) found that grazed areas contained more unpalatable shrubs and forbs, less total grass cover, shorter grass leaf heights, lower total yield, less litter, and lower water infiltration than areas inside exclosures after 15-18 years. The two dominant grasses comprised slightly higher cover on the protected areas than on grazed ranges. In southeastern Alberta, Smoliak (1965) found that protection from grazing for 33 years favored a higher proportion of needleandthread (*Stipa comata*) while light grazing favored a higher proportion of blue grama (*Bouteloua gracilis*). The same species of grasses and forbs were present on both sites. However, the lightly grazed sites produced less herbage than the exclosure.

In mixed grass prairies of the ponderosa pine (*Pinus ponderosa*) zone in southern British Columbia, McClean and Tisdale (1972) found higher production of bunch grasses in exclosures compared to grazed areas. They concluded that the mixed prairie vegetation type would improve in range condition from poor to good within 30 years after removing grazing, reaching excellent condition in 10 more years. However, in one exclosure condition improved from poor to fair in 23 years but retrogressed to poor condition in the next 10 years after extensive disturbance of soil and vegetation by pocket gophers. This is an example of potential hazards of relatively small exclosures.

Vegetation changes described in the previous studies can be explained by the current range condition model. However, not all studies can be similarly explained. Johnston (1961) reported that light grazing on rough fescue (*Festuca scabrella*) rangelands in southwestern Alberta resulted in a more diverse flora dominated by Parry danthonia (*Danthonia parryi*) while protection from grazing simplified the flora with a trend toward dominance by fescue. Total production was similar on both sites but root penetration was considerably greater on the lightly grazed site.

Watts et al. (1987) reported that rest-rotation grazing by cattle and protection from grazing were equally effective in maintaining production and ground cover of recently burned areas in the Missouri breaks of north central Montana. Season-long grazing resulted in declining forage production and increasing bare soil exposure.

On rough fescue grasslands in the mixed grass prairie of western Manitoba, further east in the mixed grass prairie, Trottier (1986) found Kentucky bluegrass communities with much lower species diversity replaced native grasses under heavy grazing. Little change occurred in these communities when protected from grazing, indicating a lower successional stable state situation. Trottier (1986) stated that recovery of this bluegrass stand does not appear possible without rehabilitative management such as fire (Curtis and Partch 1948). Singer (1987) and Merrill and Stanton (1990) reported significant differences between ungrazed and grazed areas of bluebunch wheatgrass (*Agropyron spicatum*) and Idaho fescue (*Festuca idahoensis*) on elk winter ranges in Yellowstone Park. Dry weight of standing dead grass was more than 6 times greater inside exclosures than in areas heavily grazed by elk and above ground production of vegetation was significantly reduced on the grazed area (Singer 1987). Nitrogen content in roots and above ground foliage was higher on grazed areas. Merrill and Stanton (1990) stated that "These studies concluded that winter grazing by ungulates had little negative effect on plants and that plants may even benefit by a shift in the flow of carbon from decomposers to ungulates with possibly an enhanced rate of nitrogen recycling back to herbivores." Similar responses to heavy grazing by livestock would likely be described as "overgrazing."

Shortgrass Steppe

Blue grama is the major species in shortgrass steppe areas in the Central Great Plains. This vegetation type appears to be the most resistant to grazing in the U.S., probably because it evolved under rather heavy grazing by bison and other animals (Larson 1940). England and DeVos (1969) examined diaries and journals of early explorers in Canada and cited numerous entries describing extremely heavy grazing and widespread disturbance by bison. Because this area historically supported large herds of bison, introduction of livestock had considerably less effect on community structure compared to areas west of the Rocky Mountains (Mack and Thompson 1982, Lauenroth et al. 1993).

In an early study comparing differences between grazed and ungrazed areas on shortgrass steppe, Costello and Turner (1941) found that moderately grazed areas had a higher percentage ground cover than exclosures. Protection from grazing results in thinning of blue grama cover (Hyder et al. 1975). Studies on shortgrass steppe in northeastern Colorado indicated that species composition changed little under long-term light, moderate and heavy grazing but total above ground production decreased with increased grazing intensities (Klipple and Costello 1960).

On upland areas, continued heavy grazing, especially summer grazing, results in a reduction of cool season grasses as blue grama increases (Hyder et al. 1975). Extremely heavy grazing over many years, heavy grazing combined with prolonged drought (Costello and Turner 1944) or a drastic disturbance such as plowing is required to remove blue grama once this stable state is achieved. In upland exclosures established in 1939 in northeastern Colorado, vegetation composition has changed little compared to grazed areas. In drier portions of the shortgrass steppe, blue grama dominated areas appear to be in a steady state condition and are resistant to change caused by different grazing intensities.

Hyder et al. (1975) noted that annuals (native and introduced) fluctuate with the weather on shortgrass rangelands. Thus, range condition standards that include annuals as a typical "invader" will provide different range condition readings in years where precipitation amount and distribution favors annuals than in years when no annuals exist. Likewise, plains pricklypear (*Opuntia polyacantha*), long thought of as a typical "increaser" on shortgrass ranges in the traditional range condition standards, changes in response to weather but not to different grazing intensities or complete protection from grazing (Bement 1968, Houston 1963). Because some species do not respond to grazing according to current range condition standards, Hyder et al. (1975) stated that "there is no ecological basis for visual recognition and separation of vegetation changes induced by grazing from those induced by weather" and "conventional range condition classification on shortgrass plains serves no useful purpose."

A second stable state occurs on some shortgrass steppe areas that have been cultivated and allowed to recover. Many abandoned fields pass through fairly well described stages of early succession (Costello 1944) which includes the return of blue grama (Coffin et al. 1991). However, in some areas with less than 38 cm of precipitation, succession sometimes stops with a community that includes most of the native species but without blue grama (Samuel 1985, Samuel and Hart 1990, Hyder et al. 1975, Wilson and Briske 1979). An area farmed by Indians in New Mexico and abandoned approximately 800 years ago still has no blue grama even though it is the dominant species on surrounding areas (Sandor 1983). Reasons for the absence or presence of blue grama on different fields, even in the same area, are not known. Where they occur, the communities without blue grama are stable and quite resistant to change.

Southwestern Desert Grasslands

When we consider the shrub-dominated arid and semi-arid vegetation types west of the Great Plains, our present range condition model ceases to be effective in explaining changes. Gardner (1950), in a study of the effect of protection from grazing for 28 years (1918-1946) on a desert grassland in southwestern New Mexico found little difference in species composition in 1946. Both protected and grazed areas were still dominated by blue grama with black grama (*Bouteloua eriopoda*), hairy grama (*Bouteloua hirsuta*) and other species present. However total basal cover of grasses was more than twice as high in the exclosure as outside. Gullies were still raw and eroding outside the exclosure but they supported some vegetative cover inside the exclosure.

In another area in southern New Mexico, Potter and Krenetsky (1967) reported increases in black grama inside one exclosure but the stand was floristically less complex than outside the exclosure. They reported a decrease in honey mesquite (*Prosopis glandulosa*) and other shrubs inside the exclosure and increases outside. However, no data were presented and the photographs showed no visible change in shrub species density or size inside or outside the exclosure during the study period. In an exclosure protected from cattle and deer grazing for 25 years on the Edwards Plateau in Texas, only minor changes in relative dominance of species took place and the primary changes followed precipitation variation (Smeins et al. 1976). No grazed area was sampled for comparison.

Once shrubby species invade desert grassland, they maintain themselves for long periods. According to Paulsen and Ares (1962), creosote bush (*Larrea tridentata*) and tarbush (*Flourensia cernua*) are the primary invaders on tobosa (*Hilaria mutica*) grasslands while velvet mesquite (*Prosopis juliflora*) or honey mesquite invade black grama grasslands in New Mexico. Beck and Tober (1985) found that exclusion of rabbits or cattle for 22 years had no consistent effect on herbaceous species and that "the concept that removal of these factors will result in range improvement was not consistently shown by the study."

On Rothrock grama (*Bouteloua rothrockii*) and black grama grasslands in southern Arizona, velvet mesquite density more than doubled from 1932 to 1949 on all grazing treatments with the largest increase occurring on the plots protected from grazing (Glendening 1952). His conclusion was "mesquite, once seed trees are present, may rapidly increase regardless of grazing treatment." The same increase in velvet mesquite in areas protected from grazing was noted by Smith and Schmutz (1975) in Arizona.

Changes from desert grasslands to shrublands in southern New Mexico have occurred during a changing period of a warmer, drier climate which started 100-150 years ago at the end of thc "Little Ice Age" (Neilson 1986). During this warming period, black grama seedlings were found in only seven years between 1915 and 1968 at the Jornada Experimental Range in southern New Mexico (Herbel et al. 1970). Livestock grazing during this warming period may have caused spatial and temporal heterogeneity of water, nitrogen and other soil resources. This may have initiated the invasion of desert shrubs (Schlesinger et al. 1990) that might have taken place over a longer period of time without grazing disturbance.

Possible reasons for the continued dominance of woody species once established were presented by Archer (1989). He concluded that the conversion of mesquite savannas to woodlands in southern Texas has been recent and coincident with both heavy livestock grazing and shifts in precipitation. In his conceptual diagram (Fig. 1), graminoid driven succession predominates within the original grassland

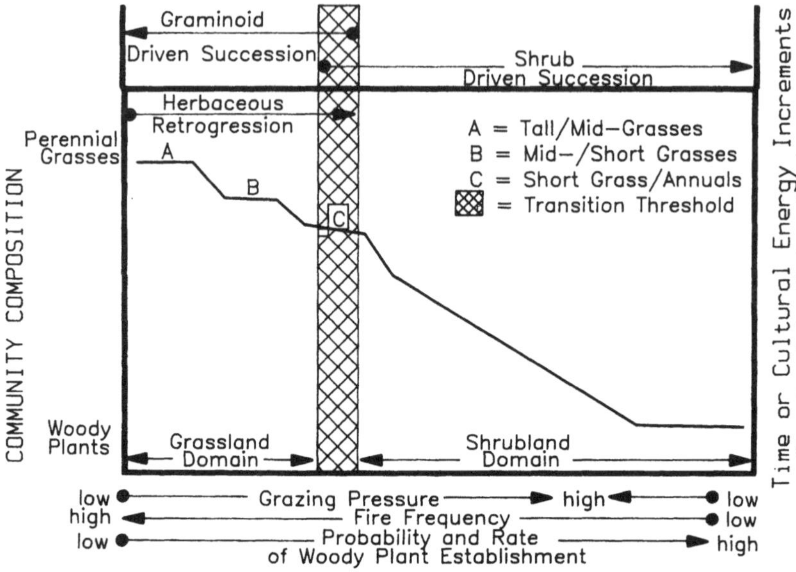

Figure 1 Conceptual diagram of threshold changes in community structure from a grassland or savannah to a mesquite woodland as a function of grazing pressure. From Archer (1989). Copyright © 1989 by the University of Chicago. Reproduced by permission of the University of Chicago.

domain. Heavy grazing alters the composition and productivity of the herbaceous species, spreads mesquite seeds and decreases fires, thereby increasing the probability of woody plant establishment. A dominance of woody plants causes the community to cross a threshold (Point C) where shrub-driven successional processes begin to predominate and the site moves toward a new woodland steady-state configuration. Once in this woodland domain, the site does not revert to grassland even if grazing is stopped. Manipulations of the shrubs (usually mesquite) will temporarily alter the grass-shrub mixture, but subsequent shrub-driven succession results in a rapid return of the shrubs. This woodland is a new steady state.

Sagebrush-Grass

Lower successional steady states are common in the sagebrush-grass vegetation type, which covers almost 50 million ha in the Great Basin due to the historical effects of grazing (Young et al. 1979 and Young 1993). Once sagebrush (especially various subspecies of big sagebrush, *Artemisia tridentata*) becomes dense with a reduced understory, it can dominate a site for a very long time with little change in range condition.

West et al. (1984) found no significant vegetative changes in a big sagebrush type in Utah after 14 years of livestock exclusion. Robertson (1971) reported that 30 years of protection from grazing on an eroded sagebrush-grass site in Nevada resulted in increased vegetal cover of all life forms, including sagebrush. Sagebrush comprised 68% of the total plant cover after 30 years compared to 64% at the start. Tisdale et al. (1969) stated that vigorous stands of sagebrush in Idaho may indefinitely delay the recovery of the herbaceous understory even after 20-25 years of protection from grazing. Sanders and Voth (1983) found no improvement over a 45-year period in three exclosures dominated by big sagebrush in southwestern Idaho. In southeastern Idaho, Anderson and Holte (1981) found that both big sagebrush and grasses increased substantially when protected from grazing for 25 years. There were no differences in species composition between grazed and ungrazed plots after 25 years and they concluded that "no evidence of seral replacement, as predicted by classical succession, was found."

In northern New Mexico, exclusion of cattle grazing for 22 years had little effect on sagebrush or understory species in either an upland or a lowland site (Holechek and Stephenson 1983). Potter and Krenetsky (1967) found that ground cover of grass, forbs and browse decreased in both protected and grazed sagebrush plots in New Mexico, and stated that grazing management alone had "poor possibilities" for improvement of dense sagebrush stands.

Sagebrush dominance represents a stable state which resists changes caused by livestock management that encourages change toward a more open grass/sagebrush state. Protection from grazing sometimes results in an increase in total cover or production of grasses, forbs and shrubs without any change in species composition or range condition. Some intervention, by man or nature, that removes the mature, dominant sagebrush plants is required to push the community out of the stable state dominated by sagebrush.

Protection from grazing in the sagebrush-grass type sometimes can result in undesirable changes in some shrub species in exclosures. Partial removal of some of the current year's growth has been shown to stimulate growth of bitterbrush (*Purshia tridentata*), which is a highly desirable shrub (Ferguson and Basile 1966). In Nevada, Tueller and Tower (1979) reported growth stagnation of several shrub species in areas protected from browsing by both livestock and deer. Bitterbrush protected from browsing for up to 6 years averaged 70% less production than those plants which were browsed annually but protected for one year for sampling. In some exclosures, both big sagebrush and rabbitbrush (*Chrysothamnus viscidiflorus*) also declined in production compared to grazed areas. Martinson (1960) reported a 71% reduction in bitterbrush production after protection from browsing for 9 years in north central Washington. These differences will be found only where the browsing pressure outside the exclosure is moderate to heavy (35 to 50%) (Lay 1965, Tueller and Tower 1979). Moderate browsing stimulates twig growth the following year. Lighter browsing has no stimulation effect and very heavy browsing over a period of years reduces production.

Other Shrub-Dominated Vegetation Types in the Great Basin

Other shrub-dominated vegetation types react similar to the sagebrush-grass type. In northern Utah, Rice and Westoby (1978) examined vegetation inside and outside 12 exclosures in communities dominated by winterfat (*Ceratoides lanata*), Nuttall saltbush (*Atriplex nuttallii*), shadscale (*Artiplex confertifolia*), big sagebrush and black sagebrush (*Artemisia nova*). Protection from grazing by sheep and/or jackrabbits for 6 to 15 years did not change the community composition. One conclusion by Rice and Westoby (1978) was that the concept of grazing succession in these semi-arid shrublands is not meaningful.

Turner (1971) found that exclusion of livestock for 10 years had little effect on shrub communities dominated by big sagebrush, shadscale and Nuttall saltbush in western Colorado. Sharp et al. (1990) documented cyclic changes over 40 years in a shadscale community in Idaho. The same changes occurred in all areas due to weather, insect infestations and other factors not related to grazing.

On a salt desert shrub rangeland in Utah, vegetation changes over a 40-year period were the same under heavy sheep grazing and protection from grazing (Norton 1978). Shadscale, the least palatable shrub, exhibited a short-term rise in cover followed by a steady decline in both grazed and ungrazed areas. The more palatable winterfat consistently increased in cover on both areas. Norton (1978) concluded that, contrary to accepted range management theory, the vegetation changes in dominant palatable and unpalatable species were not a function of grazing pressure. However, in some situations, heavy grazing during the wrong season can cause severely reduced yields or cause death of many shrubs on salt desert shrub rangelands (Cook 1971). Whisenant and Wagstaff (1991) concluded that season of grazing had more effect on vegetation composition than intensity of grazing and that spring grazing was particularly damaging.

Annual Grasslands

Introduced annual grasses have invaded many areas of western rangelands. While heavy livestock grazing and other disturbances have created niches for these annuals in many situations, alien annuals also occur in areas never grazed by livestock or otherwise seriously disturbed. These include isolated "kipukas" in southeastern Idaho (West 1991) which were never grazed and Anaho Island in Nevada (Svejcar and Tausch 1991) which has not been grazed. West (1991) suggested that subtle changes in climate may be the cause of the invasion of alien species into undisturbed habitats. Magnuson (1990) stressed the importance of long-term research because many changes in vegetation either occur slowly or the community changes lag many years behind causes (such as climate changes).

Two large areas in the U.S., once perennial communities are now occupied by apparently stable annual grasslands. It is interesting to note that Weaver and Clements (1938) recognized the occurrence and apparent stability of these areas and stated: "In some cases disturbance and the introduction of alien species act together to produce a community with the general character of the original climax. This is well illustrated by the almost complete replacement of the bunchgrass prairie of California by wild oats and brome grasses (*Avena-Bromus* disclimax). A similar replacement by downy bromegrass (*Bromus tectorum*) has taken place over large areas of the Great Basin." Clements and Shelford (1939) called these annual grasslands "proclimaxes" which includes "all communities that suggest something of the permanence or extent of a climax, but are not controlled by climate."

The California annual grassland originally was dominated by perennial bunch grasses (Biswell 1956). After settlement by the Spanish with their livestock in the late 18th century, and the introduction of a number of well-adapted annuals from the Mediterranean and other regions of the world, the perennial vegetation was almost completely replaced by annual grass communities (Heady 1958). Only remnants of the original vegetation remain and there seems to be little potential for the rather complex annual communities to return to their original condition.

Cheatgrass brome (*Bromus tectorum*) and, in some areas, medusahead (*Taeniatherum asperum*) now dominate more than 2.5 million ha of former sagebrush-grass rangelands in southern Idaho and surrounding areas (Murray et al. 1978, Miller et al. 1993). Cultivation and land abandonment caused some of this condition. However, heavy grazing, repeated fires, and the presence of highly adapted annuals account for most of this vegetation type change (Hull and Pechanec 1947). Large areas have few perennial grasses and no sagebrush close enough to provide seed for re-establishment. In addition, fire-return frequency is now less than five years due to cheatgrass abundance and flammability in contrast to pre-settlement fire frequency of 60-110 years (Whisenant 1990). Melgoza et al. (1990) reported that cheatgrass competes with native species for soil water, negatively affecting their water status for at least 12 years after a fire. Thus, frequent fires do not allow perennial grasses, sagebrush, or other shrubs to establish and set seed, even if a seed source is close by. Both these annual communities consist of recognizable, stable vegetation states where current range condition criteria do not apply.

Riparian and More Mesic Mountain Communities

Most riparian areas and mesic communities of the mountainous West react somewhat predictably when grazed heavily, followed by predictable recovery when grazing pressure is reduced or stopped. In these communities, availability of water is the key to rapid recovery.

Riparian Areas

Another paper in this volume (Elmore 1993) presents detailed information about reaction of riparian areas to grazing. Therefore, only a small amount of information will be presented here. Kauffman and Krueger (1984) presented a thorough review of the impacts of livestock grazing on riparian ecosystems. The vegetation of many riparian areas has been depleted by heavy and prolonged grazing, especially by cattle, resulting in decreased vigor, biomass and change in species composition. Since additional water usually is available from run-on or capillary sources, the herbaceous layer of depleted riparian areas often responds rather quickly to a reduction or removal of grazing.

Recovery time for woody vegetation depends upon the extent of the degradation. Willows recovered after 10 years of protection from grazing in south central Washington (Rickard and Cushing 1982). If woody vegetation has been completely eliminated by heavy grazing, natural regeneration may take a long time. The different herbaceous community may represent a stable state requiring intervention in the form of plantings to restore woody species. Recent studies of the effects of grazing on wildlife diversity in riparian areas will be reviewed later.

Aspen

Throughout the Rocky Mountains, aspen (*Populus tremuloides*) commonly occupies more mesic sites, including draws, north slopes and some riparian areas. On many sites, aspen forms a rather long-lived successional community, eventually being replaced by conifers. The leaves and shoots of aspen are quite palatable. Heavy prolonged browsing by cattle, sheep, elk, deer or moose suppresses vegetative reproduction of aspen, eventually causing stands to degenerate and disappear.

Browsing by a relatively large elk population severely limited aspen reproduction in Grand Teton National Park in Wyoming (Beetle 1974). In southern Utah, deer browsing effectively prevented aspen reproduction on a small clear cut area as evidenced by abundant reproduction in an adjacent exclosure (Mueggler and Bartos 1977, Young 1958). Larger burned areas had ample aspen reproduction, indicating that the small clearcut areas may concentrate browsing which suppresses reproduction. In contrast, Jones (1975) found that widespread heavy elk browsing in northern Arizona did not prevent establishment of a new aspen stand.

Smith et al. (1972) reported that proper livestock management is essential for regeneration of aspen following clearcutting or burning. Medin and Clary (1991) found that lightly grazed and ungrazed aspen areas had similar vegetation. However, heavy cattle grazing had a major impact on aspen regeneration and stand structure.

Vegetation composition changes in aspen caused by grazing are somewhat predictable. Heavy, prolonged grazing and browsing causes a decline in sprouting success and depletion of the understory. Aspen and understory vegetation recover when grazing is reduced due to water availability during much of the growing season in high precipitation zones or in moist sites. However, if heavy browsing of aspen reproduction continues long enough, mature aspen trees die and the stand becomes an aspen-shrub community. With continued heavy grazing, the aspen will eventually stop growing. Aspen would not return to this converted type even if grazing and browsing were eliminated. If this occurs before a substantial number of conifers become established, a new herbaceous steady state community would be created, i.e., one without aspen.

Mountain Shrub Communities

In a Douglas fir forest in central Washington, a one-acre deer and elk-proof exclosure was established in 1939 in a thick stand of snowbrush ceanothus (*Ceanothus velutinus*). Thirty years later, the exclosure had twice as much vegetative cover and crown cover of shrubs (mainly ceanothus) was 5 times greater than the outside area. Heavy browsing by deer and elk converted the outside area to a stand dominated by grasses, sedges and forbs with only scattered, decadent ceanothus plants (Tiedemann and Berndt 1972). This community shows a predictable response to overbrowsing by native ungulates. If continued heavy browsing outside the exclosure eventually eliminated ceanothus, the resulting community could be quite stable and subsequent protection from browsing might not encourage the return of ceanothus without fire or logging with ground disturbance.

Subalpine Spruce Forests

Baker (1991) found that burning a forest dominated by Engelmann spruce (*Picea engelmannii*) in Colorado resulted in bristlecone pine (*Pinus aristata*) becoming established on an area grazed by livestock. Engelmann spruce became re-established in an adjacent ungrazed area. Even though the grazed area is not now grazed as heavily as it was after the fire, the bristlecone pine is firmly established and would not be affected by removal of grazing, at least during the rather long life span of the existing trees.

High Elevation Mountain Parks

Some high elevation mountain parks in the Rocky Mountains have been invaded by relatively pure stands of Kentucky bluegrass. Meadows were formerly dominated by mountain muhly (*Muhlenbergia montana*), fescue (*Festuca* spp.) and timber danthonia (*Danthonia intermedia*). Forage production was compared on ungrazed, moderately grazed and heavily grazed bluegrass stands at 2500 m in Colorado (Schwan et al. 1949). Bluegrass production diminished as grazing intensity increased but presence or absence of grazing had little effect on species composition. This indicates that bluegrass stands are in a stable state condition, requiring intervention, such as fire or seeding, to initiate change. Application of heavy and light mulches on the heavily grazed range increased total forage production 42% and 50%, respectively, without changing species composition.

Effects Of Grazing On Biodiversity

Previous volume papers have shown that heavy livestock grazing of western rangelands occurred in the last half of the 19th century and continued into the first part of this century. A lack of knowledge about the effects of livestock grazing led to heavy use of these arid and semi-arid systems. Another reason for heavy grazing was due to the "tragedy of the commons" (Hardin 1968). When public rangelands were open to anyone and grazing was uncontrolled, everyone tried to get his/her share of the grazing before someone else did. In addition to large local livestock herds, roving bands of sheep ranged over large areas, contributing to heavy forage utilization. This overgrazing caused some serious declines in range condition, especially in the shrub-dominated ecosystems of the intermountain and southwestern U.S.

With the formation of Forest Reserves (now National Forests) around the turn of the century and passage of the Taylor Grazing Act in 1934, grazing on public lands was restricted to ranchers with local base property, with controls on animal numbers and seasons of use. Roving bands of livestock were also eliminated. As a result, range conditions began to improve. According to Box and Malachek (1987), most western rangelands are now in the best condition of this century. However, many vegetation types are in a "steady state" situation. Further improvement probably will be limited and slow, even if livestock grazing were removed altogether (Beck and Tober 1985, Laycock 1991).

Has Biodiversity Been Reduced by Grazing?

Ehrlich (1990), Wuerthner (1990) and the popular environmental press claim that continued overgrazing on public lands has resulted in serious reduction of species diversity. The pertinent questions are: 1) has overgrazing that occurred 80-100 years ago been curtailed? and 2) has past heavy grazing and the current level of more moderate grazing affected species diversity? The answer to the first question, as discussed previously, is an obvious "yes." The type of abusive grazing which occurred 100 years ago on some of the land has not existed on the majority of public lands for at least 50 years. With regard to the second question, the range management literature is filled with evidence that prolonged heavy grazing can change the productivity and/or composition of most rangeland ecosystems. However, moderate grazing the last half century probably has not decreased species diversity of either plants or animals and may have increased diversity.

Clements (1905) seems to have been the first to point out that diversity, in terms of number of species, is not linearly correlated with stage of succession. "The number of species is small in the initial stages, it attains a maximum in the intermediate stages; and again decreases in the ultimate formation, on account of the dominance of a few species." This relationship between diversity and stage of succession has been ignored by most of the critics of grazing on public rangelands and others.

Milchunas et al. (1988) found that patterns of plant species diversity in grasslands caused by different grazing intensities were a function of both the evolutionary history of grazing of the vegetation and the moisture available. In the tall grass prairie (long history of grazing and subhumid), diversity is low at low grazing intensities because a few species dominate the canopy. Diversity is high at moderate grazing intensities because of large differences in growth forms. With increasingly heavy grazing, diversity declines as short grasses dominate a greater proportion of the community. Johnson (1961) reported similar relationships concerning diversity on grazed and ungrazed rough fescue grasslands in Alberta. On a tall grass prairie site in Oklahoma, Collins (1987) found that plant species diversity was lowest on ungrazed, burned treatments, highest on grazed plus burned treatments and intermediate on both the grazed and ungrazed, unburned treatments.

Discussions on the effects of grazing on biodiversity often ignored long-accepted principles in ecology and wildlife habitat biology, including: 1) plants are distributed in patches (Watt 1947), 2) grazing usually increases patchiness of plants and 3) diversity of habitats (patchiness) and ecotones or edges where different vegetation types meet are important as wildlife habitat (Leopold 1933) and create diversity on a landscape level and 4) grazing generally reduces a relatively few dominants in a system and they are replaced by more numerous secondary species.

Thomas et al. (1979) stated, "Wildlife species richness should be approaching the maximum where the average habitat size is approximately 81 hectares. Pay special attention to the emphasis on 'average.' This indicates the existence of habitats both larger and smaller than 81 hectares. The larger habitats will accommodate those relatively few species that require blocks larger than the average while smaller habitats will increase the edge effect." Heavy livestock grazing, unless extremely widespread, increases the variety of habitats on a landscape level, improving both habitat diversity *and* edge. Thus, even if plant and animal species diversity is reduced in smaller, heavily grazed areas, this grazing could *increase* habitat diversity on a larger total landscape level, theoretically increasing overall bird and animal diversity.

At the local level, grazing affects spatial distribution and plant species composition, creating patchiness that affects avian habitat selection (Wiens 1976), and bird and animal density and diversity (Severson and Urness 1993). Ryder (1980) found that the effects of grazing on avian habitats varied from area to area. In areas of higher precipitation, grazing may be desirable to open up areas and provide more patchiness and diversity. Effects of soils, slope and exposure along with amount and distribution of precipitation may be more important than the effects of grazing on food, cover and water for birds (Ryder 1980).

On restored rangeland areas, such as reclaimed surface mined lands, many studies have indicated that plant species diversity of seeded native or introduced species, is higher on grazed areas compared to ungrazed areas (DePuit and Coenenberg 1978, Kleinman and Layton 1981, Laycock 1989b, Williamson 1981, Young and Rennick 1983). Graul (1980) recommended against planting exotic species such as crested wheatgrass (*Agropyron cristatum* and *Agropyron desertorum*) due to the negative effect on both species diversity and relative density of vertebrates. This conclusion resulted from studies of large (several thousand hectares) blocks of crested wheatgrass compared to native vegetation. Currently, the common practice is to plant much smaller (up to several hundred hectares) areas with exotic species. These smaller areas provide increased forage production but, on a landscape scale, increase habitat diversity and edge, thereby stimulating an *increased* diversity of birds and animals (Laycock 1980).

Moderate cattle grazing on a polder in the Netherlands, seeded to introduced grasses and used as a meadow bird reserve, resulted in increased numbers of breeding birds compared to an ungrazed area (de Jong 1977). In Western Australia, Majer (1985) found that pastured ex-sandmines produced an invertebrate fauna more than twice as high in numbers of taxa as grazed farmland.

Even prolonged heavy grazing may not be detrimental to every species of plant or animal. The mountain plover nests only in relatively heavily grazed shortgrass steppe areas. Originally, they nested where the stature of grass had been reduced by buffalo grazing. Now, they successfully nest only in areas heavily grazed by cattle (Graul 1973, 1975; Ryder 1980). Complete removal or drastic reduction of cattle grazing could lead to further reductions of mountain plover populations because of reductions in nesting habitat. Each altered state serves as habitat for some organisms. Whether change to that altered state is good or bad depends upon our viewpoint and whether it meets our goals and objectives.

Effects of Grazing on Diversity in Specific Habitats

This section is not intended to be a complete review of the effects of grazing on diversity. Examples will be given of studies where grazing has had little effect on diversity and others where grazing has affected diversity.

A number of recent studies have focused on the diversity of plant, bird and animal species in adjacent grazed and ungrazed riparian areas. In riparian areas of eastern Idaho, little difference in total breeding bird density was found between an area grazed by cattle and a 114 ha area ungrazed for 14 years (Medin and Clary 1990). However, total bird biomass, bird species richness and bird species diversity were 1.6 to 1.8 times higher in the grazed habitat. Small mammal populations were almost a third higher on grazed areas compared to ungrazed area. However, both species richness and species diversity of small mammals were higher in ungrazed habitat.

In riparian areas of northern Nevada, vegetation biomass and structure (Clary and Medin 1990) and breeding bird biomass and diversity (Medin and Clary 1991) were evaluated on a riparian area moderately grazed by cattle compared to a similar area (40 ha) protected from grazing for 11 years. The grazed area had less graminoid biomass and height but more nonwillow large shrub biomass than the exclosure. There were no differences in bird density, species richness, species composition, total standing crop biomass or other attributes of breeding bird communities between the grazed and protected areas.

Krueger (1985) found that cattle grazing in thick willow stands created "tunnels" by browsing, trailing and rubbing. These tunnels increased structural diversity and amount of grassland habitat within the stand which increased diversity of insects. Tunnels also provided avian flight corridors affording cover and protection from aerial predators. For these reasons, the moderately grazed areas supported a greater diversity of birds than did ungrazed areas.

In high elevation willow communities in Colorado, population densities of habitat generalist birds differed little between vigorous (winter grazed) and decadent (summer grazed) stands (Knopf et al. 1988). However, many habitat specialist bird species were scarce or absent in the decadent stands.

In northeastern Colorado, prolonged grazing of a riparian area changed the habitat structure by removing most of the willow plants (Schulz and Leininger 1990). Bird species richness was higher in grazed than in ungrazed exclosures. However, bird species diversity (evenness) was slightly higher in grazed than in the exclosures (Schulz and Leininger 1991).

In Oregon, Kauffman and Krueger (1984) demonstrated that a seasonal short-duration grazing system had no impact on breeding densities of selected migratory birds. In another study, Kauffman et al. (1982) found that total avian densities did not differ between grazed and ungrazed areas. Wiens (1973) concluded that season of grazing influences avian communities on grasslands more than intensity of grazing.

On a cottonwood bottomland in northeastern Colorado, three years of moderate fall cattle grazing had little effect on densities of 6 species of birds using ground shrub resources. In the same area, small mammal habitat use and seasonal habitat shifts on cottonwood bottomlands were similar on pastures grazed by cattle in the winter and on ungrazed areas (Sedgewick and Knopf 1987).

On a desert grassland site in southeastern Arizona where cattle grazing had been excluded for 14 years, there were significant differences in vegetative composition and bird and mammal populations compared to areas of continued grazing (Bock et al. 1984). Short grass species were equally common inside and outside the exclosure with several taller grass species more abundant on the protected site. However, the grazed area supported significantly higher numbers of birds in the summer, but densities did not differ in winter. Rodents were significantly more abundant in the exclosure. Bock et al. (1984) concluded that grazing appears to favor birds over rodents. Some small mammals may prefer ungrazed areas due to protection against avian predators.

Severe grazing had a negative effect on passerine bird abundance in riparian areas in southeastern Oregon (Taylor 1986). Since 1940, bird counts were 11-13 times higher on ungrazed areas than on areas severely disturbed by dredging and heavy grazing. Differences were highly correlated with the volume and height of willows, which were greater with increased intervals since grazing.

In general it appears that prolonged, heavy grazing can cause substantial changes in wildlife habitats and subsequent impacts on bird populations and diversity, mainly by impacts on the structural aspects of woody vegetation (Knopf and Cannon 1982). Moderate or proper grazing usually has little or no negative effects on bird populations and often has positive effects on vegetation and bird biomass, density and diversity. Numbers, biomass or diversity of rodents may be reduced in some grazed areas, probably as a result of heavier avian predation in more open conditions.

Almost every study that compared diversity between grazed and ungrazed areas failed to calculate the diversity (richness or evenness) on the combined grazed/ungrazed area, even when the study sites were adjacent. Only when this is done can some analysis of these small-scale studies, such as those summarized above, begin to indicate what impact grazing might have on diversity on a larger, landscape scale. A mosaic or combination of ungrazed, lightly grazed, moderately grazed and heavily grazed areas probably would maximize diversity in most landscapes. In most cases, moderate grazing with good management usually results in this mix of different intensities of grazing. Kellner and Bosch (1992) reported selective grazing in the semi-arid grasslands of southern Africa leads to the development of a mosaic of patches representing palatable and unpalatable grasses.

Effect of Scale on Biodiversity

One factor that usually has been ignored in discussions of the effects of grazing on diversity is that of scale. Claims that diversity has been affected by any factor, including grazing, are meaningless unless the scale is identified. Even if heavy grazing were to reduce species diversity on smaller, heavily grazed areas, this grazing should increase patchiness and increase the variety of habitats, thus *increasing* both habitat diversity and edge on a total landscape level. This, in turn should increase overall bird and animal diversity on that landscape unless grazing has been so severe and prolonged to simplify the landscape. Generally this has not occurred on most North American rangelands and the current grazing pressure is moderate.

New Conceptual Stable State Models

The current range condition model used in the United States was discussed earlier in this paper and by Pieper (1993). I have indicated which vegetation types this model seems to fit and those where it does not apply. Briske (1993) discussed some of the physiological reasons why the model does not work in some vegetation types. For the rangeland vegetation types with multiple stable states, and where the current model does not seem to apply, we need now conceptual models to interpret range condition.

The concepts of relatively stable states of vegetation is not new in the ecological literature (Holling 1973, May 1977, Wissel 1984). Malin (1955) seems to have been first to raise the issue related to rangelands. He suggested the terms "steady state" because it is not laden with the same unwarranted overtures, implications and inferences that accompany the term "climax." However, only recently has the concept of state states received much attention in the range management literature. Friedel (1988, 1991) used the concept of thresholds of environmental change to help describe and explain anomalies in condition assessment of central Australia's arid rangelands. Laycock (1991) described examples of stable states on North American rangelands, presenting several possible conceptual models to clarify these conditions. These will be summarized below. One of these is Archer's (1989) conceptual diagram (Fig. 1) describing the changes from a grassland to savanna to a mesquite woodland.

Another type of model from the ecological literature is the "domain of attraction" (Lewontin 1969) which depicts a community as a ball or marble in a cup or trough. The boundaries of each cup represent the range of environmental conditions under which that community is stable (Hurd and Wolf 1974, Forman and Godron 1986, Krebs 1985).

Krebs (1985) presented an analogy of different types of communities with different numbers of stable states. Figure 2A represents a community that is *globally* stable because, after all disturbances, it will return to its original configuration. This may be equivalent to the equilibrial model described by Ellis and Swift (1988). This model seems to represent the "climax" range condition concept because the disturbed rangeland community will *always* return to the "climax" condition (the bottom of the cup) via internal feedback mechanisms.

Figure 2B represents a community which has multiple stable states, only three of which are shown. It is locally stable at configuration I. If this represents "excellent" range condition, minor grazing perturbations may change the composition of the community somewhat (i.e., move the ball up the side of the cup) but, once this disturbance is stopped, the community returns to its original state (climax or excellent condition). While the community is within the bounds of configuration I, current concepts about range condition and reversibility of change may fit quite well. However, if the community is perturbed beyond the bounds of configuration I, it will cross a threshold (A) between configurations I and II and move to a new locally stable configuration at II. For a rangeland in this new stable state, conventional range condition standards *no longer apply* either to describe the state or identify the forces required to move the community out of this state. If stressed beyond the limits of this new stable state, the community can then be forced across threshold B to configuration III *or* to some other configuration not shown.

The most useful multiple stable state model may be the state-and-transition concept of Westoby et al. (1989). A state is a recognizable, relatively stable assemblage of species occupying a site. Forces that cause vegetation to cross a threshold and move toward another state are shown as transitions. The sagebrush-grass system and its various stable states, described previously, including its transformation into a stable cheatgrass community, were depicted in a state-and-transition model by Laycock (1991). Similar descriptions of changes in sagebrush vegetation were presented by West (1988).

In Laycock's (1991) state-and-transitional model (Figure 3), states I, II, III occur in areas without annuals. State I is the "climax" or condition undisturbed

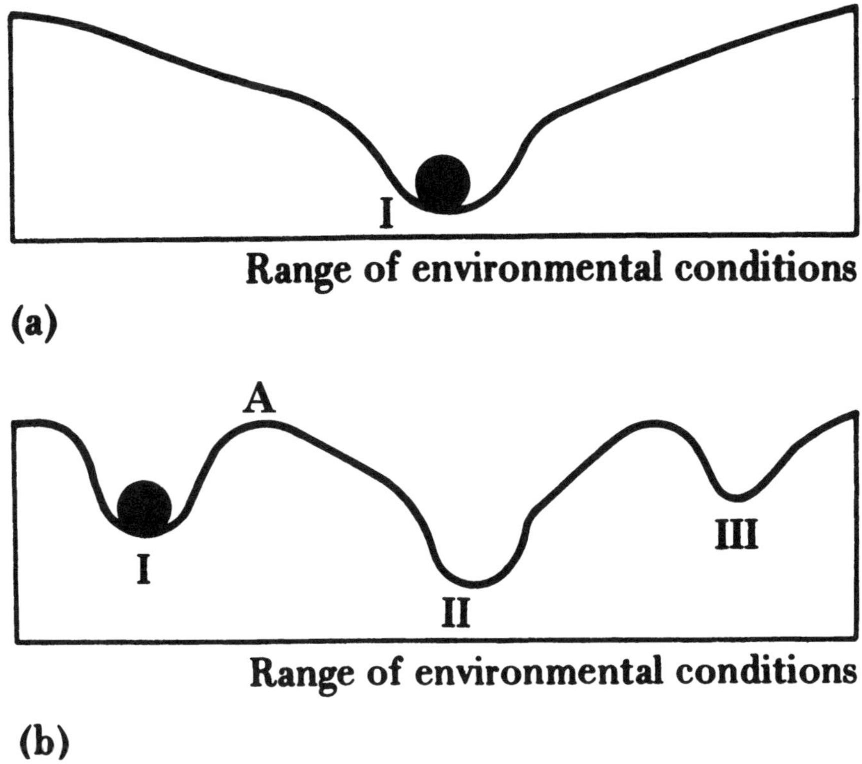

Figure 2 Diagrams illustrating the ball and cup or trough analogy to illustrate global and local stability concepts. The community is represented as a black ball on a topographic surface (cup or trough) which represents the range of environmental conditions under which the community is stable. In (a) the community is both locally and globally stable because, after *all* disturbances or perturbations, it will return to configuration I. In (b) the community is locally stable, but if perturbed beyond a certain critical range, it will cross threshold A and move to a new locally stable configuration II. From Ecology: The Experimental Analysis of Distribution and Abundance. Third Edition by Charles J. Krebs. Copyright © 1985 by Harper and Row Publishers. Inc. Reproduced by permission of Harper Collins Publishers, Inc.

by grazing. Transition arrow T-1 represents heavy grazing which causes deterioration of the understory and increased density and vigor of sagebrush. State II is dominated by sagebrush and, as previously indicated, will remain stable for long periods of time. Transition 3 is fire or some other force (insects, disease, or an herbivore that eats sagebrush) that reduces the sagebrush, which releases the understory (State III). With proper grazing management (T-5) it *may* move back to a state resembling State I. With heavy grazing (T-4) sagebrush will again dominate the stand (State II). From a biodiversity standpoint, a mosaic of States I, II, and III in a landscape would have higher plant and animal diversity than any of the states individually.

State IV represents the situation in a heavily grazed area where a well-adapted annual like cheatgrass occurs. Continued heavy grazing (T-6) causes replacement of perennials in the understory with annuals. The transitions to State V (T-8) and then to State VI (T-10) represents the role of fires in the conversion to a stable

STATE-AND-TRANSITION MODEL FOR A SAGEBRUSH GRASS ECOSYSTEM (After Westoby, Walker and Noy-Meir, 1989)

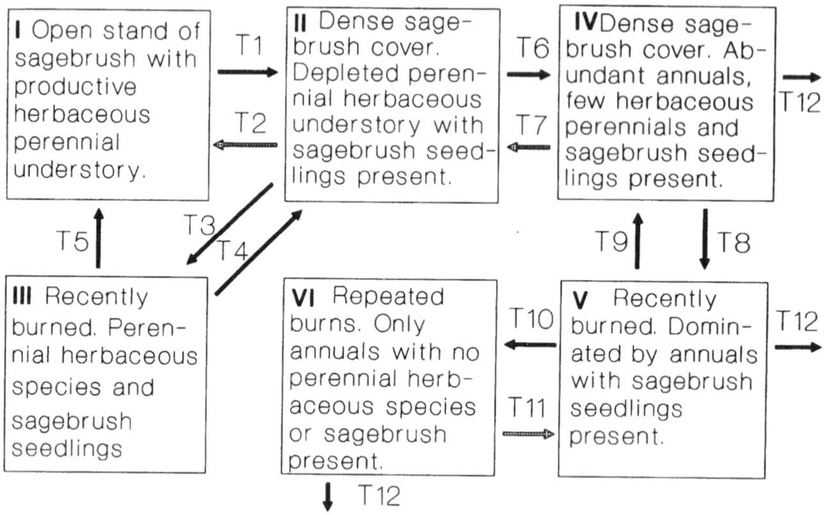

Figure 3 State-and-transition diagram for sagebrush-grass vegetation. From Laycock (1991).

cheatgrass community. T-12 represents intervention by man such as seeding an adapted perennial like crested wheatgrass. The Bureau of Land Management is planting strips of vegetative fuel breaks consisting of crested wheatgrass, other grasses, forbs and shrubs to slow the spread of fires (Pellant 1990).

In terms of biodiversity, State I undoubtedly has higher diversity of plants and animals than State VI. However, a mosaic of several of these states shown probably would have higher diversity than any of the states taken individually. The "greenstripping" i.e., planting species other than cheatgrass in a grid pattern to slow fires (Pellant 1990) will increase diversity in the cheatgrass areas.

In South Africa, Bosch and Kellner (1991) combined the elements of both the "domain of attraction" (ball and cup) concept and the state-and-transition model in the development of what they termed a "quantitative, descriptive degradation model." This model described various states with reversible transitions to other states within a given "domain of attraction" but with irreversible transitions which move the community into a "new" domain with no possibility for return.

Management Implications

Heavy grazing or browsing by ungulates usually is defined as that amount or pattern of grazing that results in a reduction of the most palatable plant species, a change in species composition, and serious erosion and site damage. The current range condition model and range condition standards or scorecards, describe the stages of deterioration fairly well for most rangeland communities. However, these changes may not be linear, as predicted by the model.

The main problem with the current range condition model is that it predicts that reduction or removal of grazing pressure will allow a deteriorated community to improve through secondary succession and move through the same stages observed during retrogression toward "climax" (Westoby et al. 1989). Only the more mesic rangeland vegetation types follow this pattern, including tall grass and mixed grass prairies, mesic mountain communities such as aspen, and riparian communities. Almost all of the arid and semi-arid rangeland types remain stable at one or more lower successional states for long periods of time, even when grazing is removed. Even when improvement does take place, the recognizable stages are not necessarily predictable. These stable states are well documented in the literature but have not been taken into consideration in our range condition and management concepts.

The existence of these steady state situations probably is pertinent only in a time frame meaningful to management (up to several decades), but may last somewhat longer in some situations. In a much longer time frame, climatic cycles or other changes may make the concept meaningless. However, the need for changes in range management concepts has prompted the search for new ideas, concepts and models.

It is important for managers to recognize that stable states exist on a great many rangeland vegetation types. For example, many allotment management plans on sagebrush-grass rangelands are designed to change range condition from fair (mid-seral) to good (upper seral) in 2 or 3 cycles of a grazing system. A much larger percentage of sagebrush than allowed by the site guides usually is the cause for the lower condition rating. The only way to significantly improve range condition score is to reduce the amount of sagebrush. On ranges grazed by cattle, no known grazing system will accomplish that goal although heavy winter browsing by deer may achieve this result (Urness 1990). Yet, our present model and range condition standards indicate that the reduction of grazing and imposition of a grazing system *will* result in such improvement.

Some managers are beginning to realize this, Winward (1991) stated, "There are more acres of sagebrush-grass lands in the western United States being held in a low ecological status the past decade due to abnormally high sagebrush cover and density than currently is occurring due to livestock grazing." He stated that an intensive fire program could help recovery of more natural sagebrush densities and that, in areas where fire is inappropriate, other approaches to thinning sagebrush should be considered. These actions, in conjunction with good grazing management, should result in the desired improvement in range condition over time.

Similar examples could be cited for other range vegetation types. If the manager does not recognize a stable state, he/she will assume that changes in grazing management (stocking rate, system, etc.) *will* improve range condition. If the current stable state is highly resistant to changes in grazing management and requires some other action, managers, livestock owners and others have been misled by our current range condition concepts and will be disappointed by the lack of response to different grazing management systems or stocking rates.

Conclusions

1) The current call for removal of livestock grazing from public lands apparently is based on the assumption that this would result in rapid increases in both range condition and species diversity. The review of the literature indicate that:
   a) Many vegetation types on public land are in a stable state condition and would change little, if at all, if livestock were removed.
   b) Very heavy grazing on small areas can decrease biodiversity in that limited area but moderate grazing often is beneficial to diversity.
   c) Grazing by herbivores increases patchiness of vegetation which should increase diversity of both plants and animals on a *landscape* level. Studies of effects of grazing on diversity, on this landscape level, are badly needed but are difficult to find in the literature. It is my opinion that, if grazing were completely removed, diversity of both plants and animals on the landscape level probably would decrease on most areas of public land.
2) Many new ecological ideas and concepts have been proposed and discussed in the last 40 years. We need to incorporate the appropriate concepts and go beyond the "conventional wisdom" of our current concepts and philosophies about range condition. Much theoretical work needs to be done before the newer models and ideas can be incorporated into range condition standards. Dialogue among scientists and managers must be initiated about the new ideas, concepts and models so new range condition standards, where needed, can be developed as soon as possible.

Literature Cited

Anderson, J.E. and K.E. Holte. 1981. Vegetation development over 25 years without grazing on sagebrush dominated rangeland in southeastern Idaho. J. Range Manage. 34:25-29.

Archer, S. and F.E. Smeins. 1992. Non-linear dynamics in grazed ecosystems: thresholds, multiple steady states and positive feedbacks. p. 4-6. *In*: T. Svejcar and J. Brown (eds). Symposium Proc.: Is the range condition concept compatible with ecosystem dynamics? 1992 Annual Meeting, Soc. for Range Manage.

Archer, S. 1989. Have southern Texas savannas been converted to woodlands in recent history? Amer. Nat. 134:545-561.

Baker, 1991. Livestock grazing alters succession after fire in a Colorado subalpine forest. p. 84-90 *In*: S. Nodvin and T. Waldrop (eds). Proceeding of the fire and the environment symposium. USDA Forest Service General Tech. Rept. SE-69. Southeastern Forest Experiment Station, Asheville, NC.

Beck, R.F. and D.A. Tober. 1985. Vegetational changes on creosotebush sites after removal of shrubs, cattle and rabbits. New Mexico State Univ. Agric. Exp. Sta. Bull. 717.

Beetle, A.A. 1974. The zootic disclimax concept. J. Range Manage. 27:30-32.

Bement, R.E. 1968. Plains pricklypear: relation to grazing intensity and blue grama yield on Central Great Plains. J. Range Manage. 21:83-86.

Biswell, H.H. 1956. Ecology of California grasslands. J. Range Manage. 9:19-24.

Bock, C.E., J.H. Bock, W.R. Kenney and V.M Hawthorne. 1984. Responses of birds, rodents, and vegetation to livestock exclosure in a semidesert grassland site. J. Range Manage. 37:239-242.

Box, T. W. and J.C. Malechek. 1987. Grazing on American rangelands. Proc. Western Sect. Amer. Soc. Animal Sci. 38:107-115.

Briske, D.D. 1993. Physiological responses of individual plants: current status and ecological significance. (This symposium).

Bosch, O.J.H.. and K. Kellner. 1991. The use of a degradation gradient for the ecological interpretation of condition assessment in the western grassland biome of southern Africa. J. Arid Environ. 21:21-29.

Brown, J.H. and E.J. Heske. 1990. Control of a desert-grassland transition by a keystone rodent guild. Science 250:1705-1707.

Brown, J.R. 1992. Introduction. p. 3. *In*: T. Svejcar and J. Brown (eds). Symposium Proc.: Is the range concept compatible with ecosystem dynamics? 1992 Annual Meeting, Soc. for Range Manage.

Clary, W.P. and D.E. Medin. 1990. Differences in vegetation biomass and structure due to cattle grazing in a northern Nevada riparian ecosystem. U.S. Forest Service Res. Pap. INT-127. Intermountain Res. Sta., Ogden, UT.

Clements, F.E. 1905. Research methods in ecology. University of Nebraska Publishing Co., Lincoln, NE.

Clements, F.E. 1916. Plant succession. Carnegie Inst. Wash. Pub. 242.

Clements, F.E. and V.E. Shelford. 1939. Bio-ecology. John Wiley and Sons, New York.

Coffin, D.P., W.K. Lauenroth and I.C. Burke. 1991. Recovery of shortgrass steppe plant communities in northeastern Colorado. No. 82. Abstracts, 44th Ann. Mtg., Soc. Range Manage., Washington D.C.

Cole, G.F. 1971. An ecological rationale for the natural or artificial regulation of native ungulates in parks. p. 417-412. *In*: Trans. 36th North Amer. Wildl. and Natur. Resources Conf. p. 417-425.

Collins, S.L. 1987. Interaction of disturbances in tallgrass prairie: a field experiment. Ecology 68:1243-1250.

Cook, C.W. 1971. Effects of season and intensity of use on desert vegetation. Utah. Agric. Exp. Sta. Bull. 483.

Cosby, H.E. Undated (circa 1975). Range ecosystem management for natural areas. U.S. Fish and Wildlife Service, Denver, C0.

Costello, D.F. 1944. Natural revegetation of abandoned plowed land in the mixed prairie association of northeastern Colorado. Ecology 25:312-326.

Costello, D.F. and G.T. Turner. 1941. Vegetation changes following exclusion of livestock from grazed ranges. J. For. 39:310-315.

Costello, D.F. and G.T. Turner. 1944. Judging condition and utilization of short-grass ranges on the central Great Plains. U.S. Dept. Agric. Farmers Bull. 1949.

Curtis, J.T. and M.L. Partch. 1948. Effect of fire on competition between bluegrass and certain prairie plants. Amer. Midl. Nat. 39:437-443.

Daubenmire, R.F. 1940. Exclosure technique in ecology. Ecology 21:514-515.

de Jong, H. 1977. Experiences with man-made meadow bird reserve "Kievitslander" in Flevoland (The Netherlands). Biol. Conserv. 12:13-31.

DePuit, E.J. and J.G. Coenenberg. 1978. Plant response and forage quality for livestock grazing on coal mine spoils in Montana. p. 892-898. *In*: M.K. Wali (ed). Ecology and Coal Resource Development, Pergamon Press, New York.

Dyksterhuis, E. J. 1949. Condition and management of rangeland based on quantitative ecology. J. Range Manage. 2:104-115.

Ehrlich, P. R. 1990. Habitats in crisis: why should we care about the loss of species? For. Ecol. and Manage. 35:5-11.

Ellis, J.E., and D.M. Swift. 1988. Stability of African pastoral ecosystems: Alternate paradigms and implications for development. J. Range Manage. 41:450-459.

Elmore, W. 1993. Riparian and watershed system: yesterday or today. (this symposium).

England and DeVos. 1969. Influence of animals on pristine conditions on the Canadian grasslands. J. Range Manage. 22:87-94.

Evanko, A.B. and R.A. Peterson. 1955. Comparisons of protected and grazed mountain rangelands in southwestern Montana. Ecology 36:71-82.

Federal Committee on Ecological Reserves. 1977. A directory of research natural areas on federal lands of the United States of America. U.S. Department of Agriculture, Forest Service.

Ferguson, R.B. and J.V. Basile. 1966. Topping stimulates bitterbrush growth. J. Wildl. Manage. 30:839-841.

Fisser, H.G., K.L. Johnson, K.S. Moore, and G.E. Plumb. 1989. 51-year change in the shortgrass prairie of eastern Wyoming. p. 29-31. *In*: Proc. 11th N. Amer. Prairie Conf.

Foran, B.D., N.M. Tainton, and P. de V. Booysen. 1978. The development of a method for assessing veld condition in three grassland types in Natal. Proc. Grassl. Soc. So. Africa 13:27-33.

Forman, R.T.T., and M. Godron. 1986. Landscape ecology. John Wiley & Sons, New York.

Friedel, M.H. 1988. Range condition and the concept of thresholds. p. 1-3. *In*: Vol I, Abstracts, 3rd Intern. Rangeland Congr. Range Manage. Soc. India.

Friedel, M.H. 1991. Range condition assessment and the concept of thresholds. A viewpoint. J. Range Manage. 44:422-426.

Gardner, J.L. 1950. Effects of thirty years of protection from grazing in desert grassland. Ecology 31:44-50.

Gillis, A.M. 1991. Should cows chew cheatgrass on commonlands? Bioscience 41:668-675.

Glendening, G.E. 1952. Some quantitative data on the increase of mesquite and cactus on a desert range in southern Arizona. Ecology 33:319-328.

Graul, W.D. 1973. Adaptive aspects of the mountain plover social system. Living Bird 12:69-94.

Graul, W.D. 1975. Breeding biology of the mountain plover. Wilson Bull. 87:6-31.

Graul, W.D. 1980. Grassland management practices and bird communities. p. 38-47. *In*: R.M. DeGraff and N.G. Tilghman (eds). Management of western forests and grasslands for nongame birds. U.S. Forest Service Gen. Tech. Rept. INT-86. Intermountain Forest and Range Exp. Sta., Ogden, UT.

Hardin, G. 1968. The tragedy of the commons. Science 162:1243-1248.

Hart, R.H. and B.E. Norton. 1988. Grazing management and vegetation response. p. 494-525. *In*: P.T. Tueller (ed). Vegetation science applications for rangeland analysis and management. Kluwer Academic Publisher, Boston, MA.

Heady, H.F. 1957. Effect of cages on yield and composition in the California annual type. J. Range Manage. 10:175-177.

Heady, H.F. 1958. Vegetation changes in the California annual type. Ecology 39:402-415

Heady, H.F. 1973. Structure and function of climax. pp. 73-80. *In:* D.N. Hyder (ed). Arid shrublands—proceedings of the third workshop, U.S. Australia Rangelands Panel. Soc. Range Management, Denver, CO.

Herbel, C.H., P.L. Dittberner, and T.S. Bickle. 1970. A quantitative ecology of the Jornada Experimental Range. p. I-133-178. *In*: R.G. Wright and G.M. Van Dyne (eds). Colo. St. Univ, Range Sci. Dept. Rep. 6, Ft. Collins, CO.

Holechek, J.L. and T. Stephenson. 1983. Comparison of big sagebrush vegetation in north-central New Mexico under moderately grazed and grazing excluded conditions. J. Range Manage. 36:455-456.

Holling, C.S. 1973. Resilience and stability of ecological systems. Ann. Rev. Ecol. Systematics 4:1-23.

Houston, W.R. 1963. Plains pricklypear, weather, and grazing in the Northern Great Plains. Ecology. 44:569-574.

Hull, A.C. Jr., and J.F. Pechanec. 1947. Cheatgrass—a challenge to range research. J. For. 45:555-564.

Hurd, L.E., and L.L. Wolf. 1974. Stability in relation to nutrient enrichment in arthropod consumers of old-field successional ecosystems. Ecol. Monogr. 44:465-482.

Hyder, D.N., R.E. Bement, E.E. Remmenga, and D.F. Hervey. 1975. Ecological responses of native plants and guidelines for management of shortgrass range. U.S. Dept. Agric. Tech. Bull. 1503.

Jacobs, L. 1991. Waste of the West: Public lands ranching. Arizona Lithographers, Tucson, AZ.

Johnston, A. 1961. Comparison of lightly grazed and ungrazed range in the fescue grassland of southwestern Alberta. Can. J. Plant Sci. 41:615-622.

Jones, R.J. 1975. Regeneration on an aspen clearcut in Arizona. U.S. Forest Service Res. Note RM-285. Rocky Mountain Forest and Range Exp. Sta., Ft. Collins, CO.

Kauffman, J. B., and W.C. Krueger. 1984. Livestock impacts on riparian ecosystems and streamside management implications—a review. J. Range Manage. 37:430-438.

Kauffman, J.B., W.C. Krueger, M. Vavra. 1982. Impacts of a late season grazing scheme on nongame wildlife in a Wallowa Mountain riparian ecosystem. p. 208-220. *In*: Peek, J.M. and P.D. Dalke, (eds). Proc. of the wildlife-livestock relationships symposium, University of Idaho, Forest, Wildlife and Range Exp. Sta., Moscow, ID.

Kellner, K. and O.J.H. Bosch. 1992. Influence of patch formation in determining the stocking rate for southern African grasslands. J. Arid Environments 22:99-105.

Kieth, J.O. 1961. An efficient and economical pocket gopher exclosure. J. Range Manage. 14:332-334.

Kleinman, L.H. and E.E. Layton. 1981. Livestock operation is born on reclaimed mined lands. No. 80. Abstracts of Papers, 34th Ann. Mtg., Soc. Range Manage., Rapid City SD.

Klipple, G.E. and D.F. Costello. 1960. Vegetation and cattle responses to different intensities of grazing on short-grass ranges in the Central Great Plains. U. S. Dept. Agric. Tech. Bull. 1216.

Knopf, F.L. and R.W. Cannon. 1982. Structural resilience of a willow riparian community to changes in grazing practices. Pp. 198-207. *In:* J.M. Peek and P.D. Dalke (eds). Proceedings: wildlife-livestock relationships symposium. University of Idaho, Forest Wildlife and Range Exp. Sta., Moscow, ID.

Knopf, F.L., J.A. Sedgewick and R.W. Cannon. 1988. Guild structure of a riparian avifauna relative to seasonal cattle grazing. J. Wildl. Manage. 52:280-290.

Krebs, C.J. 1985. Ecology: The experimental analysis of distribution and abundance. 3rd Ed. Harper and Row. New York.

Krueger, H.O. 1985. Avian response to mountainous shrub-willow riparian systems in southeastern Wyoming. Unpublished PhD Dissertation, Department of Zoology and Physiology, University of Wyoming, Laramie, WY.

Larson, F. 1940. The role of the bison in maintaining the short grass plains. Ecology 21:113-121.

Lauenroth, W.K., D.G. Milchunas, J.L. Dodd, R.H. Hart, R.K. Heitschmidt, and L.R. Rittenhouse. 1994. Grazing in the Great Plains of the United States, pages 69-100 *in:* Ecological Implications of Livestock Herbivory in the West. Society for Range Management, Denver, Colorado.

Lay, D.W. 1965. Effects of periodic clipping on yield of some common browse species. J. Range Manage. 18:181-184.

Laycock, W. A. 1975. Rangeland Reference Areas. Society for Range Management, Range Science Series No. 3.

Laycock, W.A. 1980. What is successful reclamation?—a look at the concepts of adaptability, productivity, cover and diversity of seeded species. p. 1-17 *In*: Proc. Northwest Colorado land reclamation seminar II. Steamboat Springs, C0.

Laycock, W.A. 1989a. Secondary succession and range condition criteria: Introduction to the problem. p. 1-15. *In*: W.K. Lauenroth and W.A. Laycock (eds). Secondary succession and the evaluation of range condition. Westview Press, Boulder, CO.

Laycock, W.A. 1989b. The reclamation of lands for agricultural grazing. p. 245-268. *In*: J.D. Majer (ed). Animals in primary succession—The role of fauna in reclaimed lands. Cambridge Univ. Press, London.

Laycock, W.A. 1991. Stable states and thresholds of range condition on North American rangelands—a viewpoint. J. Range Manage. 44:427-433.

Leopold, A. 1933. Game management. Charles Scribner & Sons, New York.

Lewontin, R.C. 1969. The meaning of stability. p. 13-14. *In*: Diversity and stability in ecological systems. Brookhaven Symp. in Biology 22. Springfield, VA.

Mack, R.N. and J.N. Thompson. 1982. Evolution in steppe with few large hooved mammals. Amer. Nat. 119:757-773.

Magnuson, J.J. 1990. Long-term ecological research and the invisible present. Bioscience 40:495-501.

Malin, J.C. 1955. The grassland of North America: its occupance and the challenge of continuous reappraisals. p. 350-366. *In*: W.L. Thomas, (ed). International Symposium: Man's Role in Changing the Race of the Earth. Univ. of Chicago Press, Chicago, IL.

Martinson, C.F. 1960. the effects of summer utilization of bitterbrush in north central Washington. M.S. Thesis, Univ. of Idaho, Moscow, ID.

Majer, J.D. 1985. Restoration of pre-mining ecosystems in forested and agricultural areas. Landline 11:4.

May, R.M. 1977. Thresholds and breakpoints in ecosystems with a multiplicity of stable states. Nature 269:471-477.

McClean, A. and E.W. Tisdale. 1972. Recovery rate of depleted range sites under protection from grazing. J. Range Manage. 25:178-184.

Medin, D.E. and W.P. Clary. 1990. Bird and small mammal populations in a grazed and ungrazed riparian habitat in Idaho. U.S. Forest Service Res. Pap. INT-425. Intermountain Re. Sta., Ogden, UT.

Medin, D.E. and W.P. Clary. 1991. Breeding bird populations in a grazed and ungrazed riparian habitat in Nevada. U.S. Forest Service Res. Pap. INT-441. Intermountain Research Sta., Ogden, UT.

Meeker, D.O. and D.L. Merkel. 1984. Climax theories and a recommendation for vegetation classification—a viewpoint. J. Range Manage. 37:427-430.

Melgoza, T., R.S. Nowak and R.J. Tausch. 1990. Soil water exploitation after fire: competition between *Bromus tectorum* (cheatgrass) and two native species. Oecologia 83:7-13.

Merrill, E. and N. Stanton. 1990. Plant responses to spring grazing by elk in Yellowstone National Park. p. 147-150. *In*: M.S. Boyce and G.E. Plumb. University of Wyoming, National Park Service, 14th Annual Report.

Michalk, D.L. and B.E. Norton. 1980. The value of reference areas in the study and management of rangelands. Aust. Rangel. J. 2:201-207.

Milchunas, D.G., O.E. Sala, and W.K. Lauenroth. 1988. A generalized model of the effects of grazing by large herbivores on grassland community structure. Amer. Nat. 132:87-106.

Miller, R.F., T. Svejcar and N. West. 1994. Implications of Livestock Grazing in the Intermountain Sagebrush Region: Plant Composition, pages 101-146 *in:* Ecological Implications of Livestock Herbivory in the West. Society for Range Management, Denver, Colorado.

Mueggler, W.F. and D.L. Bartos. 1977. Grindstone Flat and Big Flat enclosures—41-year record of changes in clearcut aspen communities. U.S. Forest Service Res. Pap. INT-195. Intermountain Forest and Range Exp. Sta., Ogden, UT 84401.

Murray, R.B., H.F. Mayland, and P.J. Van Soest. 1978. Growth and nutritional value to cattle of grasses on cheatgrass range in southern Idaho. U.S. Forest Service Res. Pap. INT-199. Intermountain Forest and Range Exp. Sta., Ogden UT.

Neilson, R.P. 1986. High-resolution climatic analysis and southwestern biogeography. Science 232:27-34.

Norton, B.E. 1978. The impact of sheep grazing on long-term successional trends in salt desert shrub vegetation of southwestern Utah. p. 610-612. *In*: D.N. Hyder. (ed). Proc. First Intl. Rangeland Cong., Soc. Range Manage., Denver, CO.

Osterheld, M., O.E. Sala and S.J. McNaughton. 1992. Effect of animal husbandry on herbivore-carrying capacity at a regional scale. Nature 356:234-236.

Owensby, C.E. 1969. Effect of cages on herbage yield in true prairie vegetation. J. Range Manage. 22:131-132.

Paulsen, H.A. Jr. and F.N. Ares. 1962. Grazing values and management of black grama and associated shrub ranges of the southwest. U.S. Dept. Agric., Forest Service Tech. Bull. 1270.

Pellant, M. 1990. The cheatgrass-wildfire cycle—are there any solutions? p. 11-18. *In*: E.D. McArthur, E.M. Romney, S.D. Smith and P.T. Tueller (eds). Proceedings—a symposium on cheatgrass invasion, shrub die-off, and other aspects of shrub biology and management. U.S. Forest Serv. Gen. Tech. Rep. INT-276. Intermountain Res. Sta., Ogden, Utah.

Pendleton, D.T. 1989. Range condition as used in the Soil Conservation Service. p. 17-34. *In*: W.K. Lauenroth and W.A. Laycock (eds). Secondary succession and the evaluation of rangeland condition. Westview Press, Boulder, C0.

Penfound, W.T. 1964. The relation of grazing to plant succession in the tall grass prairie. J. Range Manage 7:256-260.

Pieper, R.D. 1994. Ecological Implications of Livestock Grazing, pages 177-211 *in:* Ecological Implications of Livestock Herbivory in the West. Society for Range Management, Denver, Colorado.

Potter, L.D. and J.C. Krenetsky. 1967. Plant succession with released grazing on New Mexico range lands. J. Range Manage. 20:145-151.

Rice, B. and M. Westoby. 1978. Vegetative responses of some Great Basin shrub communities protected against jackrabbits or domestic stock. J. Range Manage. 31:28-34.

Rickard, W.H. and C.E. Cushing. 1982. Recovery of streamside woody vegetation after exclusion of livestock grazing. J. Range Manage. 35:360-361.

Robertson, J.H. 1971. Changes on a grass-shrub range in Nevada ungrazed for 30 years. J. Range Manage. 24:397-400.

Ryder, R.A. 1980. Effects of grazing on bird habitats. p. 51-66. *In*: R.M. DeGraff and N.G. Tilghman (eds). Management of western forests and grasslands for nongame birds. U.S. Forest Service Gen. Tech. Rep. INT-86. Intermountain Forest and Range Exp. Sta., Ogden, UT.

Sampson, A.W. 1919. Plant succession in relation to range management. U.S. Dept. Agric. Bull. 791.

Samuel, M.J. 1985. Growth parameter differences between populations of blue grama. J. Range Manage. 38:339-342.

Samuel, M.J. and R.H. Hart. 1990. Sixty years of succession on the Wyoming high plains. No. 217. Abstracts, 43rd Annual Meeting, Soc. for Range Manage., Reno, NV.

Sanders, K.D. and A.A. Voth. 1983. Ecological changes of grazed and ungrazed plant communities. p. 176-179. *In*: Managing Intermountain rangelands—improvement of range and wildlife habitats. U.S. Forest Service Gen. Tech. Rep. INT-157. Intermountain Forest and Range Experiment Station, Ogden, UT.

Sandor, J.A. 1983. Soils at prehistoric agricultural terracing sites in New Mexico. Ph.D. Dissertation, University of California, Berkeley, CA.

Schlatterer, E.F. 1989. Toward a user-friendly ecosystem: myth or mirth? p.223-227. *In*: D.E. Ferguson, P. Moergan, and F.D. Johnson (eds). Proceedings: Land classifications based on vegetation: applications for resource management. U.S. Forest Service Gen. Tech. Rep. INT 257. Intermountain Forest and Range Exp. Sta., Ogden, UT.

Schlesinger, W.G., J.F. Reynolds, G.L. Cunningham, L.F. Huenneke, W.W. Jarrell, R.A. Virginia, and W.G. Whitford. 1990. Biological feedbacks in global desertification. Science 247:1043-1048.

Schulz, T.T. and W.C. Leininger. 1990. Differences in riparian vegetation structure between grazed areas and exclosures. J. Range Manage. 43:295-299.

Schulz, T.T. and W.C. Leininger. 1991. Nongame wildlife communities in grazed and ungrazed montane riparian sites. Great Basin Nat. 51:286-292.

Schwan, H.E., D.J. Hodges and C.N. Weaver. 1949. Influence of grazing and mulch on forage growth. J. Range Manage. 2:142-148.

Sedgewick, J.A. and F.L. Knopf. 1987. Breeding bird response to cattle grazing of a cottonwood bottomland. J. Wildl. Manage. 51:230-237.

Severson, K. and P. Urness. 1994. Livestock Grazing: A Tool To Improve Wildlife Habitat, pages 232-249 *in:* Ecological Implications of Livestock Herbivory in the West. Society for Range Management, Denver, Colorado

Sharp, L.A., K. Sanders and N. Rimbey. 1990. Forty years of change in a shadscale stand in Idaho. Rangelands 12:313-328.

Singer, F.J. 1987. Grazing effects of elk on Yellowstone's northern winter range. Yellowstone National Park Progress Report, Mammoth, WY.

Smeins, F.E., T.W. Taylor and L.B. Merrill. 1976. Vegetation of a 25-year exclosure on the Edwards Plateau, Texas. J. Range Manage. 29:24-29.

Smith, A.D., P.A. Lucas, C.O. Baker, and G.W. Scotter. 1972. The effects of deer and domestic livestock on aspen regeneration in Utah. Utah Div. Wildl. Resour. Publ. 72-1.

Smith, D.A. and E.M. Schmutz. 1975. Vegetative changes on protected versus grazed desert grassland ranges in Arizona. J. Range Manage. 28:453-458.

Smith, E.L. 1978. A critical evaluation of the range condition concept. p. 226-267. *In*: D.N. Hyder (ed). Proc. First Intl. Rangel. Cong., Soc. Range Manage., Denver, CO.

Smith, E.L. 1989. Range condition and secondary succession: a critique. p. 103-141. *In*: W.K. Lauenroth and W.A. Laycock (eds). Secondary succession and the evaluation of rangeland condition. Westview Press, Boulder, CO.

Smoliak, S. 1965. A comparison of ungrazed and lightly grazed *Stipa-Bouteloua* prairie in southeastern Alberta. Can. J. Plant Sci. 45:270-275.

Svejcar, T. and R. Tausch. 1991. Anaho Island, Nevada: A relict area dominated by annual invader species. Rangelands 13:233-236.

Svejcar, T. and J.R. Brown. 1991. Failures in the assumptions of the condition and trend concept for management of natural ecosystems. Rangelands 13:165-167.

Taylor, D.M. 1986. Effects of cattle grazing on passerine birds nesting in riparian habitat. J. Range Manage. 39:254-258.

Thomas, J.W., C. Master, and J.E. Rodiek. 1979. Wildlife habitats managed rangelands—the Great Basin of southeastern Oregon. Edges. U.S. Forest Service Gen. Tech. Rep. PNW-85, Portland, OR.

Tiedemann, A.R. and H.W. Berndt. 1972. Vegetation and soils of a 30-year deer and elk exclosure in central Washington. Northwest Sci. 46:59-66.

Tisdale, E.W., M. Hironaka, and M.A. Fosberg. 1969. The sagebrush region in Idaho. Univ. Idaho Agr. Exp. Sta. Bull 512.

Trottier, G.C. 1976. Big game exclosures in the National Parks of western Canada. Canadian Wildlife Service, Edmonton, Alberta.

Trottier, G.C. 1986. Disruption of rough fescue, *Festuca hallii*, grassland by livestock grazing in Riding Mountain National Park, Manitoba. Canad. Field-Natural. 100:488-495.

Tueller, P.T. and J.D. Tower. 1979. Vegetation stagnation in three-phase big game exclosures. J. Range Manage. 32:258-263.

Turner, G.T. 1971. Soil and grazing influences on a salt-desert shrub range in western Colorado. J. Range Manage. 24:31-37.

Urness, P.J. 1990. Livestock as manipulators of mule deer winter habitats in northern Utah. p. 25-40. *In*: K.E. Severson (ed). Can livestock be used as a enhance wildlife habitat? U.S. Forest Service Gen. Tech. Rep. RM-194. Rocky Mountain Forest and Range Experiment Station, Ft. Collins, Colo.

Vogel, W.G. and G.M. Van Dyne. 1966. Vegetation responses to grazing management on a foothill sheep range. J. Range Manage. 19:80-85.

Watt, A.D. 1947. Pattern and process in the plant community. J. Ecol. 35:1-22.

Watts, C.R., L.C. Eichhorn, and R.J. Mackie. 1987. Vegetation trends within rest-rotation and season-long grazing systems in the Missouri River Breaks, Montana. J. Range Manage. 40:393-396.

Weaver, J.E. and F.E. Clements. 1938. Plant ecology, 2nd ed., McGraw-Hill Book Co., New York.

Weaver, J.E. and T.J. Fitzpatrick. 1934. The prairie. Ecol. Monographs 4:108-295.

West, N. 1991. Grazing may be unfairly blamed for changes in pristine areas. Utah Sci. 53:50.

West, N.E. 1988. Intermountain deserts, shrub steppes, and woodlands. p. 210-230. *In:* M.G. Barbour and W.D. Billings (eds). North Amer. Terrestrial Vegetation. Cambridge Univ. Press, New York.

West, N.E., F.D.. Provenza, P.S. Johnson, and M.K. Owens. 1984. Vegetation change after 13 years of livestock grazing exclusion on sagebrush semidesert in central Utah. J. Range Manage. 37:262-264.

Westoby, M., B. Walker, and I. Noy-Meir. 1989. Opportunistic management for rangelands not at equilibrium. J. Range Manage. 42:266-274

Wiens, J.A. 1973. Pattern and process in grassland bird communities. Ecol. Monogr. 43:237-270.

Whisenant, S.G. and F.J. Wagstaff. 1991. Successional trajectories of a grazed salt desert shrubland. Vegetatio 94:133-140.

Whisenant, S.G. 1990. Changing fire frequencies on Idaho's Snake River Plains: ecological and management implications. p. 4-10. *In*: E.D. McArthur, E.M. Romney, S.D. Smith and P.T. Tueller (eds). Proceedings—a symposium on cheatgrass invasion, shrub die-off, and other aspects of shrub biology and management. U.S. Forest Service Gen. Tech. Rep. INT-276. Intermountain Res. Sta., Ogden, UT.

Wiens, J.A. 1976. Population response to patchy environments. Ann. Rev. Ecol. Syst. 7:81-120.

Williams, S.S. 1951. Microenvironment in relation to experimental techniques. J. Brit. Grassland Soc. 6:207-217.

Williamson, R.L. 1981. Improving species composition and seasonal variety on reclaimed strip mined grasslands through selective grazing intensities and time periods. Proc. N. Dak. Acad. Sci. 35:5.

Wilson, A.D. 1989. The development of systems of assessing the condition of rangeland in Australia. p. 77-102. *In*: W.K. Lauenroth, and W.A. Laycock (eds). Secondary succession and the evaluation of rangeland condition. Westview Press, Boulder, CO.

Wilson, A.D. and G.J. Tupper. 1982. Concepts and factors applicable to the measurement of range condition. J. Range Manage. 35:684-689.

Wilson, A.M. and D.D. Briske. 1979. Seminal and adventitious root growth of blue grama seedlings on the Central Plains. J. Range Manage. 32:209-213.

Winward, A.H. 1991. A renewed commitment to management sagebrush grasslands. The Grazier. 273:1-5, Oregon State University. Corvallis, OR.

Wissel, C. 1984. A universal law of the characteristic return time near thresholds. Oecologia 65:101-107.

Wuerthner, G. 1990. The price is wrong. Sierra (September): 38-43.

Young, S. 1958. Exclosures in big game management in Utah. J. Range Manage. 11:186-190.

Young, J.A. 1994. Historical and Evolutionary Perspectives on Grazing of Western Rangelands, pages 1-12 *in:* Ecological Implications of Livestock Herbivory in the West. Society for Range Management, Denver, Colorado.

Young, J.A., R.E. Eckert and R.A. Evans. 1979. Historical perspectives regarding the sagebrush ecosystem. p. 1-13. *In*: The sagebrush ecosystem—a symposium. Utah State Univ., Logan, UT.

Young, S.A. and R.B. Rennick. 1983. Using grazing animals to alter vegetation composition of reseeded minesoils. p. 406. Abstracts, 36th Ann. Mtg., Soc. Range Manage., Albuquerque, NM.

Changing Institutional Arrangements for Setting Natural-Resources Policy

Frederic H. Wagner

Abstract

Although natural-resources professionals have traditionally operated on the premise that their primary commitment should be to the resources, there is a growing view that their obligation is to society, and that we set policy and manage natural resources to satisfy social values.

We are ending an era when natural resources were viewed largely as commodities, and their consumptive use was the overriding social value. That view was promoted by the prevailing social values, a political and governmental sector pressured by commercial users, and by professional schools which promoted these values in students soon to be agency employees.

In recent decades, an increasingly urban public has broadened its values to emphasize esthetic and nonconsumptive uses, preservation, and animal-welfare attitudes. Natural-resources professionals have lost public trust because their values and policy setting have not kept pace with changing public values. Consequently policy-setting is often taken out of the professionals' hands and carried out by ad hoc combinations of interest groups variously termed interest-group pluralism and constituency-based multi-resource planning. In this mode, professionals (1) continue to be the hands-on managers, (2) provide technical information for interest-group deliberations, (3) mediate the deliberations, and (4) should avoid advocacy in order to regain public trust and credibility.

Introduction

This symposium was organized in part because of the range profession's concerns about the future of livestock grazing, particularly uncertain on public lands in the western U.S. Range professionals are somewhat caught in the middle between the stockmen's reluctance to change and demands for policy change by other interest groups, including demands for complete elimination of livestock use on public lands.

Rangeland policies are not the only natural-resources public policies under tension and likely to change. Timber harvest, predator control to protect livestock, mineral exploration and extraction, hunting, fishing, and trapping are all under challenges of varying intensities, and from a variety of different groups. These challenges reflect changing public attitudes regarding natural resources which, in turn, are being and will be translated into political action and altered policies.

The changing mechanics of policy setting and the role of resource professionals in this process are receiving intensive study by policy analysts. The purpose of this paper is to review some of the results of these analyses.

A Question of Values

Many natural-resources professionals have traditionally operated on the premise that their primary obligation was to the resources. Peek (1986:4) stated, "It should go without saying that a wildlife biologist's primary obligation is to the resource rather than to the agency or to the people who pay his or her salary." A recent Logan, Utah newspaper quoted a U.S.D.A. Forest Service Branch Chief on a nearby national forest as saying, "Essentially the Forest Service is in the business of managing vegetation rather than managing cattle." Kennedy (1985) has commented on ". . . foresters' traditional view of themselves managing and protecting resource things (i.e. objects such as trees or game animals), rather than managing these resources as objects of changing social value. . . ."

However, there is a growing view, reflected by the latter half of Kennedy's comment, that our obligation is in fact to society, and that we manage natural resources to satisfy social values (Hendee 1974, Giles 1978, Wagner 1989, Kania and Conover 1991). This can raise the issue of whether resources have intrinsic value, a debate over which there is not enough space here for review. In brief, I do not think they do and would use the same logic I have used elsewhere (Wagner 1988) to argue that animals do not have intrinsic rights. And the point is implicit in the textbook definition of natural resources as objects and substances in nature that have value to humans.

Policies, as we all know, are plans or statements of intent to achieve goals. Natural-resources policies are, therefore, plans to achieve goals for the management of natural-resources. Given our emerging realization that our natural-resources goals are social goals, it follows that natural-resources policies are really set to satisfy social values.

All of this may be trivially self evident to many. But it becomes the basis for understanding the changing mechanics of policy formulation, and the changing role of resource professionals in the policy process, about which there is considerable perplexity in professional circles these days. Let us review these changes over the last century or so.

The Commodity-Use Era

We are ending an era when natural resources were viewed largely as commodities, and their consumptive use was by far the overriding value orientation. That orientation dominated the policy process, and was reinforced by at least three sectors in American society.

One sector was society itself and its values. The American public was much more rural than it is today, and more involved in deriving agricultural commodities from the land. It naturally looked on such other resources as timber and range forage as commodities to be harvested like agricultural crops.

The second was the political and governmental sector, formulating policies under pressure from commercial users of natural resources, and operating in a sanguine public atmosphere. Federal agencies were formed, and legislation passed, to facilitate consumptive use. The Fish and Wildlife Service, established in 1871, was given, and continues today to have, the regulation of hunting and fishing as a major part of its charter (cf. Yaffee 1982).

Dana and Fairfax (1980) have reviewed in detail the close association between timber interests and the Forest Service, essentially formed in 1891 with passage of the Forest Reserve Act, and between livestock interests and the mission of the Bureau of Land Management established in 1934. To this day, timber policy is significantly influenced by timber-oriented legislation like the Knutson-Vandenburg Act of 1930, and by legislators from timber-producing states on the House and Senate Appropriations Committees (O'Toole 1991, Ribe 1991a, b). These broad legislative mandates are then translated into operational policies by the agencies.

A third reinforcing influence has been the professional, natural-resources schools in the universities. The faculties, many from rural backgrounds, have themselves been oriented toward commodity and consumptive-use values. And they have found receptive student bodies for teaching these values as their protégés were often from rural origins, strongly drawn to the out-of-doors, and motivated by the dream of becoming rangers, foresters, or conservation officers.

Given this orientation, the graduates readily assumed the missions and ethics of the agencies. As Ribe (1991b) put it, "The Forest Service has long been composed primarily of foresters trained in the American pioneer ethic and the timber production and tree farming school of thought that dominated most forestry schools over the last half century. Through the 1974 clearcutting crisis and before, these USFS foresters have allied themselves with timber industry interests to persuade Congress to keep logging as the dominant use of the national forests."

In sum, broad policy at the level of legislation has been set by Congress, significantly led by senators and representatives who had influential user groups among their constituencies. Operational policies within the framework of such legislation has been set by agency professionals whose values were close to those of the user groups. All took place in a social climate that was sanguine about these procedures because most of the American public had values that were, if not similarly inclined, not opposed. The commodity-use era continues to the present, but somewhat in decline due to pressures exerted on it by changing societal values.

The New Age

Incongruency Between Public and Professional Values

During the first half of the 20th century, American society moved to the cities. By the latter decades of the century, 90 percent of the people were no longer extracting a living from the land or handling livestock to produce agricultural commodities. Society was also becoming more environmentally concerned, and becoming aware of such issues as population growth and resource depletion. Hays

(1988) also stresses the importance of a great increase in outdoor recreation after World War II.

In consequence, public values assigned to natural resources have broadened and changed (Kellert 1985). Esthetic and nonconsumptive uses of resources, a preservation orientation, and concerns for animal welfare now permeate much of social attitudes toward nature. Much of the American public now values noncommodity entities such as spotted owls, snail darters, and red-cockaded woodpeckers. And it assigns noncommodity values to such things as old growth timber, whales, and arid-land vegetation which were previously valued largely as commodities. Hays (1988) comments on what he calls "The New Environmental Forest":

> While the old view emphasized [national] forests as sources of commodities such as timber, water, minerals, and rangeland, the new emphasizes forests as human environments that can be managed for the enjoyment of human life. Although commodities still play an important role in that environment, a new set of values, emphasizing the benefits of natural environments, now play an increasingly important role in defining the American standard of living.

Values of natural-resources professionals have not kept pace with the changing societal values. The public sees a continuing association between the agencies and user groups, and continuation of exploitive policies. Yaffee (1982) encountered something less than enthusiasm in the Fish and Wildlife Service over carrying out the mandate of the Endangered Species Act. Indeed, the Association of Forest Service Employees for Environmental Ethics (AFSEEE) arose as a protest against the Service's reluctant responsiveness to changing values.

The public also sees what it takes, right or wrong, to be malpractice by the agencies. In unsightly clear cuts, degraded riparian zones, and overexploited animal resources it sees what it considers a violation of the public trust to manage resources properly.

The result has been a decline of public trust in the agency professionals. In the past year, I have heard three different Forest Service employees from two regions state "I have zero credibility with the public." In this same period, an official of the Utah Division of Wildlife Resources was incredulous when an environmental group questioned his figures on a sandhill crane population. The Division had been advocating a crane hunt which the environmental group was opposing.

University faculty in professional natural-resources programs have also not moved fully with the changes in social values (Wagner 1989). The preservation and nonconsumptive-oriented conservation biology movement developed in university biology departments and not in fisheries and wildlife departments. And though painful to say, this symposium, organized and largely contributed to by university range-management professors, has a defensive tone over the public challenges to continued livestock grazing.

In addition to value diversification, the public is also becoming more assertive and politically active, frequently challenging the professionals' decisions. The plight of most of the national forest plans is one of the best testimonies to this changing public activism.

Evolution of New Policy-setting Protocols

Because of the incongruence between public and professional values, loss of trust in the professionals, and growing political activism by the public, policy setting in many cases is being taken out of the professionals' hands and assumed by various, ad hoc combinations of interest groups. We are in a period of change, and there is no consistent pattern to these groups. In general, society is groping for new institutional mechanisms to deal with changing social concerns. However, the political scientists on my campus classify these arrangements generically under the term "interest group pluralism."

Some examples illustrate the pattern. A process called Coordinated Resource Management Planning (CRMP) determines the number of livestock, wild horses, and native wildlife to be carried on each BLM Herd Management Area in Nevada. In each locale, representatives of the livestock interests, environmental groups, wild horse advocacy groups, and wildlife organizations meet with a BLM representative to present their views and desires, and arrive at a compromise on the allowable number of animals of each class. The proceedings may be moderated by a county agent, or someone else who is familiar with the issue but has no vested interest in it.

In 1989, a local environmental group challenged the Forest Service's estimate on the allowable timber cut in the Tahoe National Forest. The group, which contained individuals who were knowledgeable in such calculations, demanded the data on which the Service's estimates were based. It called in industry representatives to participate in the recalculations. The result, agreed upon by both the citizen's group and industry officials, was an allowable cut only about two-thirds of that calculated by the Forest's personnel.

As in many cities, deer populations in recent years have increased in four Minnesota cities bordering the Minnesota Valley National Wildlife Refuge and Fort Snelling State Park in the lower Minnesota River Valley. These populations have risen to levels where deer-vehicle accidents and damage to landscape plantings have become a serious concern to local residents. However, attempts at controlling deer numbers, including a state-advocated public hunt, were highly controversial, and the city of Burnsville asked the state Department of Natural Resources (DNR) to convene a task force that would recommend an alternative deer management policy. The resulting Deer Management Task Force (DMTF) was comprised of 1 representative each from the federal, state, and county resource agencies, 1 from each city, 1 from each of 3 pro- and 3 anti-hunting organizations, and 1 from a non-hunting conservation group. The DMTF held 18 meetings and 1 field trip from April, 1989 through June, 1990 and issued a set of recommendations which were unanimously agreed upon by the DMTF members and adopted by the cities (McAninch and Parker 1991).

Behan (1990) has termed these kinds of solutions to controversial policy issues "constituency-based, multi-resource management":

> Constituency-based, multi-resource management is an emerging paradigm of professional forestry. It deals with complex systems, both biological and social, but the problems it encounters can be kept empirical, substantive and 'close to the ground.' Solutions to these problems can be negotiated, on-site and within the constituency and should prove both durable and economical.

Behan sees these approaches as superior to solving natural-resources problems with national legislation and administrative policies. These tend to be abstract, ambiguous, and impersonal in order to apply to the entire country — the National Forest Management Act is his prime example. They thus invite challenge and explain why 98 percent of forest plans so far completed are under formal appeal.

Changing Role of Professionals

If operational policy setting is being taken out of the hands of the natural-resources professionals, what role will they play under the new policy procedures? They will, of course, continue to be the hands-on managers in the resource agencies. Beyond this, they have two roles, according to Behan (1978).

One is the provision of technical information for the group policy deliberations. If policy formulation is to be a rational process based on facts it must be illuminated with technical information which predicts the consequences of the different policy options. Resource professionals, by virtue of their education, contact with other professionals, and access to the technical literature and data are in a better position than anyone else to serve this role.

A second role which Behan (1978) urges on the professionals is to serve as mediators for the interest-group deliberations. This procedure was followed in developing the Minnesota deer management policy. The state DNR provided a "facilitator" to conduct the meetings, take meeting minutes, and write the DMTF's final decisions.

In order to succeed in the two roles, it becomes crucial that the professionals have trust and credibility. Their technical data and judgments must be considered objective by all group participants in order for them to have a significant influence in policy decisions. The professionals' mediation of group deliberations cannot appear to favor any of the participants. As McAninch and Parker (1991) comment on what they call "professional facilitation," the procedure ". . . is based on respecting the legitimacy of conflicting views and honoring the dignity, intelligence, and sincerity of antagonists."

The critical importance of trust and credibility requires that the professionals avoid advocating any particular policy option, in Behan's (1978) view. To do otherwise would erode the image of objectivity if not in fact taint it. It was, after all, the advocacy of consumptive use which diminished the professionals' public trust. If that trust is to be regained, the professionals are going to have to abandon advocacy. We professionals are a tiny fraction of society, and there is no reason why our values should weigh disproportionately in policy decisions made to satisfy total societal values (Wagner 1977, 1978).

There are uncertainties and potential risks with the interest-group pluralism model and neutral role of the professionals. Susan Giannettino, Forest Supervisor of the Wasatch-Cache National Forest in Utah, and Ulysses Seal, of IUCN's Species Survival Commission, have both commented to me that a potential problem with the newer policy-setting procedures is the public's insufficient technical knowledge needed to understand the consequences of actions which it demands. There is the distinct possibility that certain management directions mandated by a public exercising its newly freed, democratic rights in natural-resources policy may ultimately lead to irreparable damage to the resources.

Lucy A. Jordan (Pers. Comm.), wetland specialist for the Utah Division of Wildlife Resources, concurs with the above, and goes one step further. In her view, natural-resources professionals, by virtue of their technical education, "have the best understanding of the short and long term value and contribution of natural resources to society" and can best predict the consequences of different use and management alternatives. Agency professionals (but not academics) should therefore advocate for the resources: "Someone must argue for the 'best' alternative from an ecosystem perspective. So called environmental groups cannot be relied upon to do this." Crowfoot (1990) extends this reasoning further to include academics, maintaining that the environmental hour is late and the professoriate has no choice but to advocate.

Giannettino suggests that an important role for resource professionals in The New Age is to assume leadership in informing the public of the consequences of alternative policy options. That is really part of Behan's technical-information-providing role, but now expanded to a larger, more aggressive public-information dimension. It is also the function addressed in part by this symposium.

Literature Cited

Behan, R.W. 1978. Political dynamics of wildlife management: The Grand Canyon burros. Trans. No. Amer. Wildl. and Nat. Res. Conf. 43:424-433.

Behan, R.W. 1990. The RPA/NFMA: Solution to a nonexistent problem. West. Wildlands Winter 1990:32-36.

Crowfoot, J.E. 1990. Academia's future in the conservation movement. Renew. Res. J. 8:5-9.

Dana, S.T. and S.K. Fairfax. 1980. Forest and Range Policy/ Its Development in the United States. McGraw-Hill Book Co., New York.

Hays, S.P. 1988. The new environmental forest. Univ. Colo. Law Rev. 59:517-550.

Giles, R.H., Jr. 1978. Wildlife Management. W.H. Freeman, San Francisco.

Hendee, J.C. 1974. A multiple-satisfactions approach to game management. Wildl. Soc. Bull. 2:104-113.

Kania, G.S. and M.R. Conover. 1991. How governmental wildlife agencies should respond to local governments that pass antihunting legislation — A response. Wildl. Soc. Bull. 19:224-225.

Kellert, S.R. 1985. Historical trends in perceptions and uses of animals in 20th century America. Environ. Rev. 9:19-33.

Kennedy, J.J. 1985. Conceiving forest management as providing for current and future social value. For. Ecol. and Mgt. 13:121-132.

McAninch, J.B. and J.M. Parker. 1991. Urban deer management programs: A facilitated approach. Trans. No. Amer. Wildl. and Nat. Res. Conf. 56:428-436.

O'Toole, R. 1991. Incentives to mismanage/Obscure laws encourage development, discourage amenity management. Inner Voice 3(2):6.

Peek, J.M. 1984. A review of wildlife management. Prentice Hall, Englewood Cliffs, NJ.

Ribe, T. 1991a. Pork barreling our national forests/Timber management by political decree. Inner Voice 3(2):1,5.

Ribe T. 1991b. Rangering from capitol hill/Politics in the timber target stew. Inner Voice 3(2):5,7.

Wagner, F.H. 1977. Species vs. ecosystem management: Concepts and practices. Trans. No. Amer. Wildl. and Nat. Res. Conf. 42:453-461.

Wagner, F.H. 1978. Livestock grazing and the livestock industry. Pp. 121-145 *In*: H.P. Brokaw (ed.) Wildlife and America/ Contributions to an Understanding of American Wildlife and its Conservation. U.S. Govt. Printing Off., Washington, D.C.

Wagner, F.H. 1988. Predator Control and the Sheep Industry/ The Role of Science in Policy Formation. Regina books, Claremont, CA.

Wagner, F.H. 1989. American wildlife management at the crossroads. Wildl. Soc. Bull. 17:354-360.

Yaffee, S.L. 1982. Prohibitive Policy/Implementing the Endangered Species Act. M.I.T. Press, Cambridge, MA.

Summary: Ecological Implications of Livestock Herbivory in the West

Harold F. Heady

This symposium examined facts that underlie the question: "Should grazing be allowed on public land?" and its corollary "Removal of livestock from the public land will return rangeland ecosystems to near-pristine conditions." The controversy expressed in these questions, extending over several decades, has included elements of biology, sociology, economics, decision making in the courts, national policy, and political maneuvering. Emphasis in this symposium is biological and largely restricted to the effects of livestock grazing on rangeland vegetation with attention given to wildlife, riparian conditions, and natural resource policy. The magnitude and completeness of the subjects addressed herein is indicated by references to over 900 published studies.

Our primary objective was to present the known effects of grazing on rangeland vegetation. We did not cover all aspects of grazing as a major ecological factor and even less of the social/policy/political aspects of our obligation to the resources and to society. Readers will find this summary borrows liberally from the papers but interpretations are mine alone.

The New Age of Rangeland Resource Policy

By 1960 public values assigned to natural rangeland resources were broadening away from livestock grazing and toward greater esthetic uses, preservation, and animal welfare. Coincident with increasing interest in rangeland resources, the urban public has become more scientifically knowledgeable and aware of environmental values. Non-commodity values are now assigned to spotted owls, old growth timber, rangeland forage, and many other elements. Another trend is that public trust of rangeland professionals has diminished because of deteriorated or eroding riparian zones, abandoned and broken fences, the belief that all grazing is overgrazing, and what appears to be malpractice.

Beyond the obvious and continuing role of hands-on rangeland management, two possible roles for professionals exist in rangeland policy. One is to provide technical and expert information for institutional policy deliberations. The second is to serve as mediators for multi-interest group compromise. Coordinated Resource Management Planning by local multi-interest groups pooling efforts on their specific problems is a proven and expanding vehicle for these roles. Increasingly, resource professionals and scientists are active in the new age.

Overgrazing or Grazing

Grazing is overgrazing only when it causes retrogressive vegetational and soil changes from a stated objective. Overgrazing is usually overstated in the public press. To say that "this area is overgrazed" indicates that the speaker knows the vegetational objectives of management and the grazing intensity and tools needed to reach it. For example, if the objective is removal of all animals, any grazing by domestic animals is overgrazing. This seminar has not used grazing and overgrazing as synonyms and considers the latter unacceptable on all but specially managed rangeland. To us overgrazing causes habitat deterioration; grazing does not.

Ecologists often refer to the carrying capacity as the population density when mortality equals recruitment. When populations reach that level there is often damage to habitat and die-offs. To range professionals, carrying capacity is optimum production of any herbivore on a longterm basis. It is well below levels that cause die-offs and destruction of habitat. Self-interest, lack of knowledge or an unclear objective can make any herbivore damaging or productive of the needs and pleasures of mankind without being environmentally damaging.

Everyone recognizes that much of the western public rangeland was stocked above carrying capacity and severely overgrazed beginning after the Civil War. Livestock stripped off the vegetation, trampled the soil, and unacceptable erosion occurred. Until the early 1900s overuse of the land was fostered by society and approved by government; it was benevolent and necessary policy for westward expansion of the United States. Now, six decades after the first steps were initiated to correct the overuse on rangeland outside national forests, improved vegetation and stabilization of many habitats have resulted; however, the job is not finished. Unfortunately, livestock grazing is still blamed for the results of formerly acceptable practices.

Grazing Factors

Herbivores affect their habitats and therefore the habitat of other land users directly by defoliation and by trampling vegetation and soil. Herbivores exert indirect effects through redistribution of minerals and plants, changes in microenvironments, altering soil and hydrologic properties, influencing energy flow, and unbalancing competition among plants through selective grazing.

Direct Effects by Defoliation

After the turn of the century, hundreds of studies gave us hypotheses about the importance of carbohydrate reserves to plant survival after defoliation. The findings were the underpinnings for management recommendations on critical frequencies and intensities of grazing and guides for proper degree of herbage use. Identification of critical frequencies and intensities of defoliation to ensure plant survival, relationship of root growth to severity of shoot defoliation, and appreciation of shoot morphology in grazing resistance are three of those traditional physiological principles. They and others that addressed individual plant responses to defoliation are oversimplifications and have been altered since 1975.

Carbohydrate reserves in roots and stem bases are of less importance to regrowth and survival than photosynthesis in the quickly developing new green leaves and nearby carbohydrate pools. Germination of buds at the nodes of perennial grasses does not appear to be entirely controlled by hormones produced in the apical meristem. However, the meristematic stimulants are not fully understood. Compensatory or increased photosynthetic rates by residual foliage have been observed in a number of important rangeland species.

Highly important to rangeland use is increased understanding that a defoliated plant may react more to competition from surrounding nondefoliated individuals than to the biomass removed from it. Partial reduction of competition can offset adverse effects of heavy defoliation. These revisions of a few hypotheses on the effects of defoliation support moderate grazing, especially rotation grazing, by indicating that plants can withstand some variable phenological degree of defoliation without critical damage.

The grazing optimization hypothesis states that primary production increases with an increasing intensity of grazing to an optimal level and then decreases with greater grazing intensity. This is theory suggested from container studies. The hypothesis lacks vegetational or ecosystem testing. Individual plant research is essential but field research must also be done. The lack of understanding of competition as affected by selective grazing is but one of the synecological items needing attention.

Evolutionary Perspectives on Rangeland Systems

The beginning development of the present rangeland animals and plants might have been when ancestors of vascular plants and modern animals arose from the primeval seas to colonize terrestrial habitats. Herbivores have evolved to allow digestion of herbaceous materials through energy release from organic compounds by microorganisms. Intestinal tracts have evolved into many different anaerobic digestive vats such as rumen and cecum.

The herbage on offer is in part the result of longterm evolution against herbivory through morphological escapes and chemical compounds, such as terpenes and alkaloids. Through time herbivory has been a sorting process of both animal and plant evolution. Other factors in the sorting process are climate and fire. These three factors have operated during eons of time, operate today and will continue to operate depending upon intensity, temporal, and spatial alterations. There have been extinctions and new species since life began and that will continue.

The Great Plains in central North America illustrate the close evolutionary development of large grazing animals and their forage supply. Based on many studies, grazed and ungrazed shortgrass steppes are similar and large shifts in community composition are not evident. Farther east in the tallgrass, grazing and/or burning can change the vegetation. Exotic species appear to be enhanced by grazing in tallgrass but not in shortgrass. A major conclusion is that grazing impacts of large native animals have been replaced by domestic animals. However, any grassland can be overgrazed by a species of animal whereby negative effects are produced on the vegetation.

Vegetational Changes on a Geological Scale

Let us illustrate vegetational changes on a geological or shifting climatic scale by examining the landscape history in the Great Basin since the last Ice Age, some 12,000 years before present (BP). The cool and moist conditions lasted until 8,000 BP. That was a time of pine, water birds, and low elevation pinyon-juniper. A warmer period until 4,500 BP is indicated by the abundance of *Artemisia.* During the last 3,000 years aridity and temperatures have been increasing but with moist/cool or dry/warm cycling on a 200 to 500 year basis. During several periods charcoal in packrat middens increased directly with *Gramineae* pollen and indirectly with *Artemisia* pollen. When large and small animals are considered, herbivory was probably high. Pre-European settlement vegetation is estimated to be 10 to 20 percent *Artemisia* with perennial grasses under and between the shrubs. The European settlement era brought many new stress factors; among them introduction of domestic animal grazing, alien plants that invade both grazed and ungrazed areas, cultivation, water diversion, unregulated hunting, seeding of perennial grasses and either lengthened or reduced fire frequencies. Elimination of any one of these stress factors (several cannot be eliminated) will cause ecosystem changes, but pristine systems will never return.

Two other examples illustrate the same theme. The Great Plains' bison and other large herbivores grazed on shortgrasses which became highly resistant to grazing. The herbaceous vegetation in the Intermountain Region farther west was less resistant to grazing, permitting exotics to flourish and native shrub stands to thicken and expand at the expense of understory after introduction of livestock. The shrubs of arid and semiarid areas are long-lived and will remain dominant for decades—grazing or no grazing. Results from many studies in widely different areas indicate that grazing alone is not sufficient to explain changes in vegetational composition.

Climate, grazing, and fire select different species, determine the space between individuals, control establishment and extinction from a place, and influence species interaction. Establishment in one climate and living into a different one is called biological inertia. Vegetationally propagating clones of grasses, for example, may have lived through changing climates since the ice age. It is important to emphasize that in the late 1990s, the land manager has little or no control over climate, large and small native herbivores, pathogens, alien species, fire, and carbon dioxide in the atmosphere. Control lies mostly with domestic herbivores, pest control chemicals, and prescribed fire. As wild space decreases, the need for more knowledgeable and intensive use of herbivores, chemicals, and fire as tools for vegetational management will increase.

Other Types of Vegetational Change

A common "grand illusion" is the simplistic idea that pristine ecosystems rapidly reestablish when herbivory is reduced, referring mainly to livestock. The principal tool to gather data on this subject has been comparison over time of vegetation that is grazed with that in nearby ungrazed exclosures and relict areas. Hundreds of these comparisons, many extending over 30 years, have given variable and unpredictable results. Obviously, factors other than grazing are dominant. Some of these in addition to climatic extremes are directed genetic change of some native species, introduction of exotic plant and animal species, changes in fire patterns, and reduced populations of many native herbivores. Reemergence of pristine vegetation of two centuries ago, or even two decades earlier, was seldom found. It is well to remember in rangeland vegetation management that pristine ecosystems supported a large measure of herbivory along with many instances of environmental variability, and inconsistencies in vegetational patterns.

Five processes result in the variability: An extreme climatic event may bring long-lasting change. Herbivory is discontinuous, variable, and may not be reversible. Some species promote and benefit from periodic burning. The soil may be permanently changed. Initial abundance of the competitors may determine the outcome of any impacting factor.

Ecosystem reestablishment is a developmental process that requires an understanding of all types of vegetational change. Already described are changes on a geological scale and those that occur to individual plants due to defoliation. On a time scale of a few decades, succession of species and stable end points were discussed with respect to inventory and interpretation of range condition and trend.

The proposition that plant succession occurs is accepted. No one argues that divergent species composition over space reflects variability in climate, soil, topography, and available organisms. On a time scale, there are daily, seasonal, and year-to-year changes reflecting competition and short-term abiotic variability. These are overlaid with perturbations caused by herbivory, fire, and other disturbances that cause vegetational destruction after which it responds by the succession of species. Distinguishing the roles of climate, space, landscape, and available biology, from impacts on rangeland ecosystems associated with pollution, timber harvesting, livestock, human occupancy and other consumptive uses is of major importance.

A mentioned description of the processes of succession was that of Clements and included migration, establishment, internal system effects and more or less stabilization. That system works well in the mesic middle United States grasslands where species compositional change can be rapid. Disagreement was expressed where aridity increases and the vegetation becomes dominated by shrub species with long lifespans and the speed of succession slows. If shrubs gain initial dominance they remain so and their stability may exceed the human lifetime whether grazed or not grazed by domestic animals.

As much as 60 percent of the western rangeland may be in stable condition including areas of sagebrush, creosotebush, and mesquite. There is mounting evidence supporting this aspect of vegetational stability and the implied concepts. This contrasts to estimates in the public press that as much as 85 percent of the western rangeland is deteriorating because of overgrazing. Both of these percentages need closer definition and evaluation.

Grazing To Modify Vegetation

Most ecologists and plant physiologists agree that herbivory results in plant community changes. The combined direct and indirect effects of grazing are expressed in species dynamics. If damaging, the changes are retrogressive in the sense that the competitive advantage favors species that are less palatable over palatable and lowest over tallest. The pathways of change following relaxation or removal of grazers may differ substantially from pathways of retrogression. In addition, the probability of ecosystem recovery to previous states may be greatly reduced or nonexistent. Vegetational changes are continuous but not always reversible and consistent. For example, after disturbance the sagebrush/grass may change in three directions, complete dominance of sage with less fire, increase in Eurasian annuals with frequent fire, and return to codominance of perennial native shrubs and herbs. Any of the three can appear to be permanent on the basis of a few decades, giving rise to such concepts as multiple stable points, steady states, and state and transition that replace climax. Nearly 100 years ago Clements recognized these possibilities and called them "disclimaxes."

Exclosures fenced against livestock and wild animals as small as rodents have been used to study herbivory. The results of no herbivory have varied greatly. I once visited an exclosure in the shortgrass region of the northern plains that was tall broad-leaved herbs inside with grass all around the outside. Another one almost totally dominated by sagebrush in central Oregon had not changed much in 40 years, but still another had changed to a mixture of sagebrush and perennial bunchgrasses. For others, increase of grass under grazing and no grazing have been parallel, as for example, black grama in New Mexico and on the Audubon Ranch in Arizona. In the Four Corners area, Indian ricegrass is little different in and out of exclosures after 50 years. On the Jornada Experiment Station, vegetation has sustained 25 years of grazing with little change. Livestock certainly have their influence but fire, weather, and invertebrates play major roles; sometimes with little effect until long-lived perennials die.

Shortgrass steppe under moderate grazing produces the highest ground cover, heavy use reduces cool season grasses and bluegrama dominance becomes nearly permanent. Annuals and pricklypear respond to weather but little to grazing. Cultivated land abandoned in the 1930s may pass through pioneer stages to bluegrama and stay there or may never reach a bluegrama dominance. Introduced annual grasses in California are here to stay so pristine climax is impossible to reach. By selective observation, one can "substantiate" either the pros or cons of livestock grazing, based on species composition or production of biomass.

Grazing and Biological Diversity

The intermediate disturbance hypothesis about the role of livestock grazing in vegetational change states that maximum diversity occurs at middle successional stages where competition among a few dominant species in minimized. We define diversity on the basis of number of species and their evenness of distribution of biomass or production. The concept is also applied to number and distribution of niches, habitats, and length of edges between them. Livestock create patches (diversity) when the grazing is light to moderate, but prolonged heavy grazing reduces diversity; so does no grazing. Grazing is a tool that can maintain plant succession at a certain stage thereby maintaining biological diversity.

Grazing and Threatened-and-Endangered Species

Livestock grazing is compatible with preservation of the natural resources and more beneficial than non-use or overuse to threatened and endangered (T and E) species. No grazing can allow plant succession to proceed toward elimination of T and E species (most are forbs) because of competition from grasses and accumulation of mulch. Of course, grazing overuse can eliminate local species. However, our ability to locate and identify T and E species today indicates their survival through odd periods of major overuse in the past. While livestock graze some of the threatened species, many of them are not directly affected by livestock.

Many T and E plants are varieties or other taxonomic subspecies units that are of questionable validity. Taxonomic research may not show that they should be separated from the named species and thereby combine them out of existence.

Grazing and Fire Hazard

Fire hazard on rangeland, as anywhere else, is closely related to the amount and type of fuel. Large amounts of grass biomass, thick stands of *Artemisia*, and mature chaparral are generally considered flash fuels and conducive to rapid spread of fire. The grasslands of concern in the western United States are the exotic downy brome (*Bromus tectorum*) in the sagebrush region and stable stands of mature perennial grasses. Grazing reduces fuel and thereby lessens fire hazard. Fire frequency has increased from approximately 40 years in pre-European sagebrush/perennial grass to 3 to 5 years where *Bromus* or other annuals are in extensive stands. However, sagebrush/grass may increase when fire occurs in chenopod stands. *Artemisia* increases in absence of fire, either grazed or ungrazed. Juniper in the same region is eliminated by fire frequency of 30 to 40 years.

Grazing and Wildlife Habitat

It must be recognized that wildlife are mostly state property for use and regulation and belong to society by law. The private landowner has furnished nearly all habitat for some species and large portions of it for other species. Some species have been destructive of landowner net income without compensation. Increasing income for hunting and trespass privileges are of recent origin. Public owned wildlife on public land are a different situation, especially when the animals move on and off private land. Major concern centers on habitats, principally on the question: "Can livestock grazing be managed for the benefit of wildlife?" Some believe that wildlife habitat management depends upon the livestock grazing industry because therein lies major financial support for improving the habitat on which wildlife management depends.

Many reported vegetational changes presumed to be caused by livestock have been negative on wildlife (perhaps reported because they were damaging). The effects are labeled as being caused by overgrazing, which often implies all grazing. Conversely, livestock have been managed to create better wildlife habitats by altering species composition of vegetation, increasing productivity of selected species, improving the nutritive quality of forage, and increasing the structural diversity of habitat. Waterfowl in particular have benefited from specialized grazing management. Land care to attain one or more of these goals always requires trade-offs because the habitat that favors one species will be adverse to others.

Big game populations on public lands are increasing in the presence of regulated livestock grazing. Between 1977 and 1988, the total big game population on Bureau of Land Management lands grew by 31 percent. Pronghorn have done well on lands also grazed by livestock. Mountain plover nests only in heavily grazed short grass. Many studies have shown in a wide variety of vegetational types that more biomass and a greater variety of birds occur in the grazed than in ungrazed areas, but other studies show the opposite. Grazing appears to favor birds more than rodents.

The present state of knowledge regarding the use of prescribed grazing by livestock and other rangeland management techniques as tools to enhance wildlife habitats is far from adequate. However, enough research-determined information and experience exist to warrant livestock grazing in some situations and to investigate its applicability in others.

Riparian

Riparian vegetational types, those dependent upon additional water coming off the slopes and the water in streams, are areas of animal congregation. Many riparian areas have been abused by too many animals for too long a time. As with livestock, wild animals may congregate there and may be a factor in reducing cover and diversity of vegetation. Herbaceous cover may be a steady state after willows are gone, and aspen may be held in a steady state by wild animals. Decisions by the Bureau of Land Management concerning riparian areas are based on livestock management practices and not on livestock grazing itself. Maintaining a healthy riparian ecosystem does not require total exclusion of livestock, but rather proper management.

Riparian habitats are resilient and often give rapid response to changes in grazing. Experience and research have indicated basic management practices. Structures in the stream frequently fail without animal management. Usually reduced number of livestock is not effective without controlling their distribution. Early spring use and no grazing until winter appear to be effective seasonal plans.

The Condition of Rangeland

Definitions of range sites, range condition, and trend are of current concern. These are three parameters used in the surveys and analyses for practical management of rangeland. A widespread application of these concepts groups plant species into decreasers, increasers, and invaders; and uses soil analysis, vegetational fluctuations, succession, climax, retrogression, stability, and other vegetational processes. These concepts are in need of refinement for two principal reasons. They must be considered differently for warm and cold regions as well as for subhumid, semiarid, and arid vegetational types. Secondly, the multiple uses or management objectives require that different end points of vegetational change be recognized. It follows that various names and concepts for rangeland sites, conditions, and trends need examination.

Tall grass, mixed grass, mesic mountain meadows, aspen, and riparian range types can be managed fairly well under the long-used ecological condition system. The arid and semiarid types do not change as predictably and pioneer states often appear stable for several decades, as if climax. A steady state may be changed to a different one only by a major disturbance. The time frame of these conditions is important to the manager as well as to the basic theory.

The basic assumptions of the long-used range condition model are not wrong. Our understanding of where and how to apply the system in the large number of time and space variations in rangeland ecosystems has been wrong. An actual inventory and evaluation system or systems will determine acceptance of the changing concepts on rangeland condition and trend.

The Question of Livestock Grazing

Differences in our views signify a healthy range science and continuing development of more reliable knowledge. Of increased importance in light of rangeland public policy issues is the separation and understanding of fact and what is believed to be fact. Our role goes further; to inform the public of the consequences of land management options and increasingly to mediate the deliberations of interest-groups.

Range scientists accept that serious deterioration occurred on western rangelands from the 1850s to the mid-1900s and even today on certain sites. With the present state of our knowledge, there is no excuse for destructive grazing by livestock.

There can be no return to the "pristine" conditions as long as humans walk the face of the earth, set fires, scatter litter, carry seed, whether or not domestic animals are removed. In the longterm their removal is likely to decrease diversity of habitats, jeopardize threatened and endangered species by allowing too much competition for them, increase fire hazard and undesirable burning, and alter wildlife habitat.

Abundant evidence exists that rangeland vegetation will tolerate a certain degree of herbivory and in fact thrives on and has evolved under grazing impacts by large animals.

Based on the results of science and experience reviewed, conservative and well managed livestock grazing, not overgrazing, is sustainable on western rangelands, with due regard for social concerns and the demands for other uses of the land.